The Home Shop Machinist PROJECTS Seven

The Home Shop Machinist PROJECTS Seven

Compiled and edited by
Joe D. Rice

THE HOME SHOP MACHINIST
(a division of Village Press, Inc.)
Traverse City, Michigan

PHOTO CREDITS
Page 1: Rudy Kouhoupt, May/June '96 cover
Page 55: Frank McLean, September/October '96 cover
Page 69: D. E. Johnson, November/December '95 cover
Page 83: Ted Wright, May/June '96, page 31
Page 113: George Genevro, March/April '94, page 34
Page 157: Roland Friestad (graphic), May/June '91, page 51

The Home Shop Machinist
PROJECTS
Seven

First Printing August 1997

Village Press
2779 Aero Park Drive
Traverse City, Michigan 49686

International Standard Serial Number 0-941653-52-8
Library of Congress Catalog Card Number 97-90915

Copy editing by Clover McKinley
Design and Layout by Luana Dueweke and Ellen Johnstone

Typeset and printed at Village Press, Traverse City, Michigan USA

INTRODUCTION

PROJECTS SEVEN is a collection of many articles and series from *The Home Shop Machinist* primarily from 1994 through 1996. However, this volume is unique among those Projects books published. We have included the very excellent columns of Roland Friestad, titled "Computers in the Shop."

Roland began his column when, for most of us hobby machinists, the computer was something seemingly unrelated to machining except for industrial applications. Today, we are finding many uses for the computer closer to home that can simplify some otherwise arduous, less exciting tasks, and better yet, increase accuracy in the more ticklish areas of machining. The reader response to Roland's articles on conversion of your mill or mill-drill to CNC has been good over the years, but what surprises me most is how many people are calling and writing us to see if they can get those articles six years after the series finished. Thus, we have decided to include them in their entirety in **Projects Seven**.

But the book is not just about computers in the shop, as you will soon see. It includes some superb work by many other authors/ machinists who have offered us a wonderful variety of techniques and projects that should keep anyone busy for a long time. Even if you choose not to do all the projects, this book will be helpful. When we listen to the wisdom and experience of our peers, we're bound to learn and ultimately apply some of that to our own work. Applying it may come in the form of emulation – doing it their way – or it may be the springboard for synthesis, their ideas blended with your own creativity to form an altogether new approach to a problem or puzzle. Either way, I hope **Projects Seven** will present you with the gifts these authors have offered.

Finally, I wish to again thank each author who appears in this book. Their unselfish efforts are appreciated so very much. And I hope you, the reader, enjoy **Projects Seven**.

Joe Rice, Editor

CONTENTS

TECHNIQUES

LATHE ACCESSORIES

MILLING & DRILLING MACHINE ACCESSORIES

SHOP IMPROVEMENT PROJECTS

HOBBY PROJECTS

COMPUTERS IN THE SHOP

TECHNIQUES

Balancing a Grinding Wheel

by A. M. Christopherson
Photos and drawings by Author

I don't know what it is about grinding wheels, but every time I loosen one or take it off a grinder spindle, I can never get it back without some imbalance resulting. I recently set up a dedicated grinder for a Darex drill sharpener and went through my usual gyrations trying to get it to run smoothly, without much luck. I noticed the grinder shaft had a shoulder to position the innermost wheel washer. The only trouble was the washer hole diameter was slightly larger than the shaft diameter, which prevented the washer from seating truly on the shoulder.

Try as I might, I could not get the wheel to run true so I could proceed with the balancing operation. I decided what was needed was a thick washer with a stepped bore to permit seating up against the shaft shoulder properly and eliminate any wobble. Making that didn't present any problems, and it did exactly what I wanted it to do.

Figure 1 shows the washer section I used, but if you want to try this, keep in mind the highlighted clearance area from the shoulder to the bearing. And any housing overhang must be taken into consideration to avoid locking the wheel, washer and shaft assembly up against the housing. Because of all the different sizes of grinders being used, I have not used specific numbers for most of the dimensions in this article.

Next, I decided to utilize an outer washer design that incorporated a balancing feature I had seen in a back issue of *HSM*. After hours of delightful but fruitless looking, I gave up and designed what I call a "flywheel" balancing washer. As you can see in Photo 2, it incorporates a moderately thick (.250") rim I used to position holes for the balancing pins at 10° intervals around the periphery, parallel to the wheel axis. This was accomplished using a rotary table to space the holes properly around the rim during drilling. Since I was going to install a cup-type wheel at the other end of the grinder, I decided to make that rim a bit thicker (.450", Figure 3), to allow the use of longer (heavier) balancing pins. The thicker rim

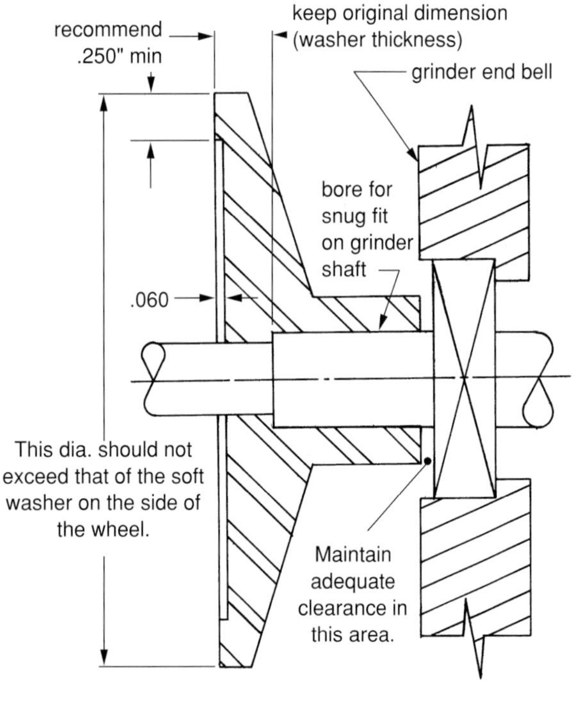

Dimension to suit your grinder.
Figure 1

2

overhang also provided a nice pocket for the retention of a rough balancing washer (Figure 4).

Final fine balance is achieved by the insertion of pins in the holes. During this phase, care must be exercised to retain the pins while their final positions are being determined by running the grinder. By careful drilling and use of drill rod for the pins, a snug sliding fit allowed me to position them temporarily. If you don't feel comfortable running the grinder with the pins just slid into the holes, a bit of insurance can be obtained by lightly prick punching the side of each pin to insure it stays put during this part of the balancing operation. During this trial-and-error phase, allow some of the pin end to protrude or make the pins a little longer than the hole is deep to allow you to grasp them for removal and replacement as you go through this fine balancing procedure. Once their final locations have been determined, they are removed and re-installed with *Loctite* to insure secure retention. Now I finally have a grinder that runs to my satisfaction.

When you start to balance a wheel or wheels, you have to know where you start, so index mark the shaft, the washer and the wheel to retain a known relationship to each other. Then, and only then, can you start looking for the weight locations that will reduce the vibration to the level you are seeking. Pick a spot, slide the washer off-center, or insert some pins and run the grinder. It may take a few trials to locate the light spot but it will work.

I found using .1250" diameter drill rod for the pins was satisfactory for the first wheel, which had very little initial imbalance as evidenced by the use of only six pins, as shown in Photo 2. However, for the second, cup-type wheel, I had to use a washer in the center to rough balance first; the pins alone did not have enough mass to do the job. Of course, cup wheels, due to their overhang, present additional dynamic balance problems not normally encountered with disc-type wheels. In retrospect, due to the cup wheel overhang, I should have made its washer rim at least .75" thick, which would have allowed heavier pins and probably eliminated the need for the rough balancing washer.

As an option to using the balance washer, consider using larger diameter pins, .1875" or perhaps .250". Larger pins may require a slight modification to the hole pattern depending on the diameter of the washer being fabricated but can eliminate the necessity of an initial balance washer. The design is flexible, allowing the use of pins in conjunction with or without a balance washer. Either way, if you try this method, I think you will be satisfied with the results.

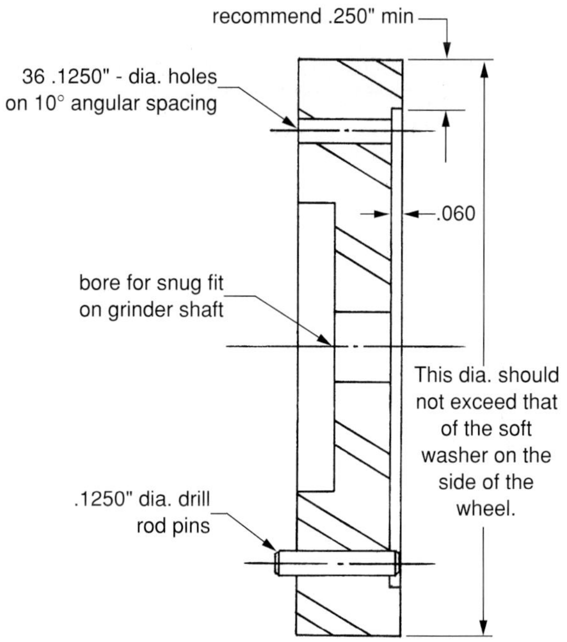

Figure 3

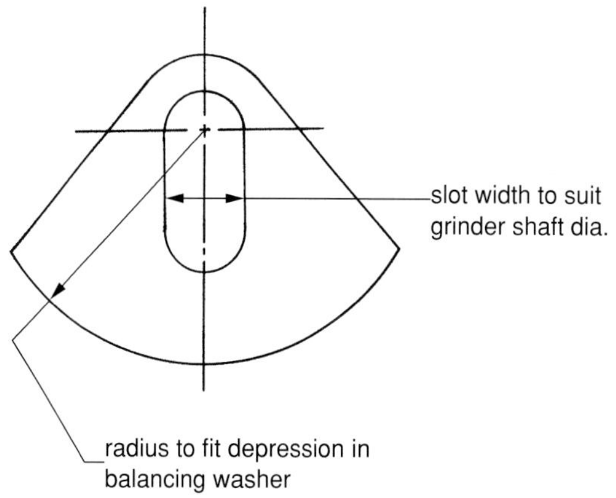

Figure 4

Clamping Tapered Material in a Vise – No Problem!

by Joe Harmon
Photos by Author

A need arises at various times to clamp tapered work in a vise. Vises are on the market that will meet this need if your pocketbook can afford them. I could not justify buying a vise that could clamp tapered work, so I decided to modify my present one.

Some rigidity will be sacrificed in order to modify the vise, but I have not experienced any problems with this tradeoff. Remember, this project is to modify, not to build a vise from scratch.

Listed is the material and equipment needed for this project.

BILL OF MATERIALS

ITEM NO.	QUANTITY	DESCRIPTION	SIZE
1	1	Vise (I used a Craftsman)	e.g., 3-1/2"
2	1	Keyway stock	1/2×6"
3	2	Mild steel	1/4×2-1/8×3"
4	4	Socket head cap screw	1/4-20
5	1	Socket head cap screw	5/16-18
6	1	Dowel pin	1/4"
7	8	Socket head cap screw	8-32
8	2	Bar stock	3/16×1×5-1/2"

EQUIPMENT

1. Band saw
2. Angle plate
3. Mill
4. Drill press
5. Jeweler's saw

First remove the movable jaw from the vise. Make sure the swivel base is tight. Mark two holes for the 1/4-20 screws on the backside of the jaw that is sloped. Using the milling machine, mill spots on the marked slope. Now center punch the milled spots for drilling the 1/4-20 clearance holes on the drill press.

After drilling the two 1/4-20 clearance holes, the next step is to clamp the vise to the angle plate. Do this such that the vise is on its side with the jaw parallel to the angle plate base. Correctly done, this will produce a cut that is parallel to the vise base. Now the jaw can be cut off on the vertical band saw. For the longest cut, I followed the casting seam, and for the short cut, I used the anvil edge as a guide.

With the back jaw cut off, I mounted the vise base on the milling machine and milled 1/4" material from where the jaw was positioned. This is important to keep the jaws in alignment.

Place the cut-off jaw and the 1/2" keyway stock onto the 1/4 × 2-1/8 × 6" as a base. Scribe the proper length onto the base and also transfer the two 1/4-20 screws holes. Drill and tap the 1/4-20 holes. Mount the jaw to the base.

Drill two clearance holes for 1/4-20 screws into the remaining 1/4 ×2-1/8" and the 1/2" keyway stock. Placing the 1/2" keyway stock onto the base with the jaw mounted, transfer the holes. Drill and tap 1/4-20 into the base, then bolt the keyway and top plate to the baseplate.

The author's modification to his vise is shown here.

Drill and ream a 1/4" hole into the top plate. Align the jaw base on the vise, leaving clearance for the base to rotate from side to side. With the base in position, transfer the 1/4" hole to the anvil. Drill and ream the 1/4" hole. Insert a dowel pin to align the jaw, and now screw it back on the movable jaw until both jaws are tight. Drill for a tap size 5/16-18 hole between the rear jaw and the dowel pin into the anvil. Remove the rear jaw, tap 5/16-18 into the anvil and drill a 5/16-18 clearance hole into the baseplate.

Scribe an arc with a 5/16" hole centered onto the base and cut out. I used a jeweler's saw. The final step is to attach side plates to the connecting bases to aid in rigidity. Some 3/16 × 1 × 5-1/2" bar stock is attached with four 8-32 screws, one plate on each side.

With the modification complete, the vise is ready to use. When clamping a tapered part, one side plate is removed and the 5/16" screw is slightly loosened. The part is inserted, and is clamped tight enough to set the angle of the rear jaw. Snug tighten the 5/16" screw, release the clamped part, then fully tighten the 5/16" screw. When not clamping tapered parts, both side plates are attached.

I did not give exact measurements because of different vises being used and working with castings. Some of the materials listed will differ with various vise sizes. This outline is more of a guide to the building concept. Modifications will vary to each vise design.

In this photo, the offset and resulting taper are clearly shown.

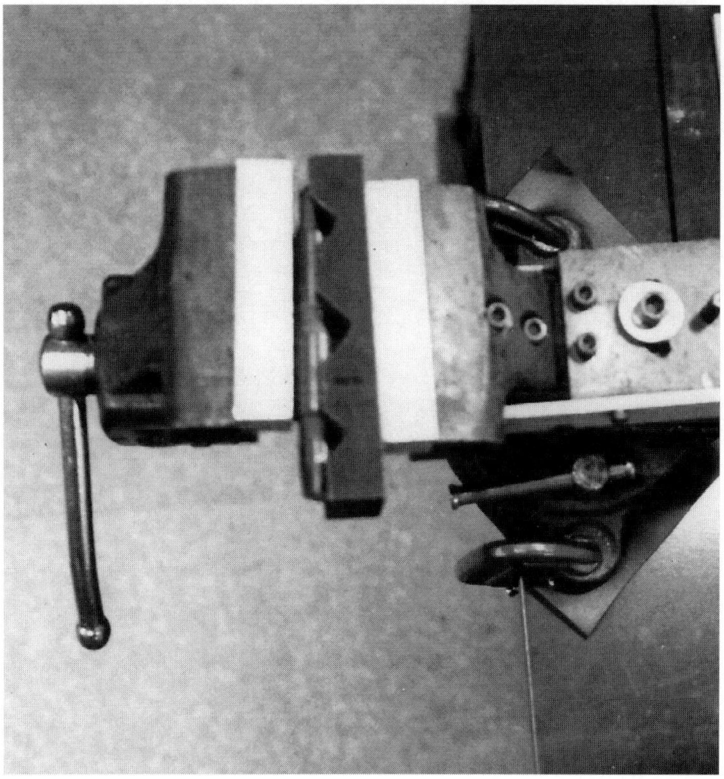

Large Radius Cylindrical Cuts on a Shaper

by Peter J. Hoijer

I had a need to make large radius (24") cylindrical grooves in steel plates recently (Figure 1) and this is my solution using a shaper.

The machining of large radius cylindrical grooves can be accomplished on a shaper by replacing the vertical feed with a vertical rod with free swiveling ends whose length is the desired radius. As the shaper table moves horizontally, the vertical rod raises and lowers the table so the cutting tool will cut a cylindrical groove with the radius of the vertical rod length (Figure 2).

While supporting the table, remove or disable the vertical feed. On my shaper, I just removed the setscrew securing the fixed nut, and moved the nut up a few turns (Figure 3).

Fabricate an adjustable link with its length in the range of the desired radii. I used a 5/8" threaded stock with a nut and 3/4" water pipe. Make both ends spherical. I machined the rod and put a large steel ball in the tube end.

Fabricate end plates to receive the ball ends and attach them to the shaper horizontal slide and base or floor. The plates may be shimmed to bring the work within range of the cutting head.

Set the rod to the desired radii and, using the end plates, install it between the horizontal

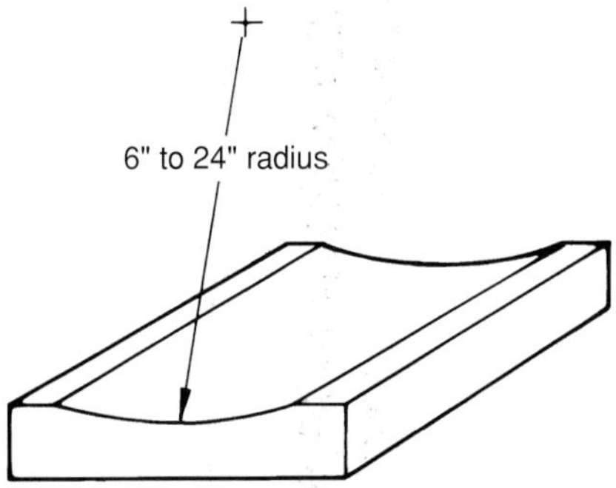

Figure 1 WORK SAMPLE

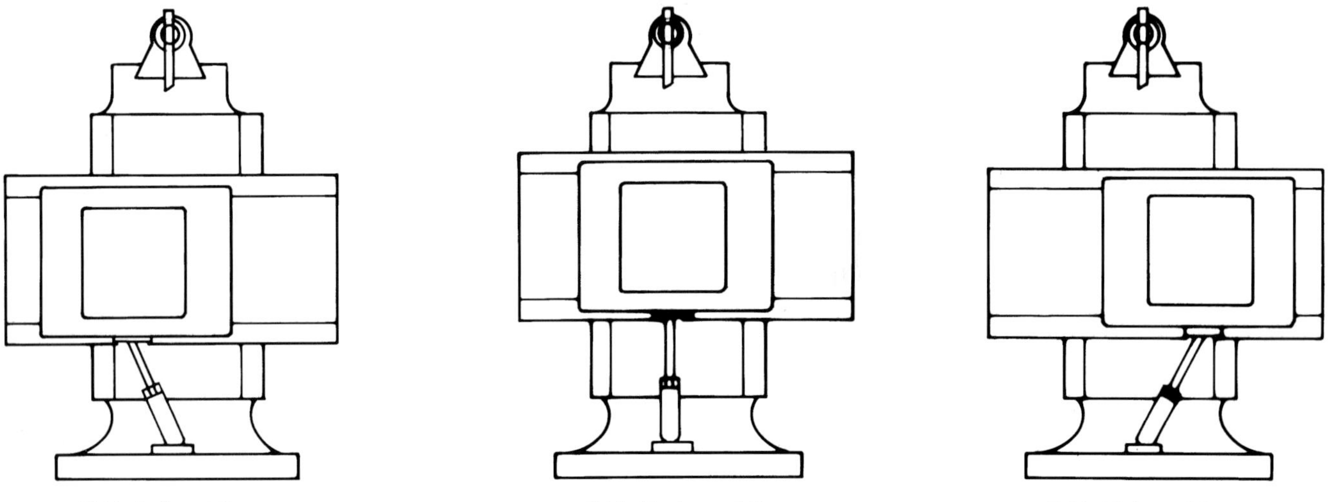

Figure 2 MOTION STUDY

slide and the base or floor. Shim the end plates to bring the work within the range of the cutting head. Adjust the horizontal feed until the rod is vertical. Then adjust the horizontal location of the piece on the table so the center of the desired groove is directly under the cutting tool (Figure 4). Operate the shaper in a normal fashion with the cutting head vertical adjustment providing the necessary depth of cut.

The work will be raised and lowered as the horizontal feed function provides a cut to the radius of the length of the adjustable rod (Figure 2).

For radii larger than the space between table bottom and the floor, the shaper could be raised off the floor or table, or a hole could be put in the floor.

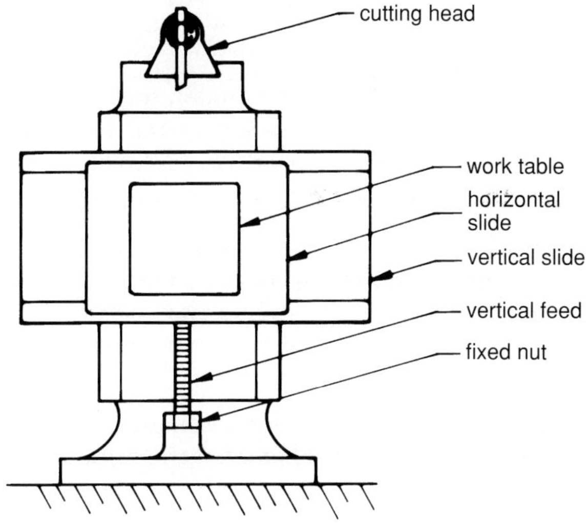

Figure 3 NORMAL SETUP

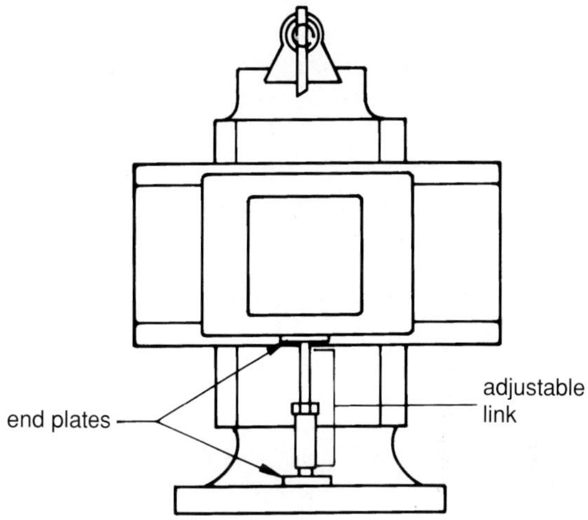

Figure 4 MODIFIED SETUP

SOME POINTERS ON

Scroll Sawing

by Guy Lautard

I've been meaning to do a Personal Use report on my Hegner scroll saw for some time. Bill pictures his Hegner in the opening installment of his serial, so perhaps all I might add about these saws is this: if you want to buy the best scroll saw available, check out Hegner. I think you will look a very long time to find anything to top them. Use will reveal differences in quality over almost any other saw that may not be apparent at first glance. And who else guarantees their saw for five years, even under eight hours per day use cutting 2" oak if that's what you want to do with it?

As it happens, I just spent some time on the phone with Ron King, technical rep for the exclusive representatives of Hegner scroll saws in the USA, and learned a few things that may be useful to others contemplating making Bill Smith's new clock, or any other project involving scroll sawing, and particularly scroll sawing in brass.

I have a Hegner scroll saw, and up to now have had a love/hate relationship of sorts with it. I love the quality, but I hate the speed at which it runs (mine is not a variable speed model, whereas several saws in the Hegner line can now be had with a variable speed motor), and I hate it when the blades break.

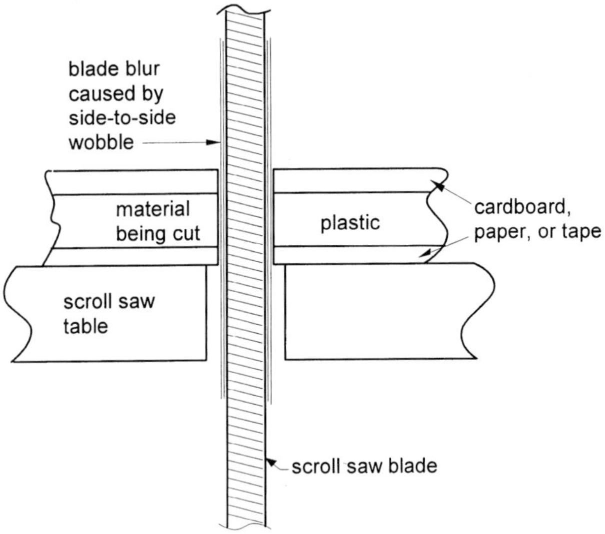

If excessive blade blur is present, resultant friction (heat) exists at the point of contact with the material being cut.

Put cardboard, tape, or paper on BOTH the top and bottom of plastics when you cut them.

Ron said the lowest speed at which my saw runs is quite normal for cutting brass, so this is just something I must come to terms with. If you've never been in a small aluminum boat, the first time you step into one it will seem awfully tippy. "Don't worry about it – it's normal," you are told. Pretty soon you realize it *is* normal, and you soon give it little thought.

Useful points Ron King made included:

Blade tension should be moderate – a turn or so of the blade tension screw after the slack is taken up. Three to five turns produces far too much tension.

The primary factor in blade breakage is friction. Friction between the sides of the cut and the sides of the blade is undesirable, and can be minimized by good sawing technique. I will come back to sawing technique in a minute.

Blade wobble – failure of the blade to travel along its own axis (blade looks like a blur when looking toward the teeth) – causes some of the friction against the side of the blade. A good saw (e.g., a Hegner) will have minimal wobble, whereas a lower quality saw may show considerable wobble, due to misalignments here and there throughout the machine.

Good sawing technique minimizes friction, hence blade breakage. The main thing to avoid is pushing sideways on the blade when you wish to saw on a curved path. The work should be turned, so the blade stays tangent to the center of the curve being cut. You don't push your car sideways to get around a curve. The idea is the same in sawing. Turn the work; don't push sideways.

Blade choice depends on the material to be sawn, the thickness of that material, and the complexity of the cut. Finer details (sharper curves) require finer blades, and will probably involve greater blade breakage. This is to be expected, so do not become discouraged by it.

Ron explained four cardinal points about blade selection: 1) For a given job – material, thickness, and complexity – there are probably three or four blades that will do the job satisfactorily; 2) the finer the blade, the thinner will be the maximum thickness of material the blade can deal with, the slower will be the feed rate, the more blades you will break, but the smoother will be the cut edge; 3) the best blade to use on a particular job is the coarsest blade that produces a satisfactory edge *for you* on that job; and 4) ultimately, blade choice is a compromise between smoothness of cut, blade durability and speed of cutting, and will be governed by personal preference and experience... which leads right back to point 1. There is no one "right answer" for a given combination of material, material thickness, and job complexity.

One trick Ron told me to try concerns sawing materials 1/8" or thinner. He suggests "sandwiching" the material from which the job is to be cut between two sheets of cardboard of the type often found on the back of a pad of paper. Stick the cardboard to the metal with double sided tape or contact cement (3M *Spraymount*, or less expensive types of similar product), because you most definitely don't want the "meat" in the sandwich to move during the sawing operation. Stick a photocopy (or an actual drawing of the job, if accuracy is critical) to the upper piece of cardboard with spray adhesive, and then go to work.

This sandwiching technique does two things: 1) It thickens the item being held, which makes it easier to turn it in the curves. (Remember, you aren't going to push it sideways any more, are you? This will help.) 2) It also helps to get rid of the heat produced by sawing out of the actual workpiece. I haven't yet come to terms with this last idea, because it seems to me the cardboard would act as an insulator, trapping heat in the part, but Ron says it works. Why argue with success? (I'm sure Ron's performance with a scroll saw would be seen as "success" compared to my efforts to date!)

So there it is: from a telephone conversation, some notes that hopefully will be helpful to some other *HSM* readers.

Postscript: Another scroll saw worth looking into is the Skil Model 3330. It's not in the same league as the Hegners, but it's a step or three up from some of the alligator-finish Taiwanese imports I've seen. I mention it because I feel it would be a pity to pay as much for a cruder saw simply because one did not know of the existence of the Skil saw. Check it out.

Facing to Accurate Length

by L. C. Melton

There are several methods for facing to length on a lathe. The best method is the one that accomplishes the task to desired accuracy with the least time and effort involved. However, not every home shop has the accessories to utilize some of the "best" methods. Here is a procedure that will get accuracy to 0.001" with a minimum of accessories.

1) Mount rough cut work in the lathe and face one end.
2) Measure from the faced end the desired length and punch or scribe a mark.
3) Face the uncut end to within 0.025" or so of the mark.
4) Measure the work length accurately and note the length to be removed. Example: Length = +0.023"
5) Set the compound at a 60° angle to the lathe center line. This gives a 2:1 ratio of the dial reading to movement of the compound, measured parallel to the lathe center line. This provides for better control and a more accurate reading of the dial.
6) Move the carriage so the facing tool is close to the end of the work, and lock the carriage to the bed.
7) Advance the facing tool toward the end of the work with the compound screw until the tool just touches the end. It is important that the tool is just touching the work and not into it a thousandth or so.

To determine this, a) note the reading on the dial; b) back the tool away, then advance it to about 0.005" short of the previous reading; c) move the tool with the cross-feed screw so it will contact the work in a different place; and d) with the compound, move the facing tool toward the work one increment at a time and observe after each movement until the tool contacts the work.

8) With the cross-feed screw, move your facing tool back off the work, ready for a facing cut.
9) Set the depth of cut with the compound screw to remove about 0.018" of the remaining 0.023", leaving about 0.005" for the finish cut. Remember the compound is set at a 60° angle from the lathe center line, so two increments on the dial advances the tool one increment into the face of the work.
10) Make the cut and measure the length to determine the amount remaining for the finish cut.
11) Set the depth of cut to remove the remainder and make the finish cut (as in 7 and 8 above).

If practical, set the facing tool so a line passing through its point and through the pivot point of the tool post is parallel with the lathe center line. Set this way, any flexing or movement of the toolholder or post due to the force of the tool feeding across the work will flex the tool out of the work and avoid cutting undersize.

By setting the compound 5-3/4° off a right angle to the lathe center line, ten increments on the dial equals one increment advance into the face of the work. Set this way, and using a proper facing tool on free machining screw stock that shears cleanly, you can obtain accuracy about as close as you can measure with a micrometer.

Turning Short Tapers on a Mill/drill

by Stephen M. Thomas

Photos by James J. Thomas

1

Here is a method for turning short tapers on a mill/drill or other vertical spindle machine. Since the setup is a bit more involved than for straight turning, I will digress for a moment to explain why anyone would bother doing this, and also to mention some accessories that can enhance production capabilities for both milling and turning operations.

After years of cutting torch, offhand grinder, and drill press machining methods, I finally thought my limited budget could handle a more sophisticated machine tool. With the funds available, the choice seemed to be either a very small capacity lathe, or a large mill/drill with some basic tooling. The mill/drills may lack the rigidity of a "real" mill, but the large models have a reasonable work size capacity and the potential for acceptable precision.

Since most of the work I intended to do involved milling, facing, and jig-type spotting, drilling, and boring, I opted for the mill/drill. Of course it very quickly became obvious that many projects were impossible to complete without some turning capabilities. Necessity and the challenge encouraged me to work out ways to accomplish many forms of lathe work on the mill/drill.

For my purposes, it was best to add tooling to the mill/drill whenever the budget allowed, rather than try to save for a lathe while getting by with less capability on the mill. I have not regretted this choice, but it clearly won't be for everyone. I hope this encourages the recognition that imaginative use of the tools at hand will often suffice to complete a task generally assumed to be the province of some entirely different machine.

Previous articles have described straight turning and facing operations, boring, and internal grinding. Within the size limitations of the machine, these operations can be nearly as quick and convenient to perform on the mill/drill as on a lathe. This is especially true if the M/D is equipped with a power up and down feed. I hope to develop a future article on building this device.

Well, as soon as straight turning became convenient, it was easy to crave taper turning abilities. Pulleys, guide wheels, expanding arbors, an R-8 shank saw arbor – all had short tapers that seemed to be within reach of machining on a mill/drill if I could come up with some sort of tool slide that would be mounted to the mill table and pivoted to any angle between vertical and horizontal. I dithered about the parameters for quite some time until, coincidentally, an information package from a company called Metal Lathe Accessories arrived in the mail. I had sent for information on a boring and facing head kit they advertised in *HSM*. As it turns out, Metal Lathe Accessories makes a range of reasonably priced kits for the type of tooling and inspection equipment that can be difficult to find or exorbitant to buy as a finished product.

The MLA-5 (Photo 1), described as a milling attachment for a lathe, was the perfect solution for my vertical mill taper turning attachment. With 3-1/8" of slide travel, it would even permit tapers up to M3. The mill/drill was used to completely machine all the components for this kit, including the revolving three-ball handle, and Acme lead screw and nut. I did modify the design to use 1/2-20 rather than the 10 tpi specified. This was to give a slower, hopefully smoother traverse while cutting tapers.

All Metal Lathe Accessories kits come with complete sets of clear elegant drawings that leave nothing to guesswork. A thorough manual of suggested procedures with photocopies of some of the tricky machining sequences is also included, though the photos are a bit murky with this style of reproduction. The castings are gems – massive and well proportioned with plenty of meat on the machinable surfaces, and a smooth finish requiring no filler on the parts to be left as cast. I have completed three different kits, as well

2

3

as parts for two others, and haven't seen any porosity, inclusions, or carbides in any so far. Having machined and scraped the 18" fitter's straightedge, I am also impressed that there is very little, if any, residual stress in the castings.

Clearly, taper turning on the mill/drill will require an attachment, to which is mounted a bit that can be traversed smoothly along a straight line for the length of the intended taper. It must also be possible to swivel the traverse mechanism so the line of travel can be arranged to any desired angle between the horizontal and the vertical. There may be reason or preference for designing one's own device, for adapting a ready-made tool such as a cross slide vise and angle block arrangement (Photo 2) or for using a lathe-style milling attachment such as the MLA-5.

Whatever the arrangement, the important points to observe are: 1) Tool must travel in a plane parallel to spindle axis. If the tool travels in a plane skewed to the center line, an external "taper" will have a concave surface, and an internal one a convex surface. 2) The tool point must be on the center line of the work. If the plane of tool traverse is parallel to spindle axis, but off-center, the taper generated will be sharper (have a smaller included angle) than the intended and measured inclination of the slide would suggest.

It is frequently convenient to obtain matching internal and external tapers by leaving the attachment at one angle setting for both operations, rather than flopping the tool slide over for the matching angle. Unless sine bar methods and equipment are available, it is likely to be more tedious to accurately match the opposite angle than it is to grind complementary tools and re-center after changing from internal to external. The following example on making an expanding arbor should help illustrate.

Many of the expanding arbors I have made were for a specific job, such as turning a set of guide wheels concentric and parallel with the metric bearing bore, and with the V-flange centered axially on the wheel (Photo 3). The need is for an accurate, quick to make (straight shank) tool that can be remachined after that job to fit the next oddball size that comes along. Check the concentricity and alignment of a collet used to make straight shank tooling. It is very disappointing to plow in and make a beautiful tool, but then find that, next time it's mounted, it won't run true. I find it worthwhile to keep a 3/4" and 7/8" collet that were ground in place in the spindle nose. These are kept solely for making and mounting straight shank tooling.

Turn the arbor blank to suit requirements for diameter, shoulders, length, etc. Nearly any mild steel will work for a temporary arbor. Perhaps some readers will explore the possibility of making hardened sets of frequently used fractional sizes.

Allen head cap screws make excellent expanding plugs. They are sufficiently hard to minimize galling in use, yet machine easily with carbide bits. Choose a screw size with a head appropriate to the arbor size. 3/8-24 is my favorite size for arbors between 5/8" and 1" in diameter. 1/2-20 will expand anything from 1" up to 3" or so in diameter. Below 5/8", my personal preference is to choose the screw with the largest head diameter that is just less than the arbor diameter. NF series threads provide greater leverage and "bite" over equivalent NC series threads.

4

5

It is preferable to bore the tap size hole for the expander screw (rather than drill it). If drilled right to tap size, keep it absolutely concentric. Next, counterbore to the major diameter of the thread, plus a few thousandths, for a depth equal to the thickness of the head of the chosen expander screw. Now thread in place with a hand tap. The counterbore will help align the tap.

Some creativity will be useful in setting the "taper attachment" and mounting the tool-holder. There will be no definitive setup in cases like this. If it works, it won't be far wrong.

The actual taper for the expander is not critical; 15° each side (30° included angle) works well and is adequate to angle the entire cap screw head without undercutting into the hex broach. Machine the internal taper as smooth and straight as possible. Actually, for work like an expanding arbor, an optimum curve could be calculated that should be superior to a straight taper, given the expected amount of expansion and the effective travel distance of the expander. But it's good practice to concentrate on accurate tool centering and straightness. For normal work such as machine mounting tapers, any curvature is a defect.

Next, remove the arbor and set up to turn the screw head. The slide angle is not disturbed; the bit and holder are changed to machine the screw at the same setting. To repeat, it is essential that this new bit be adjusted to center line.

The finished arbor can be slit by mounting in a vise. It is not difficult to arrange to cut two slots at 90° by positioning it in a vise and checking with a square. The vise bed keeps the cut parallel with the axis, and a few light trial passes will help to find center.

A spin index is relatively inexpensive from the import houses, and speeds up all kinds of semi-precision simple dividing work like cutting expander slots, bolt heads, etc.

Large diameter expanding arbors of shallow depth will need undercutting with a parting tool. This operation should be performed before slitting, since the bit can tend to hang up on slots, and fingers will start to bend or distort if any binding or heavy shock occurs. I haven't been able to find any recommendation or formula to determine amount of undercut in proportion to diameter or length. From measuring several arbors that seem to work well for me with easy tightening and minimal distortion, it seems possible to suggest that: 1) the undercut should be below the portion with the tapered bore; and 2) the diameter of the undercut portion should be approximately twice the nominal diameter of the threaded portion of the expander screw.

The photos, by the way, show the taper turning of guide wheels for a movable platen in a wood-working press. Each wheel has a ball bearing pressed in each side, separated by a spacer inside. At the left in Photo 4 is a completed wheel, and Photo 5 shows another one in progress.

1 *A well-worn example of the typical 16" leather Gladstone bag as carried by workmen circa 1930.*

2 *Enlarged view of the completed shaper. The ram is fully forward.*

Reviving a Lunch Break Shaper

by Terry Sexton
Photos by Author

The title refers to the process of breathing life back into an interesting artifact. Recently while building my own version of Dave Gingery's shaper I was discussing the job with my friend Ian. He mentioned that some years back, he had been given a fairly strange little metal shaper. It had lain around his workshop and generally got in the way for some time before eventually being stored underneath a bench out of sight. Having a long-standing interest in these things, I prevailed upon Ian to describe it to me.

I was intrigued enough by his description to suggest that we might resurrect the wee beastie with a view to putting it up for discussion during "models on the table" at a monthly meeting of the club. These one-off machines create plenty of interest when they surface among enthusiasts. Unfortunately, most get thrown out once their owner/builder has passed on.

As Ian had his hands full at the time, I offered to do the cleanup. It turned out to be a sad-looking effort, and vital parts were missing, but it was clear enough that this was one very rare animal.

The story goes that the unit was built a long time ago during lunch breaks at the railway workshops. It was put together from scrap materials, smuggled out the gate one piece at a time in a leather carry bag and assembled at home. The son of the builder inherited the shaper, and kept it for many years. He was approaching retirement when he gave it to Ian. The shaper was most likely built during the early 1930s.

I began by making a detailed inspection of the unit and taking notes prior to dismantling it. The unit had obviously been cannibalized for screws, nuts and bolts, etc., over the years. It had originally been painted with that black yukky undercarriage paint that railways use, but during years of idle storage, had collected several misty multicolored layers of overspray, dust and runs of spilled paint. A moderate amount of rust was evident on the working surfaces, but surprisingly all mechanisms actually functioned. The table was missing, no doubt cannibalized into something else or souvenired for use as an angle plate.

Making up a new table was straightforward – simply a matter of working out the maximum stroke and side travel of the unit then getting hold of a piece of heavy angle iron and cutting a gusset to fit. I managed to weld up something from the scrapbox which was subsequently machined up and fitted to the unit. Sticking to the spirit of the thing I settled on a pattern of tapped holes in the table rather than milled slots.

The front plate of the body has certainly seen the magnetic table of a vertical spindle surface grinder, because it still carries the radial score marks left by a segment-type wheel. The horizontal slide carrying the table was most likely made from bright mild steel bar sections cleaned on a shaper before being ground on a horizontal spindle surface grinder. The vertical slide and clapper box are made up the same way. However, the rest of the unit is made from offcuts of hot rolled plate and angle iron, etc. The ram is machined to a 2" square section and slides in babbitt-coated guides. (Railways were once major users of poured babbitt bearings. A ladle full of trimmings and cold splashes could be melted and poured during the lunch break without attracting attention.)

Close inspection inside the body framework revealed that a surprising amount of arc welding had been employed. There are ten fabricated sub assemblies which have been welded up, then bolted together. The critical ones have been squared up by machining on

SPECIFICATIONS:	
Maximum stroke	4.5" (power) 6" (hand lever)
Table size	4.25 x 7.25" (an accessory table can increase this)
Table vertical travel	3"
Table cross travel	7"
Cap above table	4" (an underslung table could be fabricated to give 10")
Toolslide travel	3" adjustable to 4"; finest feed via ratchet to the table is 0.004" per stroke

3 ***Rear view of the machine with handle. The ram is fully rearward.***

a shaper. The shaping was probably done covertly while watching out for the boss, because the surface finish is fairly rough.

Access to a drill press must have been easy enough because there are holes everywhere (something like 300 of them) to allow for final assembly using whatever nuts, bolts and screws were lying around. Short runs of arc weld also assembled each of the carrier bearing bodies for the countershaft and drive shaft. The steel ratchet pawl arm on the table feed apparatus has been neatly brazed onto the bronze bush which carries it, and the main drive cast iron pulley has an extension brazed on to lengthen the hub.

The table leadscrew is left-hand threaded, which shows that some thought went into it beforehand. The largest single item is the baseplate, and that just fits neatly inside the traditional 16" leather Gladstone bag that most workmen carried in those days (Photo 1). The other smaller lightweight sub assemblies would have been smuggled out the same way.

Initially the unit must have been hand-operated via the lever which fitted over the two posts visible on the top plate and ram.

In the fullness of time it would have been run under power. And in the fashion of the day, the unit would have been powered by means of a pulley held in the chuck of the home

4 ***View inside the body showing the pinion, bull wheel, trunnion and rocking arm. Note the unusual arrangement where the lower pivot floats in the slot and the top end of the rocking arm is a fixed pivot.***

workshop lathe. This procedure required a crossed round belt in order to get the direction of rotation right (slow cutting stroke, quick return stroke) with the lathe cone pulleys providing a range of speeds.

Among my treasures, I just happen to have a 1920s vintage Drummond round bed lathe mothballed for future restoration and the idea of dragging it out, firing it up and lashing up the shaper for a run was irresistible (Photo 5). What a sight to behold! The old lathe with its chipped but still glossy black enamel, and the shaper in its fresh coat of light blue enamel rocking back and forth at 70 strokes per minute. I fancy I heard the ghosts of men from the railway workshops cheering.

Before I started working on this shaper, I knew nothing at all about the young man who built it. But having examined the machine, I've drawn a few conclusions of my own. I now feel I know him slightly, at least the bit of him that went into his machine. It's a fair bet that he was a keen home shop type, intent on cutting metal in his spare time. Judging by the fairly rough finish on the shaper he was probably an apprentice, one who was popular enough with the staff to receive a certain amount of help and encouragement. He certainly managed to indulge in skill- and character-building exercises. There is no doubt in my mind that he was in a hurry to get projects going in the home shop.

We must remember that this machine has outlasted the builder and that is a mighty rare thing these days. And the question arises – did we build anything in our youth that might interest anyone when we've been dead and gone a few years? Will they hold aloft some artifact and say of us, "Old whatsisname built that when he was a lad... carved it out of the solid with his bare hands"?

One-off machines such as this shaper cannot be found preserved in science museums. Apart from their practical use in metal cutting, such artifacts serve to remind us of the way real home shop people managed to get by in days past. The unit has proved to be a fine conversation piece already, being dubbed "the cantering bedstead" by one visitor. There has been no shortage of offers to give it a good home with plenty of TLC.

I would urge all home shop people, regardless of their main interest, to keep an eye open for these "one-off" rare items. Make an effort to grab and restore them when they show up. Do a good job of it, and who knows, you may hear the ghosts cheer for you someday.

5 ***Bolted to a temporary stand and lashed up to a circa 1920s 4" Drummond round bed lathe for a run. In the style of the times, neither machine was fitted with graduated handwheels.***

Dividing on the Lathe

by James Schmidt

In the May/June 1985 edition of *The Home Shop Machinist*, John McDermott uses W.C. Grosjean's method of dividing a circle on the lathe, but laments that the only way to set up the divisions is by trial and error. Well, there is a better way.

In Mr. Grosjean's method, a paper strip is wrapped around the lathe chuck and cut off so its length forms the exact length of the chuck circumference. This strip is then divided into whatever spacing you need. After reattaching the strip to the chuck, a pointer is mounted on the headstock and used to index from one graduation to the next. The chuck is usually much larger than the work being divided, so the errors on the part are reduced to very small amounts – *if* the divisions are accurate.

To do really accurate dividing by this method, you need to do a little drafting work. You need a drafting table or board, a T-square, a triangle, and a good accurate scale. The board can be made in a pinch (I use a chopping board with one smooth face and a straight left edge), but the other items should be purchased for accuracy's sake. Check with your local vocational school bookstore. They often have good equipment for their students that you can buy at rock-bottom prices. You will also need a sheet of paper wide enough to fit the strip of paper on. Even if the strip fits only diagonally, it will work fine.

Rough-measure the length of your paper strip. If you have a 4" diameter chuck, your strip will be roughly 12.57" long. Look for the next even length shorter than your paper strip that can be easily divided into the number of segments you want.

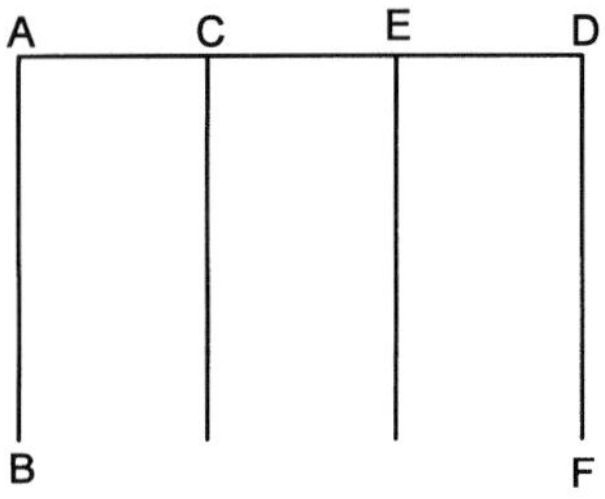

Figure 1

Let's say you want to graduate the work with three divisions. Obviously, the next smaller even length divisible by three is 12". Mount a suitable size sheet of paper on your drawing board and draw a line (A-D, Figure 1) 12" long across it near the top. This is the line you will actually be dividing.

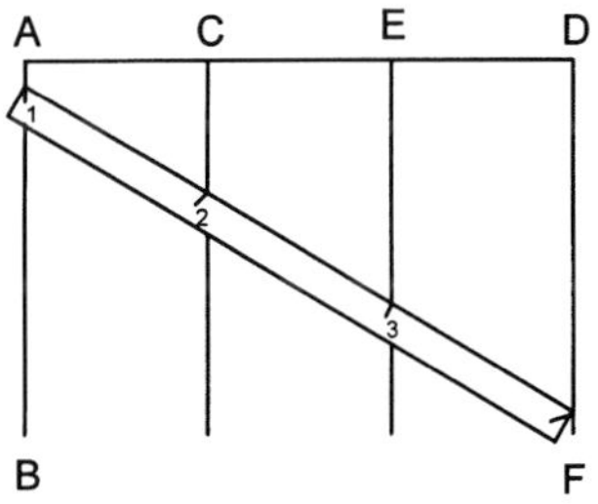

Figure 2

Now, with your scale, carefully mark off your divisions – in this case, at 4" and 8". Now drop vertical lines from the ends and these points straight down the sheet. Your 12" line is now divided into three equal segments. The problem now is to divide the odd length strip the same way. You can do this without any further measurements.

Take from the chuck that odd length strip you have been saving so patiently, and put it on your drawing board with the left end anywhere on line A-B. Holding the left end on A-B, pivot your strip until the right end lines up with line D-F. At this point you should have something like Figure 2. If your strip slopes up instead of down, don't worry about it, as long as the ends match up. Mark on your strip the points on which your dividing lines meet the strip. These points are accurate divisions of your lathe chuck circumference. You can lengthen the lines and number them for your convenience, if you wish. Tape the strip on the lathe chuck, mount a pointer that will line up with the graduation on the strip, and you have a very accurate means of dividing anything mounted in the chuck.

While a simple example was used for clarity, the divisions possible are limited only by the number of narrow pencil lines you can fit on the circumference of your chuck. I have already done 50 divisions easily on a 4" chuck. The best part of this process is that it can be used easily to put divisions on other more versatile equipment, such as the rotary table described in *HSM* in May/June and July/August 1984.

SOME FINAL SUGGESTIONS

Count your graduation lines beginning with the left edge as 1. Thus, you will have one more line than the final number of divisions you will get. This is because the final line (D-F, Figure 1) "wraps around" to line A-B when in use and isn't counted.

When you are moving from one graduation to another on the lathe, put the lathe in low speed back gear, and pull it over by hand with the drive belt or motor pulley. The very high gearing ratio this provides minimizes shifting while scribing graduations, and gives you a very accurate way of aligning the graduations and the pointer.

Operations such as reamer grinding or milling splines might require a more positive locking method than the back gear method noted above.

P.S. Just to show you there's nothing new under the sun, while writing this article I saw the same concept used to divide a board into equal parts. Some carpenter beat me to the punch by a few centuries!

Heat-treatment Processes for Engine Components

by George W. Genevro

The heat-treatment of metals, either prior to or after fabrication, forming, or machining, is a very useful and often necessary process in the manufacture of engine parts. Requirements such as light weight, resistance to deformation, long fatigue life, wear resistance, and structural integrity make the development of physical properties appropriate for the intended life and use of the part mandatory. In many cases, two or more of the common heat-treatment processes are used in sequence to achieve a desired end. A basic understanding of the commonly used processes will be valuable to those who design and build engines and make replica parts when restoring commercially built engines.

ANNEALING

Basically, annealing is the process of heat-treating a metal to bring it to its softest possible condition. Both ferrous and non-ferrous metals may be annealed but the processes vary considerably.

Steels with a carbon content above .25% may be fully annealed by being slowly heated to just beyond the critical temperature (usually 1330° to 1380°F), held at that temperature for a period of time determined by its cross section and mass, and then cooled slowly. The steel may then be formed or machined fairly easily since its tensile strength is at its lowest level and it is free of internal stresses.

Aluminum alloys may also be annealed, and the process will differ depending on whether the material is a work-hardening type of alloy (such as 3003) or a precipitation-hardening alloy (such as 2024). Aluminum castings may also be annealed to achieve dimensional stability and to relieve stresses that might have developed while the casting cooled in the mold. Depending on the annealing process used (solution heat-treatment and controlled cooling), precipitation-hardening alloys such as 2024 will remain in the annealed condition or will age-harden either at room temperature or more rapidly at elevated temperatures. For example, if the metal is cooled slowly (50°F per hour down to 500°F) from the solution temperature (usually 910 to 930°F for 2024), the material will be in 0 temper and will stay soft. If the material is quenched in cold water from the solution temperature, the precipitation-hardening reaction is temporarily stopped and may be further retarded by cooling to well below room temperature. So-called "ice box" rivets used in aircraft structures are an example. At room temperature, the aluminum will usually reach T4 temper in several days.

Annealing of copper after it has been hardened by cold-working requires heating in the 750° to 1200°F range, depending on the cross-sectional shape and thickness and how much it had been cold-worked. Heating beyond 1200°F is counter-productive since it will cause excessive grain growth and brittleness. Yellow brass may be annealed partially or wholly in the same manner using the minimum amount of heat necessary to achieve the desired end. Both copper and brass may be allowed to cool slowly or may be quenched in water.

HARDENING

Steels with more than .30% carbon are usually hardened by heating to a selected temperature (depending on alloy and carbon content) above the critical range and are then quenched in brine, water, oil, or a heated salt bath. In some cases, a blast of cold air is used as the quenching medium. In the fully hard condition, the steel is at its highest tensile strength and is brittle, with the degree of hardness and brittleness determined by the carbon content and alloying elements.

Aluminum alloys, depending on the alloying elements, may by strain-hardened (also referred to as work hardened) or precipitation-hardened. The strain hardening may be accomplished by rolling, hammering, embossing, stretching, or by any other means used to shape the material. The amount of hardening that occurs is dependent on the amount of cold working done. Precipitation hardening takes place only if solution heat-treatment has been properly done in terms of temperature control and quenching procedures. Hardening may occur at room temperature (natural aging) or at elevated temperatures (artificial aging).

NORMALIZING

Ferrous metals in cast, fabricated, or wrought form may be normalized. Basically, the process consists of heating the metal to above the critical temperature and cooling it in still air. Internal stresses that might have been caused by casting, forming, welding, or other procedures are relieved and the tensile strength and hardness are about 10 to 20% greater than for fully annealed material. For example, AISI-E-4130 steel sheet is available in N (normalized) or A (annealed) condition, with the A condition material having lower tensile and yield strengths.

TEMPERING

Tempering is the process of selectively or wholly reducing the hardness of a steel object that has been fully hardened. This is accomplished by accurately heating part or all of the object to a predetermined temperature below the critical range and then cooling by quenching in oil or water or allowing the object to cool slowly. Selection of higher tempering temperatures will result in greater softening of the part being tempered. This is usually the last in a series of heat-treating processes performed on a tool or an engine part.

CARBURIZING

Steel with a carbon content below .25% may be carburized to produce a hard, wear-resistant shell up to .060" thick while retaining a ductile, shock resistant core. The process consists of heating the part to be carburized in a closed container of carbonaceous material such as petroleum coke or some of the commercially available carburizing compounds. Generally, the depth of the hardened case is dependent on the length of time the material is kept at the carburizing temperature, but continuing the process beyond a certain point does not materially increase case depth. Hardening is accomplished by quenching the part in the appropriate medium from above the critical temperature and tempering if desired. Selective carburizing may be done by copper plating the part of an object that must be kept soft while leaving the areas to be carburized bare since the copper prevents the passage of carbon into the steel.

NITRIDING

Only specially alloyed steels that contain small amounts of aluminum, chromium, molybdenum, or vanadium may be nitrided. The steel does not have to be heated beyond the critical temperature since the process basically involves the diffusion of nitrogen from anhydrous ammonia into the surface of the part at about 950°F.

While the case produced by this process is thin (usually about .010" to .020"), it is highly resistant to wear, galling, and seizing. It is widely used to increase the wear resistance of engine cylinders, crankshafts, and similar parts in full-size aircraft engines and could have applications in model engines. Since the parts are not quenched after nitriding, there is little chance of distortion.

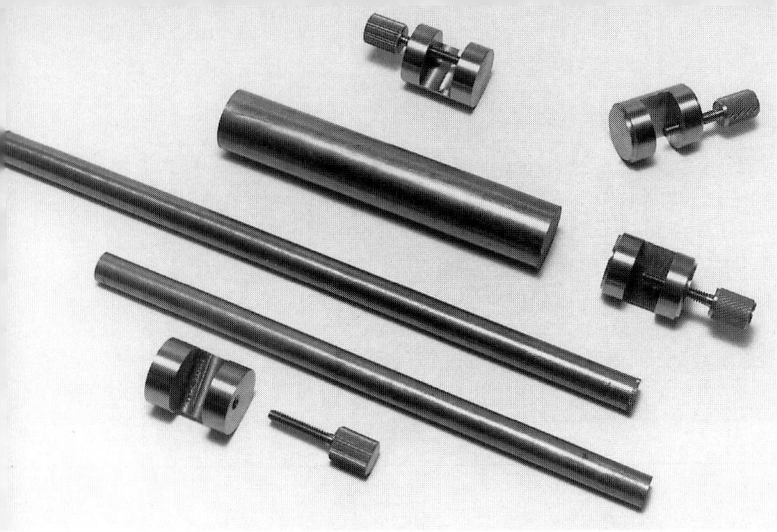

1 ***The completed mini screw clamps together with the bar stock used to make them.***

Starting Turning

MINI SCREW CLAMPS

by Audrey Mason

Photos and drawings by Harold Mason

Why is it that you always need a third hand when you're in the middle of an intricate job? And there is never anyone about to lend a hand, is there? These thoughts were in my mind as I wondered what to make that would demonstrate simple screw cutting with the use of taps and dies in the lathe. Once you start making components that fit together, you are sure to find that you need to be able to screw cut, so Harold and I put our heads together and came up with the mini screw clamps shown in Photo 1.

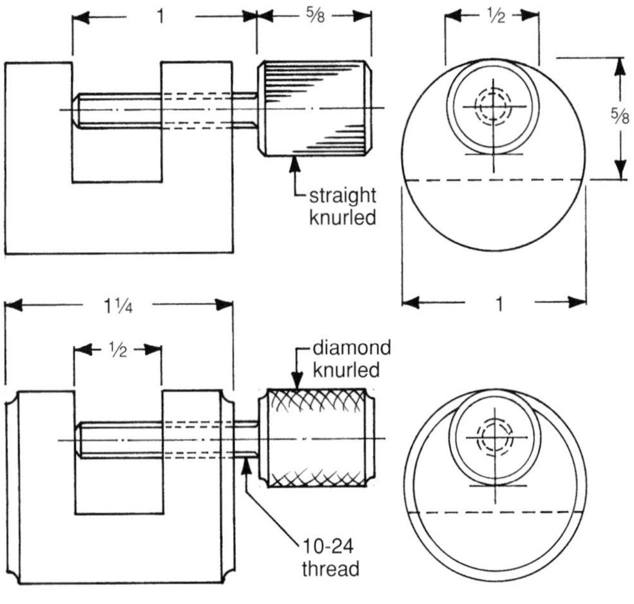

GENERAL VIEWS OF MINI SCREW CLAMPS
Figure 2

I cut a 2-3/4" long piece of 1" diameter CRS (free cutting) bar to allow for facing the ends and parting off (Figure 2). This will make a pair of clamp bodies as we are going to work one on either end of the bar.

I made two pairs of clamps – the first pair with straight sides, which were tapped in the lathe and have screws with straight knurled knobs, and the second pair with a flute top and bottom, which were tapped in the drill press and have diamond knurled knobs. I will explain in detail how I made the first pair, and then go into making the second pair giving you the differences to look out for.

Put the bar into a three-jaw, self-centering chuck, and have a turning and facing tool ready in the tool post. My spindle speed table shows that I need a speed of 410 rpm for 1" CRS bar, but our lathe doesn't have that particular speed, so I ran it at its nearest speed, which is 425 rpm. You will see from some of the photographs that we have indulged in a new quick-change tool post. This means that once the tools have been set to center height in their holders, the holders can be interchanged without the need of tool adjustment. Sheer bliss!

Make a mark with a hacksaw 1-1/4" from the face of the bar (the length of the clamp), and a second mark 1/4" towards the chuck. Check that the top slide does not overhang the carriage. Touch the tool to the face of the work, lock the carriage, zero the top slide handwheel thimble, and wind the tool out from the work with the carriage handwheel. Put a .005" cut on the top slide thimble.
I used coolant, and fed in the tool with the cross slide handwheel to face the bar.

3 ***The position of jaw 1 of the chuck marked on the bar. You can see the cleaned-up portion of the bar.***

Continue taking small cuts until you are satisfied with the finish on your bar.

Release the carriage, touch the tool point to the circumference of the bar, and zero the cross slide handwheel thimble. With the carriage handwheel, wind the tool clear of the bar and put on a .002" cut, set the auto feed to .0025" per rev, let the tool run along the bar until you reach fractionally beyond the 1-1/4" mark, and stop the lathe. This should have cleaned up the outside of the bar. DO NOT TOUCH the cross slide handwheel or thimble; wind the tool away from the work with the carriage handwheel. We are turning the work end for end, and we want to take off exactly the same amount of metal from the other end of the cylinder.

Mark on the bar the position of jaw 1 of the chuck, as shown in Photo 3, so when the bar is turned end for end and you position it again with the mark against jaw 1, this gets the work as concentric as possible.

Reverse the bar in the chuck, tighten the chuck, and remove the chuck key; skim along the circumference with the tool still at the same setting on the cross slide, and let the tool run until it reaches the cleaned-up surface. Switch off the autofeed, and face the end of the bar as before.

Remove the work from the chuck and blue both faces of the bar. Rest the work in a small V-block with a clamp (Photo 4, No. 3). Set the point of a scribing gage to the bottom of the bar and, with a 6" rule fastened to a large V-block, set the 1" mark of the rule to the height of the scriber point. Move the point of the scriber up the rule to very slightly below 1-1/2", and set the pointer there. Take the scriber to the work and pass it across the face; it should pass through the "pip" made when you faced the work. When it does, scribe a line across the face and extend the line down the sides of the bar for a short way. Repeat at the other end of the bar. Turn the small V-block round to do this. Do not remove the bar from the block to turn it to get to the other end. We want both ends to be identical (Photo 5).

Remove the work from the V-block and measure in from the edge along the scribed line 1/4". Draw a cross line here, and center pop the intersection. Repeat at the other end of the bar; be sure to measure from the same end of the scribed line as before, so a hole drilled from one end would run right through to the other, concentrically. Make a center pop mark here. Check the mark with a magnifying glass, and

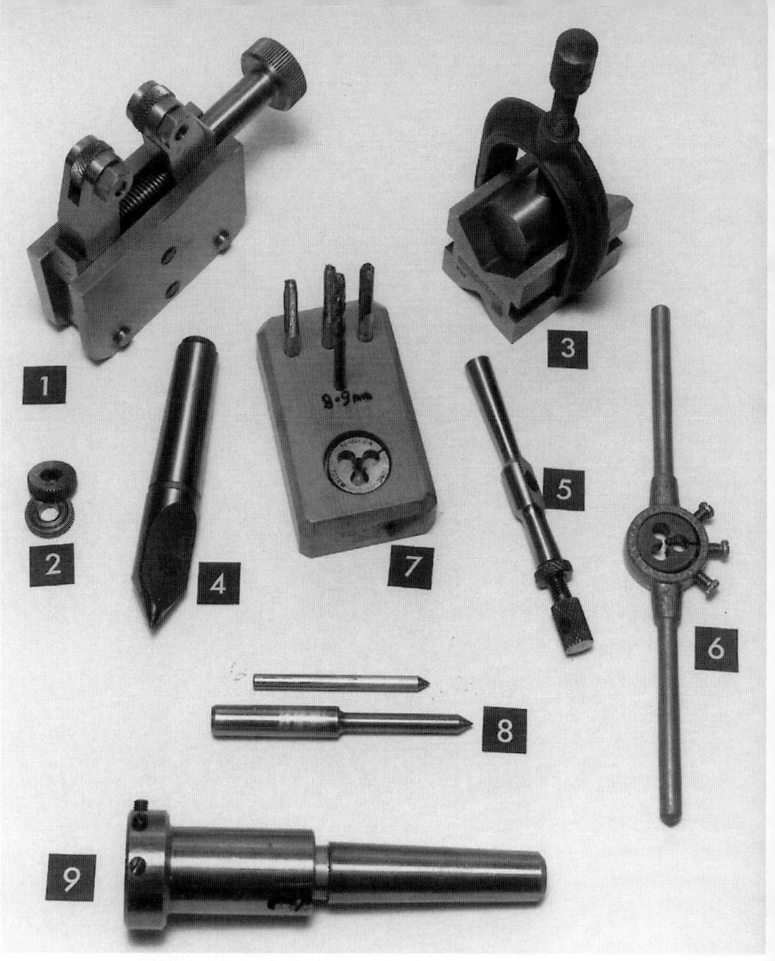

4 ***Some of the tools used: 1) a caliper knurling tool; 2) knurls; 3) a small V-block with a clamp; 4) a half center; 5) a tap wrench; 6) a die stock; 7) a homemade block to store taps and matching die; 8) centering points; and 9) a traveling die holder.***

5 ***The bar set up in the small V-block and the center line being scribed across the face.***

6 ***Setting up the pop mark in the four-jaw chuck using the DTI. See the half-center supporting the centering point.***

if necessary, gently move the mark over until it is positioned accurately on the intersection.

Put a four-jaw independent chuck on the lathe. Use shim to protect the work, if you wish, and with a centering point and a DTI (Photo 6), set up the bar using the pop mark as its center, as we did when making the circular place marker. As you will see from Photo 6, I used a half center in the tailstock, which supported the centering point in the pop mark. Take readings on the DTI, and adjust the chuck jaws until the bar is centered on the pop mark.

When the work is running true, put a center drill in the tailstock, and make a pilot hole as we have done before for drilling. I used a No. 25 drill which is the size recommended for tapping 10-24 threads. Using cutting oil and a spindle speed of 2,445 rpm, I drilled out a hole 5/8" deep, measuring the depth off on the tailstock barrel. This depth will take the hole through the top section of the screw clamp. Then with a single flute countersink bit, I countersunk the hole to help the taper tap enter the hole and stop it from raising a burr.

Put the taper tap in the drill chuck, so if you hit the bottom of the drilled hole, the tap will move in the chuck rather than snap off. Use cutting oil, and set the spindle speed on fastest back gear – 80 rpm. Leave the tailstock loose; do not lock it in position. Set the lathe running and, by hand, push the tailstock along the bed until the tap engages in the drilled hole. The tap will be pulled in by the work. If the tap slips in the chuck, stop the lathe, reverse the motor, and gently ease the tailstock away from the work. Re-engage as before and feed in the tap until it reaches the end of the bore, when the cutting oil will be forced up the side of the tap. Reverse the motor and withdraw the tap (Photo 7).

7 ***The taper tap held in a chuck in the tailstock.***

8 ***Using a tap wrench with the work held in the soft jaws of a vise.***

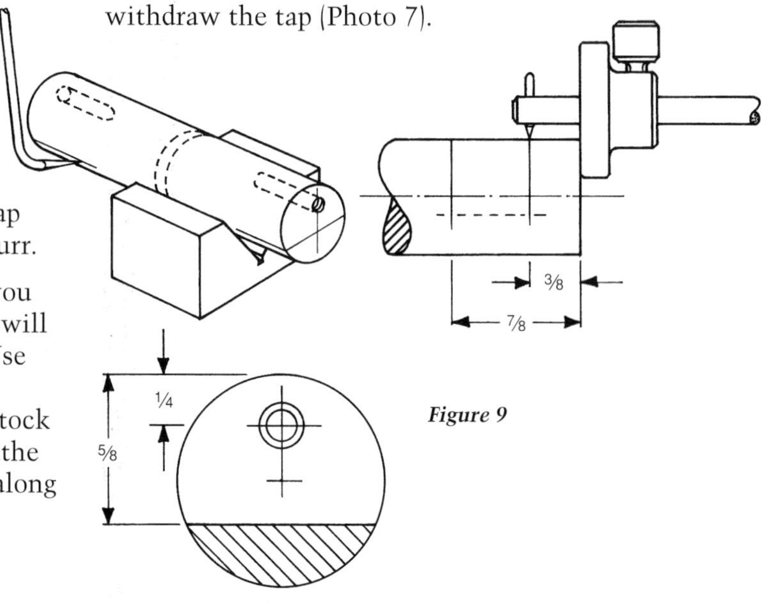

Figure 9

10 ***An engineer's square being used to set the work in the milling vise with a piece of metal on the vise to sight along to get the work parallel with the vise. Note the soft-faced hammer to persuade the work into the correct position.***

11 ***The second mouth being milled. The coolant pipe is held in place, to give maximum coverage, by a clip secured to the machine by a magnet.***

Reverse the bar in the four-jaw chuck and set up, centering on the pop mark, as before. Drill and tap a hole in this end of the bar in the same way.

Remove the bar from the lathe and hold it upright in the soft jaws of a vise (Photo 8). With the second tap in your fingers, dip the end of the tap in cutting oil; feel for the screw thread and fit a tap wrench (Photo 4, No. 5). Feed the tap into the threaded hole by giving a half-turn, then reverse the cut and you'll feel the swarf break. Cut again, repeating until you feel the tap reach the bottom of the hole. Unwind the tap.

Finish out the hole with a plug or bottoming tap using the same technique. Repeat the operation at the opposite end of the bar. Remove from the vise and blue all over.

With the bar in the small V-block use a scribing gage, and make marks 3/8" and 7/8" from each end of the bar (Figure 9) to indicate the width of the clamp mouth. Then measure 5/8" from the edge above the tapped hole on the face of the work, draw a line across the face, and continue the line along the circumference to give the depth of the mouth.

I set the bar in a milling vise with the tapped holes at the top, as shown in Photo 10, and used an engineer's square to set the line on the faces through the hole vertically. I then sighted along a flat piece of metal sitting on the milling vise to set the work horizontally parallel to the milling table.

Using lots of coolant, a 5/16" end mill and a speed of 1,100 rpm, I set the table stops to the width of the mouth, and taking .030" cuts, worked down to the depth line. I cleaned up the faces of the mouth, turned the bar end for end – setting up as before – and cut the second mouth (Photo 11).

Watch your fingers as you take the work out of the milling vise, and deburr all the edges! Scribe a line round the circumference where you need to part off the two clamps. Either saw the two clamps apart, or return the bar to a three-jaw SC chuck on the lathe and part off. Clean up the surface with emery cloth; don't forget to protect the lathe bed with a board or some kitchen tissue. Run a file around any sharp corners, and you've made your first pair of clamp bodies.

I made the second pair of bodies next. I put the 2-3/4 × 1" diameter bar in a three-jaw chuck, and worked exactly as for the first pair until I had faced the second end of the bar. At this point, I did not remove the bar from the chuck as I wanted to do more work on it. I put a parting-off tool in the tool post, found the mid-point of the bar, and positioned the tool to cut with half its width on either side of the mark (Figure 12).

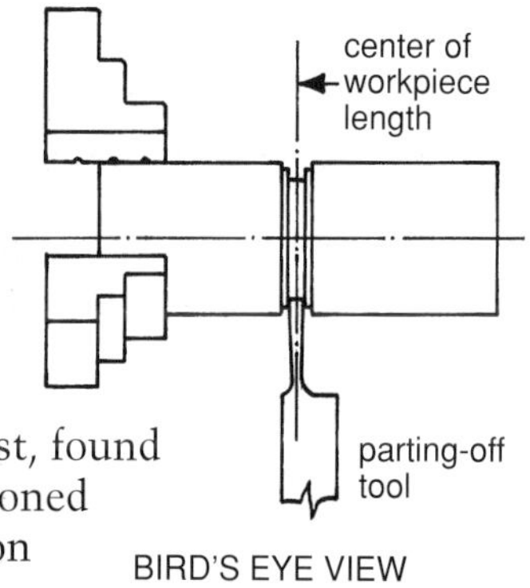

Figure 12

Set the spindle speed to 200 rpm and, using a firm, steady feed, flood coolant into the groove as you cut. Lock the carriage in position with

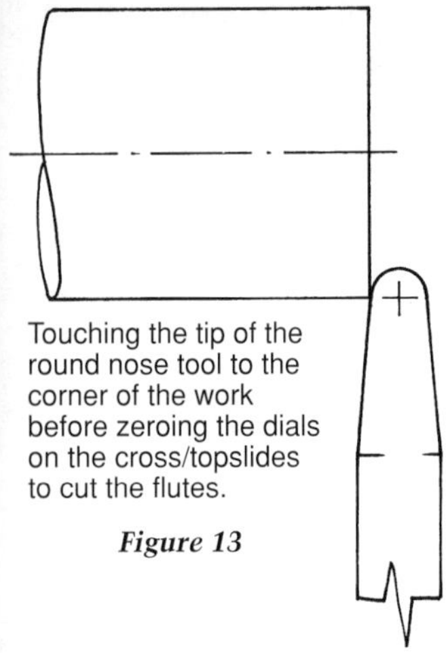

Touching the tip of the round nose tool to the corner of the work before zeroing the dials on the cross/topslides to cut the flutes.

Figure 13

the tool point on the mark. Zero the cross slide handwheel thimble with the tool point on the circumference of the bar. Feed the tool in for .150", which is one-and-one-half revolutions of my thimble, then withdraw the tool.

Replace the parting-off tool with a round nose tool. I used a 1/16" (0.0625") radius round nose. Move the tool to the face of the work, and zero the top slide handwheel thimble with the tool touching the face of the work. Figure 13 shows you how to set the tool to the face of the work for zeroing. Wind out the tool, move its point to touch the surface of the circumference of the bar, and zero the cross slide handwheel thimble.

Set the spindle speed at 425 rpm, and put a .030" cut on the cross slide thimble. Position the tool at the end of the bar, bring the top slide thimble to zero, then feed the tool in for .0625". Wind the tool out with the top slide handwheel. Then turn the cross slide thimble to .0625" and feed the tool in again with the top slide handwheel for .0625". This gives you a quarter-circle (quadrant) cut, or as we say in the UK, a flute.

Move the tool to the parting-off groove, and zero the top slide handwheel thimble on the headstock edge of the groove. The cross slide handwheel thimble should be turned back to zero, and should not need to be re-zeroed. Put a .030" cut on the cross slide handwheel thimble, and proceed as before to cut the quadrant.

Turn both thimbles back to zero; they should not need re-setting. Place the tool against the other lip of the parting-off groove, and cut the quadrant by turning the top slide handwheel backwards to cut towards the free end of the bar (Photo 14). Turn the bar end for end, and cut a quadrant on the outer edge as we did before.

Remove the bar from the lathe, blue the ends, and mark the position of the tapped hole as you did for the first pair of bodies. But this time I took the work to the drill press after making the pop marks. I held the work in a drill vise and used a centering point (Photo 4, No. 8) to center the pop mark, locking the drill table when the work was centered. I set the drill speed to 1,270 rpm, and put in a 3.9mm drill, touched the drill to the metal, and set the depth gage to 1/2". Using cutting oil, I gently drilled out the hole to a depth of 1/2". I took the V-belt off the spindle of the drill press so I could countersink the hole with a countersink bit, by hand.

I put the taper tap in the drill chuck, and used a tommy bar in the chuck key socket of our Jacobs chuck to feed in the tap to the bottom

14 ***Cutting the flutes where the parting-off tool cut a groove. Here you can see our quick-change tool post.***

15 ***A tommy bar being used in a Jacobs chuck to feed in the tap.***

of the hole (Photo 15). Release the tap from the drill press, and screw out the tap by hand. Turn the bar end for end and repeat the operation. This cuts out the four-jaw setting-up procedure. Complete the tapping and milling as for the previous clamp bodies.

Now to make the screws. My thread tables showed that the outside diameter needed for a 10-24 UNC thread is 0.190" (Figure 2). I turned the screws one on each end of 1/2" free cutting CRS bar – using one piece slightly more than 14" long and the other 13" long, as I wanted to make the two pairs together. This bar must fit through the hole in the spindle nose of your lathe. Using a three-jaw chuck with about 1-1/2" of bar projecting, a T/F tool, and a spindle speed of 615 rpm, face the bar.

Make a mark 1" from the face of the bar; zero the cross slide handwheel thimble with the T/F tool tip on the circumference of the bar. With the carriage handwheel, wind the tool clear of the work and put on a .010" cut with the cross slide thimble. Set the autofeed to .003" per rev, and run the tool to the 1" mark. I took 13 cuts of .010" and then measured the diameter of the peg. I had .080" still to remove to get down to 0.190". I needed to take off .040" from each side. From here on in I used cutting oil and not coolant. It took three more cuts of .010" and one of .011" for me to reach the correct diameter. Take a cut, and measure and adjust your cuts to your measurements. Put a slight chamfer on the end, to make it easier for the bar to enter the die.

Turn the bar end for end and repeat. Photo 16 shows one bar with the thread diameter turned on either end and one bar in the three-jaw chuck being turned. I turned all the thread diameters first.

There are various ways of using a die with the lathe, and I am going to show you two. Wind the tool post out of the way. Now you can put a traveling die holder (Photo 4, No. 9) in the tailstock with the die (lettering on the die towards you) held in it. Move the tailstock until the die holder is within 1/4" of the work, and lock the tailstock in position.

Using back gear, reduce the spindle speed of the lathe to 55 rpm. Use plenty of cutting oil, and feed the die onto the work with the tailstock handwheel; let the work pull the die along the spigot of the die holder and feed in the tailstock as the peg on the die holder reaches the end of its travel. Feed in until the die reaches the shoulder (Photo 17). Stop the lathe, reverse the motor, and allow the die to

16 ***The diameter of the bar being turned down to form the peg on which the die will cut the thread. A bar with a peg cut on each end rests on the carriage.***

17 ***The traveling die holder being withdrawn.***

slowly wind out. The die will need cleaning before it can be used on the other diameters. Repeat on the other three pegs.

Or you can use a die stock (Photo 4, No. 6). Here you push the die, held in the die stock,

18 ***The die stock held in place by the tailstock barrel.***

along the peg with the tailstock barrel (Photo 18). Feeding in the barrel at the same speed as the die is cutting, cut to the shoulder, withdraw the tailstock barrel, reverse the lathe, and wind out slowly.

Now to form the knobs, again you have many choices. You need a surface to grip to tighten the screws, and you can, if you wish, cut grooves or ridges. I chose to knurl mine.

20 ***The knurls being withdrawn from the work.***

The knurls and knurling tool I used are shown in Photo 4, Nos. 2 and 1.

To apply the knurls to the work, you need a holder of some kind, and there are several kinds. I prefer to use a caliper knurling tool as it places less stress on the lathe than some of the others.

First I set the speed to 615 rpm, and cut a chamfer on the shoulder to accommodate the knurls. Using a knife tool, I cut a chamfer and cleaned up the shoulder of all four screws. Then on the last one, I extended the length of bar protruding from the chuck so the shoulder was a good 1-1/2" from the chuck jaws. Mark off 5/8" plus the distance from the side of the knurl to the outside of the nut holding the knurl, from the shoulder. This is so you can see where you're cutting.

We have two types of knurls, straight and diamond, so I cut one pair of knobs straight and the other pair diamond pattern. With a pair of straight knurls in the knurling tool, position the tool in the tool post. Your knurling tool may be different from mine, so follow your instruction leaflet. This is the way I did it. With the cross slide handwheel, place the tool over the bar and close the knurls onto the work, with the knob on the knurling tool, until both knurls revolve. Position both knurls so they are central over the bar (Figure 19). Zero the cross slide handwheel thimble, and wind the tool out for two complete revolutions of the handwheel. (Be sure to end on zero again.)

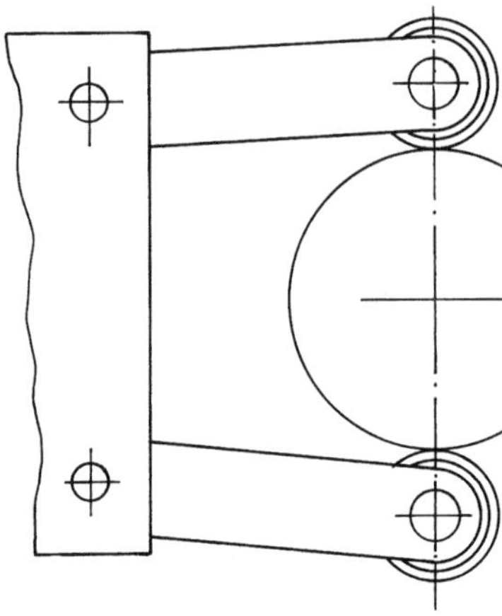

The centers of both the knurls and the axis of the work must lie on the same vertical line when knurling begins.

Figure 19

Put a cut of one division on the knob of the knurling tool. Set the spindle speed to 200 rpm. Wind the knurling tool clear of the end of the knob section of the screw; turn the cross slide handwheel through two complete revolutions to zero. This positions the knurls centrally over the work. Flood the work with coolant and, using the carriage handwheel, feed the tool along the bar to the mark (Photo 20). Withdraw the knurling tool from the work.

Repeat on the other end of the bar to make matching screw knobs. Then I changed to diamond knurls, and cut the other two knobs in the same way.

Finally, let's deal with decoration and parting-off. Mark off your preferred length of knob, and with the 1/16" radius round nose tool in the tool post, form a flute. Take the tool in for .030" on the cross slide handwheel at each end of the knob. Change to a parting-off tool and part off at the end of the flute (Photo 21).

All the parting-off pips and any other blemishes can be removed by putting the pieces back into the three-jaw chuck, using shim to protect the worked metal and making sure that the work is running true. Then skim off any excess of metal and clean up generally (if necessary).

With an eye dropper, put a few drops of kerosene through the threads in the clamp bodies, oil the screws, and fit the screws to the clamps.

These small clamps can be used to hold components together for marking out, or they can be used with other fastening devices to give extra support, as shown in Photo 22.

21 *Parting off the completed screw.*

I hope you get as much enjoyment making yours as I had from making mine, and that you find lots of uses for them. At least you now know you can make a screw to fit a threaded hole.

22 *The mini screw clamps in use.*

Hot Bluing Steel

by Steve Acker

Photos by Author

The home shop machinist can build tools, jigs, fixtures, models, engines, and myriad other items using basic machine tools, hard earned skill, and a liberal dose of imagination. Fine quality work is often the result. One thing sometimes missing, however, is a really professional finish for steel parts.

Easy-to-apply finishes, such as paint and cold bluing solutions, work cosmetic wonders but have limitations. Paint can flake off, add to the dimensions of a part, run or vary in thickness, and be time-consuming to apply. Cold blue looks good on small parts but tends to streak, lacks durability, and does not always give the best luster to a polished part.

Hot salt bluing is a derivative of commercial black oxide metal coating technology. It is widely used in the firearms and gunsmithing trade. It works beautifully on most machined mild steels. It is meant for low alloy steel and will not work for stainless steels, high alloy steels, cast iron, aluminum, brass, or other non ferrous metals. Small amounts of copper will actually kill the bluing solution. You may find that bluing hardened steels will give inconsistent results.

Welding the base.

A simplified form of hot salt bluing can be used with shop-built tanks and accessories to give your projects a professional black oxide finish very similar to that used in large manufacturing facilities. The solution is used at 292° F, gives a distinctive blue-black finish to mild steel, does not measurably add to the work's dimensions, is fairly tough as finishes go, and can be applied to a large number of small parts quickly, if desired. It is not overly expensive to get started. The solution lasts a long time, so the ongoing costs are small on a per part basis.

Before you start, be aware that such benefits do not come without some negatives. Proper safety precautions must be adhered to, as the chemical bath is both very caustic and hot enough to cause severe burns. Proper ventilation, rubber gloves, a face shield, and long-sleeved cotton shirts are absolute requirements.

The basic necessities.

Do not breathe the fumes given off by the hot bluing solution. The fumes are corrosive, attacking wool, leather, your lungs and your skin, and will cause any nearby steel or iron to rust. It is a good idea to keep some vinegar on hand to put on your skin if you come in contact with the solution. Do not work around children or pets, and ask that you not be disturbed.

This is definitely not a kitchen stove procedure. The corrosive fumes will settle on everything inside a building if this work is done indoors. Working outdoors is for me preferable to rusting the entire contents of a small garage or shop. Most gun shops blue in dedicated rooms that remove fumes by using hooded exhaust fans. The salt solution requires close to perfect temperature control, constant vigilance of the water level due to evaporation, and that all parts are perfectly clean and free of oils before treatment.

For all this, the process is relatively simple and within the capability of the home shop machinist. The part is cleaned mechanically and chemically, immersed in the 292° supersaturated salt bath for 20 to 30 minutes, cleaned of residual solution, and then immediately oiled.

The cosmetic quality of the resulting finish will vary according to the surface preparation of the part. Bead blasted or as-machined parts will tend to have a matte finish, while a polished part will appear bright and shiny with an apparent depth to the finish. Different effects can be created by finishing entire projects or only accent pieces. Blued steel headbolts or component parts on small engines, for instance, really set off fine work on other parts that are left as-machined.

Welding the bluing tank.

Most equipment required can be easily fabricated or purchased from Brownell's gunsmithing supply. The bluing salts used are *Oxynate No. 7* from Brownell's. A 40 lb. bucket of salts is approximately $40. The only other item that should be purchased is the specially made bluing thermometer – also from Brownell's. Cooking or candy thermometers are often aluminum and will be dissolved by the bluing solution.

The gunsmith's method of bluing fine firearms uses the same supplies but is more involved than the industrial oxide process. Bluing firearms is difficult because parts are made from a variety of machined, cast, and forged steels at differing hardnesses.

A great amount of experience, labor intensive surface preparation, and process familiarity are required to consistently attain the desired high quality finish. Don't use the methods outlined here and expect good results on firearms. If you would like more information on firearms bluing, Brownell's has a pamphlet entitled "Bluing Instructions and Supplies." The Brownell's catalog, by the way, is a treasure trove of hard-to-find items and is recommended reading. To get a catalog, send $3.75 to Brownell's, Inc., 200 South Front Street, Montezuma, IA 50171. The cost is refunded with a $35 order.

The finished tank ready for water and bluing salts.

A mild steel tank is required for the bluing salt solution. Do not use galvanized steel or stainless steel. The tank should be welded to avoid future problems. Use uncoated mild steel welding rod, if welding with gas or TIG. Brazing the tank would kill the bluing solution because of the copper content of the brass. The stand for the tank can incorporate the burner. You may wish to make the burner detachable for any later service.

A good size for small parts is 6" deep, 6" wide, and as long as the longest part you might blue. Note that there needs to be at least 1" of space between any immersed part and the tank wall so the solution can circulate properly and the part won't pick up heat from the bluing tank itself. My tank is 6 × 6 × 12" long and holds just over a gallon of bluing solution. It was welded up from 1/8" mild steel plate I had on hand. Make a lid with a short lip all the way around it to fit your tank. The lid design should include provisions for a gasket and a handle. The gasket can be made from rubber weather-stripping.

Make a small parts basket from expanded metal or other steel product sized to fit inside your tank. The basket allows small parts to be blued and allows parts to be agitated during the bluing process for an even color. Circulation of the bluing solution is ensured by 1" tall legs welded to the bottom of the basket.

The tank should be filled to an operating depth of 4-1/2", or 1-1/2" down from the top of your tank. The resulting volume can be determined by simply filling the tank from a quart jar and noting the volume. Another way is to multiply in inches the inside length times the inside width, and multiply the result by the desired depth of solution. The answer is then divided by 231 (the number of cubic inches in a gallon) to give the gallon capacity.

The heat source must be capable of putting out enough heat to boil and keep boiling the amount of solution you are using. The most common heat source is a gas burner, usually a pipe burner as long as the bluing tank. A gas camping stove can be used. One friend made a serviceable burner from a gas central heat unit. A gas

The thermometer face. Note the operating temperature range mark.

grill burner the right size also works well. As with any open flame, proper ventilation is a must. If you use your burner outside, you may wish to construct a wind baffle or place the whole unit in a relatively protected area.

Supersaturation is the name of the principle that allows more of a water soluble substance such as bluing salts to be dissolved in water as the temperature of the water is raised. In addition, the boiling point of the water rises as the amount of dissolved salts increases. As the supersaturated solution cools, the dissolved salts come out of solution and become solid again. When the correct amount of bluing salts is dissolved in the water the solution will boil at 292° F. If the solution is just a few degrees cooler, the steel will turn a gray color. If the solution is just a

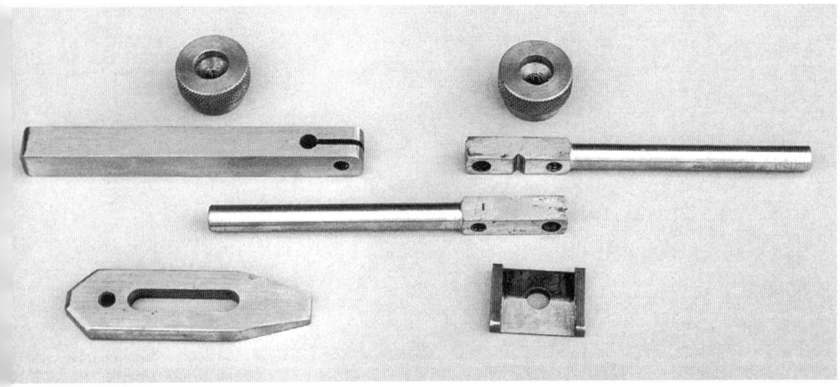

A few parts ready for bluing.

few degrees warmer, the steel will turn a reddish color. If either of these colors appear, correct the solution temperature, remove the coloring from the steel, and start over.

The desired mix will be very close to a ratio of 10 lbs. of salts to one gallon of water. When the boiling point of the solution is correct, the ratio will be correct. Weigh the salts and measure the water. You can weigh the salts by putting them in a plastic bag and using a fishing scale to get the right weight. When one gallon of water and 10 lbs. of salts are mixed in solution, you will wind up with about 1.1 gallons of solution, so plan accordingly.

The first time you mix up a batch of bluing salts, you will start by using a modified ratio of 3-1/2 quarts of water to 10 lbs. of salts. Remember, the ratio is the same for any amount of solution and will require some calculation for the capacity of your tank. Note that you are intentionally using less water for the initial mix. This will let you get to the correct temperature before boiling occurs and

Parts after bluing.

then adjust the solution ratio by adding water until boiling occurs.

Start by putting on all of your safety equipment and making sure nothing flammable is in the vicinity. Figure out where the hot solution would go if it tipped over, and decide if that is an acceptable place. If not, move to a better location for this operation.

Pour the proper amount of clean, cold water into the tank. Slowly stir in the salts. Stir constantly as you go, using a mild steel stirring rod. The initial chemical reaction will create some heat, so watch for possible bubbling or eruption. Don't forget your safety equipment. The thermometer should be mounted 1" in and centered on the long side of the tank. The bottom of the thermome-ter should be at least 1" up from the bottom of the tank, and the ring around the stem of the thermometer should be at least 1" under the top of the solution. Turn on the heat source and stir constantly until you get to 292°. Ignore the scum and foaming that appears on top of the tank.

The solution will probably not be boiling yet because using less water has effectively raised the ratio of salts to water. You have more salts in solution than needed for the solution to boil at 292°. If this is the case, dilute the solution by adding water carefully until you have a rolling boil at 292°.

CAUTION: Added water can turn to steam and violently blow hot, caustic bluing solution all over you. Do not simply pour water into the tank, as it will penetrate into the solution before it turns to steam and will cause a violent eruption. A good way to add water is to use a

Before and after knurled clamping knobs.

long-handled steel ladle. A key point here is the long handle on the ladle. Fill the ladle about half full of water, push it partially into the solution, and move it across the top of the bluing solution while slowly tilting it, allowing the water to enter the bluing solution. The bluing solution should not enter the ladle. You are, in effect, spreading the added water on top of the solution. The solution will still froth and sputter when water is added in this manner, very likely throwing some solution out of the tank. Your protective clothing must still be in place and remain in place during the entire bluing session. Now go back and read this paragraph a second time.

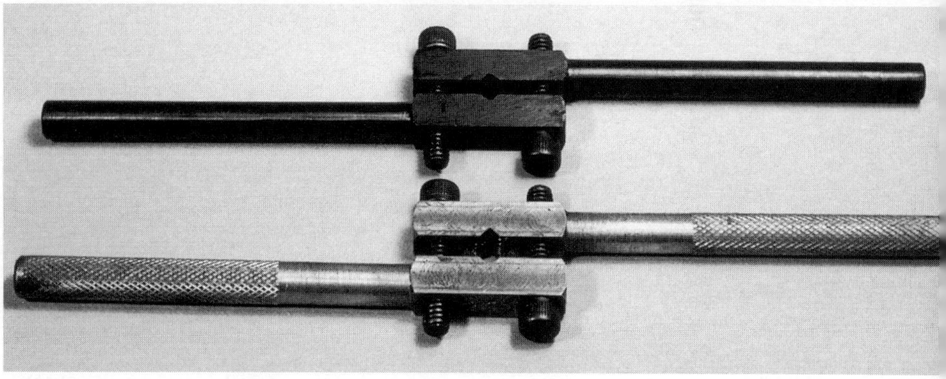

Before and after tap handles.

If you are using a very small tank, evaporation of the water is quite rapid and will affect the solution ratio quickly. Add replacement water in small amounts. Because of the small volume of bluing solution, adding water will cool the solution very quickly, possibly causing color problems. You can heat the fresh water before adding it to the bluing solution to minimize temperature variation. This will not prevent eruptions when adding water, as the heated water will still turn to steam in the 292° solution.

You may wish to turn down the heat source somewhat after boiling commences, but remember that you are actually controlling the boiling point of the solution by adjusting the ratio of salts and water. If, for instance, the solution boils at less than 292°, let some water boil off or add more salts. You are now ready to immerse your prepared steel items. Keep on looking at the black section of the thermometer face that represents the proper temperature range.

If you are reheating a previously used bluing solution, you must first break up all the salts crusted on the top and bottom of the tank. If you don't, any liquid trapped under the salts could steam up and cause eruptions. Heat slowly and stir continuously to get the salts back into solution. The mixture should boil at the same temperature as the last use.

Parts to be blued must be mechanically and chemically clean steel. Anything to be blued should be disassembled to individual component parts, since traces of bluing salts will work their way into recesses such as the minute spaces present in assembled threaded fasteners, and cause rust later. Mechanical cleaning means that any existing surface finish such as paint, rust, old oxide coatings, or associated crud must be physically removed by sanding, bead blasting, or wire brushing until bare metal is exposed.

Chemical cleaning is done primarily to remove oils and contaminants from the metal surface. Naval jelly contains phosphoric acid and works well to remove rust. Something as simple as cleaning with hot, soapy water (using a true detergent) in combination with vigorous brushing will usually give acceptable results. One brand of detergent that has been used is *Joy* dishwashing detergent. Any oil, especially silicone-based oils, left on the parts will prevent good metal coloration. Chemicals such as 1,1,1 Trichlorethane, acetone from the paint store, or even isopropyl alcohol, used with proper precautions, will remove most oils including the silicone-based oils. Brownell's sells a proprietary hot bath cleaner called *Dicro-Clean No. 909* that works well. Use a brush and cold,

A blued centering fixture clamp screw.

Salt crust on top of cooled bluing solution.

The reassembled holding fixture.

clean water to rinse and scrub the work after cleaning to remove traces of the cleaning chemicals.

The parts are now very clean and will rust in a heartbeat. Go into the bluing solution as quickly as possible. Do not handle the clean parts with your bare hands. The oil from your skin will show up as bluing defects. Use a clean steel or iron wire or gloved hands to handle the parts at this stage.

Stir your boiling bluing solution one more time and then suspend your work on mild steel wire in the solution. Use a parts basket for small items. Keep the parts immersed for 20 to 30 minutes. Move the parts around occasionally, don't let them touch the sides of the tank, and monitor the tank temperature and solution level. As water boils off, the solution density changes and the boiling temperature rises. Carefully add water as already described until the temperature is again at 292°. If the thermometer face becomes covered with salts, dip a corner of a paper towel in the solution and wipe the thermometer face with it, then immediately dry it off. Don't try to use cold water or the hot thermometer bezel will crack.

After enough time, remove the parts from the bluing solution, and wash them gently in cold water. Inspect for any bluing salts adhering to the parts and remove them. Ideally, the parts should then be boiled in clean water for ten minutes to prevent residual bluing salts from causing rust later. You might instead try slowly and carefully drying the parts with a heat source such as a hair dryer, heat gun, camp stove, or propane torch. Don't get the blued part over about 200°. Oil the part with a light non-additive, non-detergent machine oil, immersing it in the oil, if possible. Brownell's sells a water displacing oil for this purpose that is a staple of the gunsmithing trade.

Handle the freshly blued part as gently and as little as possible for a day or so to give the blued surface time to cure. If you are done with the salt bath, turn the heat source off, remove the thermometer, and cover the tank with its lid. A layer of bluing salts will come out of solution and cake up on top of the mixture as it cools. Allow the thermometer to cool and clean it with warm soapy water, then rinse in fresh water and store. The salt bath will be stored covered in its tank until the next use. Wait until the tank is cool to move it.

If you don't get the proper results, the problem is usually one of three things: 1) the bluing solution is at the wrong ratio or boiling temperature; 2) the bluing solution has been contaminated by putting something improper in it or storing it uncovered; or 3) the part to be blued is not truly clean. One possible source of contamination is the water used in the solution. Most municipal tap water is fine, but when in doubt, run the water through a charcoal filter.

The solution is worn out when it turns reddish and quits coloring the steel. When this happens, or when you wish to dispose of the solution for any other reason, check with local authorities before disposing of the salts. Brownell's sells a neutralizing kit for this purpose. Instructions for environmentally safe disposal are included with every order of bluing salts. Some fire departments have a hazardous waste disposal station.

Gearing your shop up for bluing small steel work is to me a lot like getting started in metal casting. It is not for the careless, it can bring a useful industrial process to the home shop, and it significantly expands your finished work capabilities. You might consider bluing as a joint venture with another hobbyist, or explore the possibility of trading such work for other work. I think you will appreciate the expanded capabilities of your shop, and experience the satisfaction of fully completing your projects.

Stuck Chucks, One-more-time, Didactic #17

by Henry J. Kratt

YO!! You with the 48" pipe wrench wrapped around your lathe chuck! C'mon over and let's talk a bit about stuck chucks and some things that have worked for me.

To ease the removal of a chuck, it must first be fitted – some of the more sensual among us might even say "mated" – to a specific spindle. I'm discussing the threaded spindle and chuck since these are the ones usually found in the home shop. This could apply to the threaded spindle on a dividing head as well as on your lathe.

How is this done? Well, the old saw your mother used to give you about cleanliness being next to Godliness – now is one more time to believe it! Clean the spindle threads and the land at the base of the threads and inspect both very closely. I use a magnifying glass sometimes, especially for a small lathe. The threads should be smooth and without any raised places. A small bit of a thread knocked out or ground off will not interfere with the fitting. Look at the flat space especially – that flat ring at the end of the threads that stops the chuck when it is threaded on. Look closely! You are looking for dings, bumps, embedded trash, any scars that stick up above the surface. Those below the surface don't interfere – unless they are really gross, like half the circumference, in which case this procedure is not going to fix your problems.

The object at this point is to remove all the bumps and scars – only those sticking above the surface. The spindle is hardened steel, so you will need to stone them off. For me, this means with a small stone file or whet stone and very careful strokes. Do not disturb the undamaged surface; this an essential register for the lathe. To do this could seriously damage your lathe. For the occasional wise soul who wants to mechanize this process by turning the lathe on and holding the stone there by hand, all I can say is DON'T EVEN THINK ABOUT IT. You will make a mess! Maybe a tool post grinder or such would help – but then the guy with that kind of stuff probably quit reading these notes two paragraphs back. You are trying to restore the truth of that flat surface. Over the years, small bits of scrap may get bedded in the surface and make it rough or irregular. Failure to get this surface leveled will preclude being able to fit the chuck. If you can't get it flat, don't go any farther in this fitting; it will just be destructive to the chuck backplate.

Now to the chuck! This is another thorough cleaning job. Backplates are soft, otherwise I could not machine them to fit the chuck. Again, look for dings and raised places. With the plate being soft, there is plenty of opportunity for damage. If you store unused chucks with jaws up, and do not have suitable cushioning on the shelf, the flat mating surface or land at the end of the threads will be scarred. Get this clean. Also inspect the threads to see if any small bits have been embedded in the flanks. Clean and then reclean.

Now starts the fitting! Dry the surfaces. Take a small smear of fitter's or scraper's blue (not layout blue – this stuff looks like blue grease) and apply it to the flat mating surface of the spindle. Thin it out; it should be visible enough but not in any way lumpy. Lock the back gears or else prevent the spindle from turning. Now mount the lathe chuck slowly. Turn it on, by hand, until it is just about to touch. Then turn it until it begins to be snug. Now, take it off. If you are lucky, there will be a wide blue ring all around the chuck where it touched.

So much for fantasy. The blue was most probably a thin mark about half around – or worse, a single blue pip. This is a basis of the problem of the sticking chuck – it is being bound onto the spindle by the torque of turning and, as such, is sprung on the threads. The mating surfaces are not taking the load evenly all around. Incidentally, this also pulls the chuck out of alignment. With a small scraper, remove the blued surface from the backing plate and wipe clean, re-blue the mating surface on the spindle, and retry the chuck... over and over until there is a nice wide blue ring on the chuck backplate all around. After all this effort of fitting, you will probably store your chucks with the jaws down to protect the scraped surface.

What we have done is ensure the even distribution of the load from the chuck to the spindle without distortion of the mating surfaces and the threads. Distortion of these surfaces will spring the spindle, backplate, or both, and lock them together much as a lock washer does.

None of this fitting will improve the accuracy of centering the work in the chuck. I am told there are some serious scroll chucks out there that have accuracies in the tenths across the range of the chuck, but I will never own one. Mostly chucks are good for 3-5 thou. All are more accurate at one diameter than another. Regrinding of the chuck jaws you hear about is often a "fix" that is only accurate at the specific diameter used for the grinding.

I once came upon a fellow who had set up his school shop with leather collars around all the lathe spindles to facilitate the removal of chucks – a real dodge but effective for novices. Of course, this could be fatal to the turning of accurate threads since it would be possible for the chuck to tighten further and ruin the alignment. The collar suggested by some is invariably of a compressible material ostensibly to take up misalignment between the mating surfaces at the spindle.

The Proper Use of Diamond Dressers

by John Bochert

It is sad but true that even some professional machinists do not know the proper way to use a diamond wheel dresser. As Frederick Oginz aptly remarked in his article titled "Diamond Tool Efficiency" (*HSM* July/August 1988), "Operators of grinding machines are well versed in dressing and truing the wheel but lack expertise in stretching the life of the tool."

The most common problem with such operators is that they present the diamond perpendicular to the wheel as in Figure 1, and as the diamond wears down, as it most certainly will, all they can do is keep trying to dress the wheel with a very blunted diamond (Figure 2). Consequently, they do not get a very good wheel surface with which to grind. A point or edge on a diamond is the most efficient at dressing.

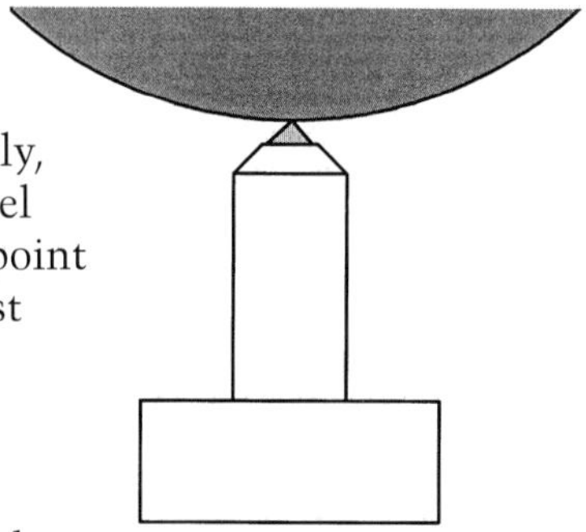

Figure 1

A diamond dresses a wheel by dislodging glazed grits and embedded metal particles from the periphery of the wheel, exposing new and sharp grit. As the diamond is very hard, it is also very brittle and fractures easily. It is the impact of the grit in the wheel that fractures off microscopic chips from the diamond, which is the source of the wear on the diamond.

To keep the diamond sharp and well pointed, the accepted practice is to mount the nib on a 25° to 30° angle (Figure 3), and as a spot or facet is worn slightly off-center (Figure 4) it can be rotated slightly to present the edge of that facet to the wheel. This can be done repeatedly, each time renewing the point on the diamond. A wheel that is dressed with a blunted diamond does not grind as well as one dressed with a sharp diamond, and needs dressing more often, thereby shortening the life of the diamond even more.

Just placing the diamond on an angle is not the entire answer. Using a surface grinder as a reference, the angle must be pointed in the direction the wheel is moving and should be placed on the vertical center line of the wheel or a fraction on the trailing or up side of center (Figure 5), never on the down side. If there is any possibility of the diamond moving, such as when the operator forgets to turn on the magnetic chuck, it could be pulled into the wheel suddenly with disastrous results. **Always wear safety glasses.**

The next items for consideration are the amount of material removal and the speed of the diamond across the surface of the wheel. Take light cuts, from .001" to .003" maximum for material removal and as little as .0002" for finish passes. The diamond should pass slowly and uniformly across the wheel, two to three seconds to cross a 3/4" wide wheel turning at 3,600 rpm should be sufficient. And avoid impacting the edge of the wheel with the diamond.

Figure 2

The same principles apply to other types of grinders, such as cylindrical grinders, centerless grinders, etc. Always position and angle the diamond so any inadvertent movement of the diamond will be away from and not into the wheel, and never use a diamond that is loose in its setting for obvious reasons.

When the diamond finally does get worn too far to be usable, it can be reset and be made like new again. This is not done by building the point back up again, but by turning the diamond over (Figure 6), as diamonds are usually double-ended. A perfect diamond crystal is shaped like two pyramids placed base to base.

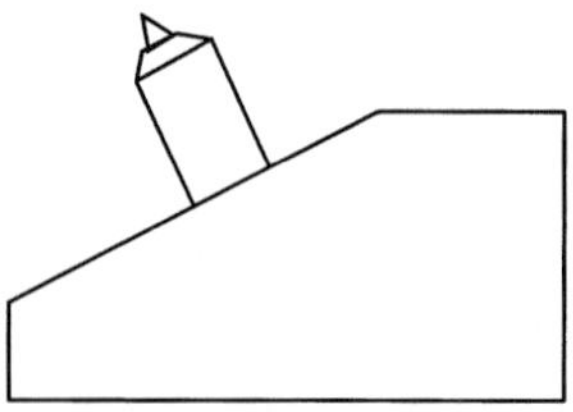

Figure 3

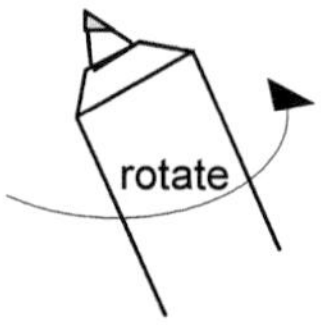

Figure 4
shaded area is worn off

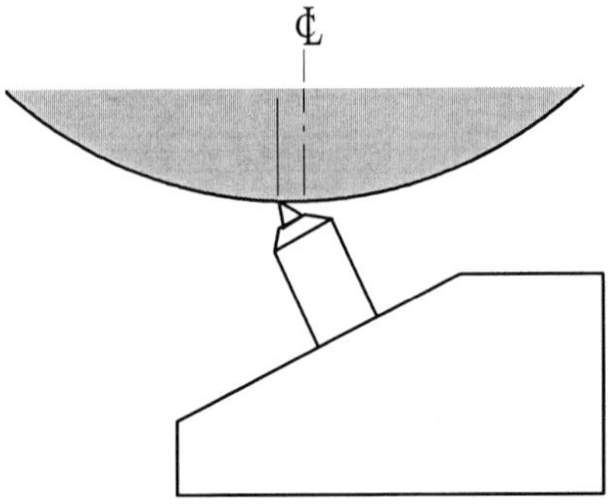

Figure 5

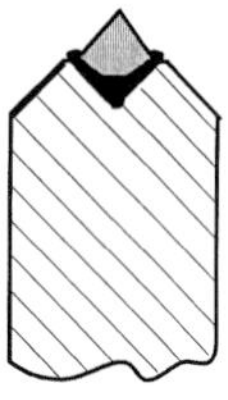

Figure 6
cross section with worn end down

The diamond is set into a conical socket in the nib and brazed or silver soldered around the edge. The brass or solder does not adhere to the diamond and so is then slightly crimped or swaged to entrap or grip it. This is best left for professionals, as the diamond is very brittle and is easily shattered.

The larger the diamond the stronger it is because of its mass; it is also more resistant to wear, so do not use too small a diamond. A .3 to .75 carat diamond is recommended for a 7" to 8" diameter wheel, and larger wheels require even larger diamonds.

Horizontal straight line dressing is not the only use of the diamond. With the proper tooling, you can dress radii, both convex and concave, and angles on the wheel singly or combined, in order to be able to form-grind complex shapes in one operation. The possibilities are limited only by the imagination and the proper tooling.

Most important of all, a little thought and a lot of common sense are always good advice in any situation.

Cutting Steel with a Circular Saw Blade

by Steve Acker

Many of my wife's relatives live in picturesque northern Arkansas. Every year, we visit the relatives and travel the area. On our last visit, my brother-in-law, knowing of my interest in machine shop topics, shared with me a story of a local man who cuts steel with a circular saw blade meant for use in wood. This of course got my attention. After a phone call, we left for the gentleman's shop.

A few miles later, we arrived in Salem, Arkansas, at the home of V. J. "Chuck" Cowell. A retired and talented machinist, Mr. Cowell graciously took time to show me his shop, and explain to me how he cuts steel in his milling machine using a 7-1/4" carbide-tipped, wood-cutting circular saw blade. It had occurred to him that, if the carbide on a wood saw blade was similar in composition to the carbide on machine tool bits, it should cut steel just as well as more expensive cutters. He showed me the arbor he made to fit the central hole of the saw blade. The arbor is mounted in a collet in the mill and runs at about 250 rpm. A slow feed lets the carbide teeth cut the steel quickly.

Once home, I duplicated the arbor and tried the process myself. The spindle speed was set at 210 rpm. I found that mild steel was cut cleanly, just as described. I cut a 1-1/2" diameter round steel bar quickly and smoothly. The chips were small and uniform and the blade did not appear to be hurt or worn by the process. The cut did tend to wander slightly on the thick stock, at least for me – possibly because of the thin blade or a too quick feed rate.

V. J. "Chuck" Cowell with his creation.

A slightly smaller diameter or thicker blade may wander less. Some chatter occurred when the work extended too far from the milling vise. Subsequent cuts in thinner work and various kinds of metals were straight and chatter-free. Like any large diameter cutter, it is potentially hazardous and requires utmost care in operation.

This technique will probably not replace the metal-cutting bandsaw in a machine shop, but I found it useful and continue to find new uses for it. I intend to experiment with saw blades having various numbers of teeth. This rig comes into its own when cutting aluminum. It is limited to cutting short pieces by the layout of a typical milling machine but might be good as an alternative for hogging out large sections of a workpiece.

Overall, the rig acts like a big slitting saw. The new smaller circular saw blades meant for trim work saws might be worth investigating as general purpose slitting saws. Given the cost of new slitting saw blades I suspect that this may be the best use for this idea. Thanks, Mr. Cowell!

Building Phil Duclos' Model Maker's Dividing Head

by Marsh Collins

Shortly after receiving my copy of Phil's book, *The Shop Wisdom of Philip Duclos,* I made up my mind to build the rascal. It was just what I'd been looking for. I have one of the smaller mill/drill machines Enco sells, as dictated by headroom (6'8") and weight limitations in my shop (wood floor and framing). When I shopped for a dividing head, I soon found what was on the market was too big and heavy. And it cost too much for what I could do with it. So build your own, Collins!

After studying Phil's plans I decided on some changes; I'm one of those birds who can't leave a good design alone, plus I had a few good reasons in mind. To maintain the level of confusion, I'll break this into segments.

THE ANGLE PLATE

Having spent a few years with a steel fabricator, I was aware that structural angles of 5/8" thickness in 4" lengths are scarce outside shipyards or bridge fabricators. I therefore built up the angle plate out of 5/8" crs plate, with three 5/16-18 Allen cap screws to tie it together. Otherwise it's as shown on page 6 in the book. I built this item first, for a good reason (Photo 1).

1

THE BODY

After taking a long think about the use to which this machine will be put, I chose to make the body from aluminum. I have a good source for remnants, and acquired a piece of 6061-T6 3" diameter × 6" long, for cheap. My decision was based on: 1) the kind of milling/cutting done on workpieces with this tool will be relatively light cuts as I'm not expecting any production demands and can afford the time; and 2) aluminum is good bearing material for the spindle, and for the few threads that will be taking any repetitive usage, such as the spindle locking bolt (page 17), I provided threaded steel inserts. Likewise, the three 1/4-20 cap screws that fasten the tilt lock plate became 1/4" studs, 20-thread on the aluminum end, and 28-thread on the nut end. The cover plate (page 14) was made of steel because of the backlash adjusting screw thread. Instead of four screws I used six 6-32 Allen cap screws.

2

3

There was a good reason why I made the angle plate first; I used it for a machining fixture for the body (Photos 2, 3 and 4). By drilling a 5/16" hole through on the center and using a 5/16" bolt and an end plate, I was able to do most of the outside shaping with a minimum of clamps and doo-dads in the way.

Note that in Photos 1, 2, 3 and 4 a feeler gage set is shown at the outer end of the workpiece to take up any deflection while milling the body. I keep a cheap set, and whenever I need a temporary shim I simply use the required leaf, or I cut off a slice to fit where needed. One gage set has lasted me years.

When it came to hollowing out the body, I installed the studs and used them along with the tilt lock plate (page 7) to clamp the body to the angle plate, again using the feeler gage to prevent deflection.

I also added a spindle bearing oil hole opposite the spindle locking bolt, provided with an Allen cap screw plug to keep trash out. The spindle locking bolt hole provides lubrication access to that side.

THE GEARS

I have specifics for the gears. Those that Phil described were Boston Gear's: 24 diametral pitch worm and gear. The gear is their No. G1036, bronze, 1.667" pitch diameter, 1/4" hole, 5/16" wide face. The worm is their model LUHB, steel-unhardened, .500" pitch diameter, single thread, 3/16" bore. For Boston's nearest jobber, call (800) 343-3353. (In Massachusetts, call (617) 328-5690).

THE SPINDLE

I had a beautiful piece of monel propeller shafting, 1-3/8" diameter, which, although harder than the hinges of you-know-where, made the cleanest spindle bearing journals ever seen. Threading the 32 tpi end was a bear, and after cobbling up the first try, I changed from 7/8" diameter to 3/4" diameter and it worked out just fine. I also, with the tougher steel, reamed it for a No. 2 Morse taper, as many of my other tools use the No. 2. I also threaded it 1"-8 instead of the non-standard 1"-10 that Phil used.

INDEX SPRING CLIP

Instead of the spring clip, with which I have had sorry experiences in the past, I threaded the spigot of Part 3 (worm shaft housing) to 9/16-18 and made a knurled nut, which gives me the tension control I wanted.

SPINDLE LOCKING BOLT

Being averse to handles sticking out in the way of later operations, I am partial to Allen screws. Since the spindle is monel (and harder than...), I used an Allen cap screw with the cup point ground smooth.

PLATES

I used 6061-T6 aluminum for these, again considering the planned usage. The segments were made from steel because of the clamping screw. I developed my own method of hole spacing, simply dividing 360° by the number of spaces. Since I have an 8" rotary table, it was

4

5

6

7

8

simple to mark off the number of degrees progressively, i.e., the angle between each hole for the 20 ring is 18°. The first hole from zero is 18, the second is 36, etc.

Some divisions get into decimals of a degree, but from a practical standpoint, with each error on the disk being divided by 40, the error becomes infinitesimal. I carried decimals to the nearest single decimal, and eyeballed the proportion between degree marks on the table scale.

I also looked up the settings for a 40-1 ratio in *Machinery's Handbook,* and copied those for up to 20 divisions (Figure 10) as I doubt any higher number of divisions will be needed by the average model builder. If they are, *Machinery's Handbook* is always handy, but with greasy fingers?! I *Varathane*-sprayed my hard-copy and tacked it to the wall.

TAILSTOCK

I made the body and base of 6061-T6 aluminum for the same reason as the divider body. The tailstock spindle is from 1/2" drill rod, and made as shown. I added a handle to the clamping nut and a clip bent to hold one face of the hex head bolt. The clamping nut was made from a 5/16-18 coupling nut (from the all thread rod section of the hardware store), and the handle was cut from 1/4×3/4" crs freehand on the grinders. Changing the position of the bolt head in the clip and/or adding a washer under the nut adjusts the handle to horizontal when clamped.

9

CHUCK

As a lot of my work is done in a chuck, I obtained a neat little 4" four-jaw independent from Enco (don't look in their catalogue – it wasn't in it) for a price you won't believe. Made in India, it had beautiful castings and was well finished all over the outside, including the jaws and screws. Like anything I've bought made in India, China or Taiwan, the inside needed cleaning of foundry sand, etc., so after a good cleaning and filing/grinding of rough spots and junk, it was as nice a chuck as I've seen in a long time. I made a backplate for the 1"-8 spindle nose (as well as a 1-1/2-8 for the lathe), and that now graces the spindle. I also plan on making a thick faceplate, probably aluminum, as soon as I can acquire a 4" square piece of 1-1/4" aluminum plate. Photos 5 through 9 show my finished dividing head at work.

Remember, the foregoing modifications were for my planned uses – not production – and materials were those I had on hand or easily came by. Your wants may be different. But I now have a dividing head of the right size for what I do and the cost to me was peanuts – $7 for the aluminum block for the body, about $30 for the gears, and about $10 for screws, bolts, etc. The rest was from my scrap bin. Beat that, if you can, for a Model Maker's Dividing Head!

Figure 10

Model Maker's Dividing Head: Disk Selection Chart

for from 2 through 20 divisions

(Excerpted from *Machinery's Handbook*, 18th Edition)

No. Divisions	Disc No.	Circle	Turns	+	Segment
	(A=Any)	(A=Any)	(full)	add	(=0)
2	A	A	20	+	–
3	3	39	13	+	13\39
4	A	A	10	+	–
5	A	A	8	+	–
6	3	39	6	+	26/39
7	3	49	5	+	35/49
8	A	A	5	+	–
9	2	27	4	+	12/27
10	A	A	4	+	–
11	2	33	3	+	21/33
12	3	39	3	+	13/39
13	3	39	3	+	3/39
14	3	49	2	+	42/49
15	3	39	2	+	26/39
16	1	20	2	+	10/20
17	1	17	2	+	6/17
18	2	27	2	+	6/27
19	1	19	2	+	2/19
20	A	A	2	+	–

Lathe Tips

by Edward P. Stone and Edward G. Hoffman

Drawings by Edward G. Hoffman

Engine lathes, in all their many variations, are generally the most frequently used machine tools in any shop. Just about every project you'll ever make will require some turned elements or components. For this reason, most of us are well acquainted with the lathes in our shop, and we've developed a variety of setup methods and techniques to suit each task. Many times these setup procedures may have been learned from others or developed through experience. Here, using the old trial-and-error approach, we usually find methods that work best by first learning what *doesn't* work. The following are a few explanations and tips to help make your lathe setups and turning operations easier.

Lathe turning tools are available in a wide range of styles and types. As with all other machine tools, the cutters you select are based on the operations to be performed. The simplest type of cutting tool used with the lathe is the single-point cutting tool. Though the design of these cutters is quite simple, the terminology used to describe the various cutter elements can be confusing. For example, simply identifying the hand of a cutting tool gives most of us fits. But if the tool is held correctly, identifying the hand is quite easy.

As shown in Figure 1, most people look at a lathe tool in the position shown at "A." Here, the cutting edges seem to be reversed. The right-hand cutter has its cutting edge on the left side and the cutting edge of the left-hand cutter is on the right. But, if the cutters are viewed from the perspective of the workpiece, as shown at "B," the cutting edges are on the correct side. To identify the hand of a lathe tool, just turn the cutter to view the end from the vantage point of the workpiece.

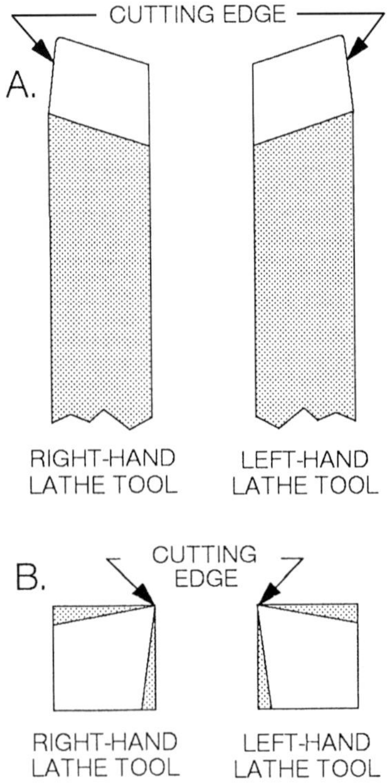

Figure 1

Cutter geometry is another topic not very well understood. As shown in Figure 2, the typical lathe tool consists of a variety of flats and angles. Each of these areas of the cutter are very important to the overall efficiency of the tool. In the simplest terms, the angles on any cutter only serve one purpose – to isolate the cutting edge from the rest of the cutter. To perform as intended, the cutting edge must be the first area of the cutter to contact the workpiece. Otherwise, the cutter will rub against the workpiece, and will not cut. Therefore, the first step in turning is understanding the cutter geometry and the terms used to identify and describe each area of the cutter.

A typical lathe cutting tool, with all the individual angles and elements identified, is shown in Figure 2. Here, you'll notice that the cutter has six different angles. These can be grouped into three general categories: cutting edge angles, relief, or clearance angles, and rake angles. As a general rule, these angles should be as large as possible to provide sufficient support for the cutter. Removing excessive amounts of material when grinding a cutter will often weaken the tool. This results in a tool that could break easily or require frequent sharpening.

The side cutting edge angle and the end cutting edge angle are ground on the top of the cutter, and form the basic shape of the cutting tool. The intersection of these angles forms the tool point, or will blend into the arc forming the nose radius on some tools. These cutting edge angles also act as clearance angles to isolate the tool point. The end relief angle and the side relief angle are ground on the flanks of the tool. These angles are intended to provide the necessary clearance that isolates the cutting edge of the cutter and allows it to penetrate the workpiece. The back rake angle and the side rake angle further define the cutting edge and also serve to direct the chip away from the cut. The intersection of the two rake angles and the two relief angles typically forms the cutting edge.

The rake angles also define the bluntness or sharpness of the cutter. All cutting tools have rake angles. Depending on the type of cutter and the material being machined, the rake may be positive, negative, or neutral. These

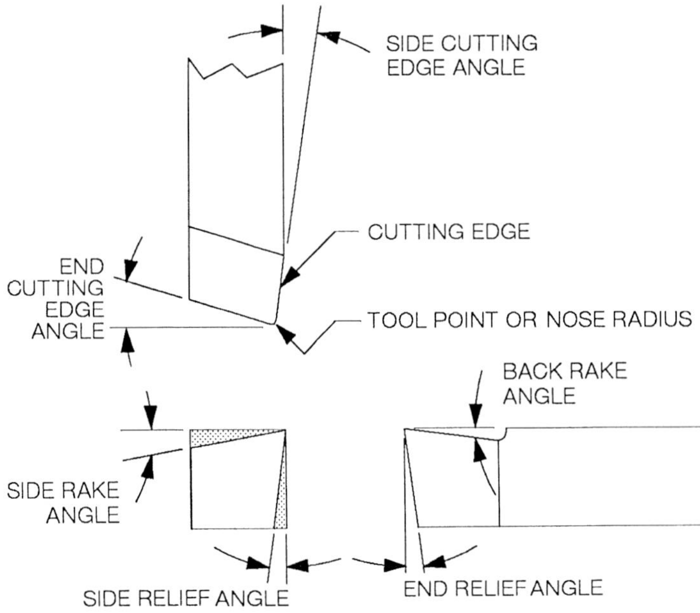

Figure 2

three forms of rake are shown in Figure 3. With positive rake angles, the cutting edge is higher than the rest of the tool face. With negative rake angles the back of the tool face is higher than the cutting edge. And with neutral rake angles, the cutting edge and tool face are on the same plane. Though positive rake cutters are the most common, there are workpiece or tool materials and processes where either neutral or negative rake angles are better suited.

Once the cutter is properly ground, the next step is aligning the cutter to the workpiece. As a general rule, the cutting edge for most turning operations should be set to the center height of the workpiece. However, for some roughing operations, where heavier feed rates are used, you may want to set the cutter slightly above center. Then as the cutter enters the workpiece, the force of the cutting action will tend to pull the cutter down to the proper center height.

Setting the cutting tool to the correct height can be accomplished with any of several different methods. The simplest way is by visually aligning the cutter to a tail-stock-mounted center or to a center reference line cut into the tailstock spindle, as shown in Figure 4. If your lathe does not have this tailstock spindle reference line, it is usually a simple process to cut one into the spindle. All you'll need to do is set a 60° cutter on its side and position the point to the exact center height. Then, using the cross-feed and the carriage handwheel, simply scribe the line into the tailstock spindle to the desired length with a series of progressively deeper cuts.

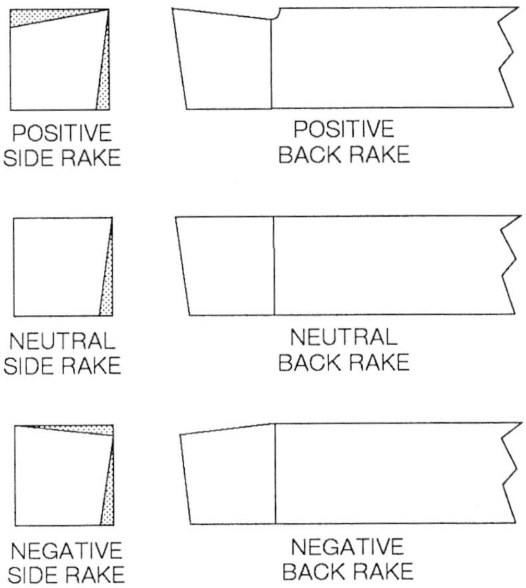

Figure 3

Another method of positioning the cutting tool is shown in Figure 5. All that is required is a 6" steel rule or a short length of flat stock approximately the same size as the steel rule. With this method, you'll first need to mount and position the cutting tool to the approximate center height. Then, advance the cutter to within 1/4" of the workpiece using the cross-feed. Position the steel rule, or flat stock, between the workpiece and the cutting tool at about the 3" mark or the middle of the bar stock. Slowly advance the cross-feed until the point of the cutter contacts the steel rule and holds it a gainst the workpiece.

If the rule stands exactly vertical, as shown at "A," the cutting tool is positioned right on the center height of the workpiece. However, if the top end of the rule tips forward, away from you, as shown at "B," the cutting tool is above the proper center height. On the other hand, if the top of the rule tips back toward you, as shown at "C," the cutting tool is below the proper center height. In either case, the exact angular position of the cutter with respect to the workpiece is equal to the amount of tilt of the rule. So, if the rule tilts 10° toward you, for example, the cutter is 10° below center. If the rule tilts 10° away from you, then the cutter is 10° above the proper center height. This fact is quite useful for those situations where you want to set the cutter slightly above the center. Rather than guessing at the cutter position, the amount of tilt of the steel rule will instantly show you how may degrees above the center the cutter is positioned.

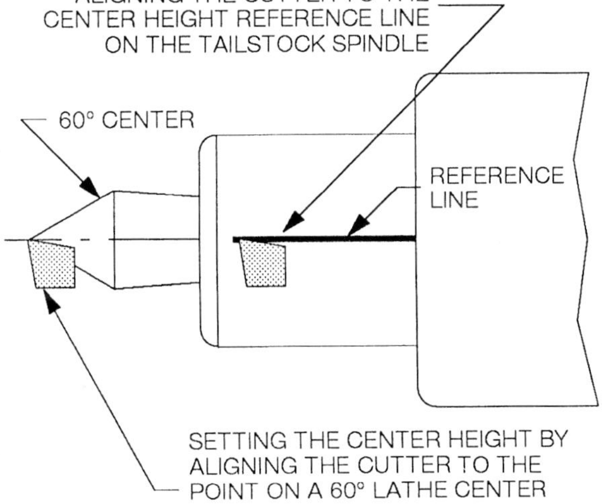

Figure 4

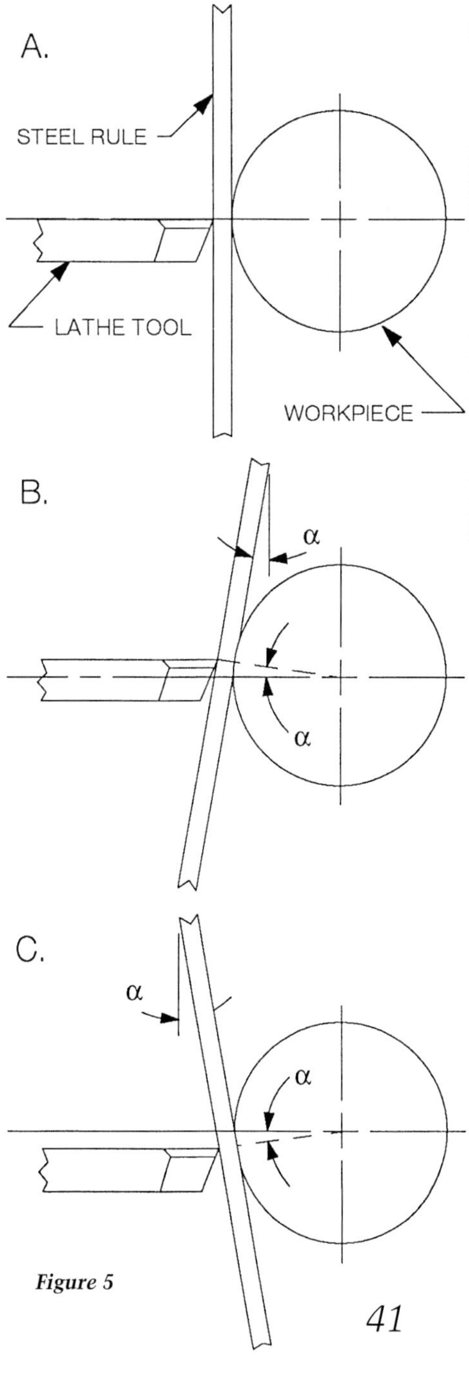

Figure 5

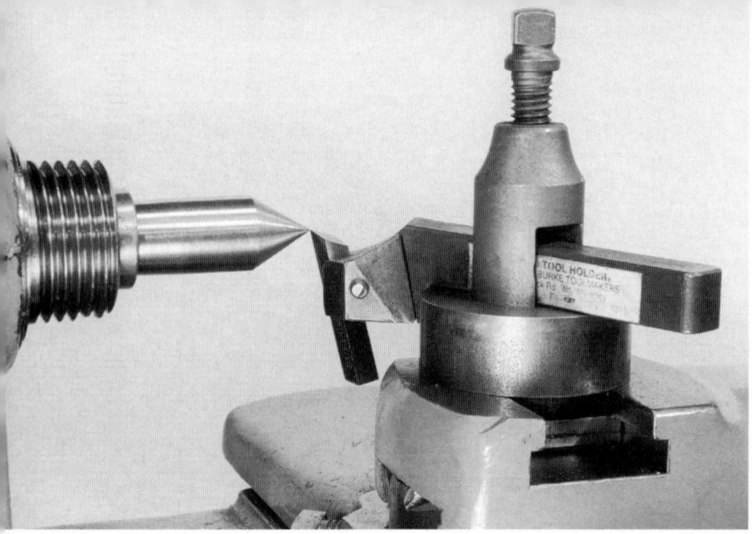

1

2

Using the Diamond Toolholder

by Rudy Kouhoupt

Photos by Author

Nearly four years have passed since the Diamond Toolholder came into my shop. My highly favorable first reaction to it prompted me to write an enthusiastic Product Review that appeared in the July/August 1993 issue of *HSM.* Since then, Diamond Toolholders have seen much service in my shop working on two lathes of distinctly different sizes.

Many readers of this magazine and users of the Diamond Toolholder have asked me questions about its features and use. This article is an attempt to present a comprehensive view of what I have learned from a great deal of hands-on experience and chip making using the toolholders on my 9" South Bend Model B and Model 4000 Sherline lathes.

Before getting into a detailed discussion of the toolholder, let me make the following disclaimer: I am in no way connected with the design, manufacture, distribution, or sale of the Diamond Toolholder. Furthermore, I have no financial interest in it and derive no profit from its manufacture and sale.

Refer to Photos 1 and 2 to get an idea of what the Diamond Toolholder is all about. In Photo 1, you see a side view of the toolholder in an American, lantern-style tool post on my South Bend lathe. The toolholder supports a tool bit of square cross-section in a tangential position. It is the tangential positioning of the tool that distinguishes this toolholder from most others. The tangentially mounted tool is capable of taking either turning cuts or facing cuts. It can also form shoulders.

Notice that the work in progress in Photo 2 was to cut a square shoulder on a flanged bearing housing. The shoulder cut is a combination of a turning cut and a facing cut made simultaneously. The acute angle of the tip of the tool bit permits the tool to cut the shoulder in a single pass, while the tangential mounting provides both front and side clearance angles. In this overhead view, the cutting surface of the properly ground and sharpened tool bit is the white diamond-shaped object in the shoulder. The name Diamond Toolholder is derived from the fact that the end of the tool bit assumes a four-sided diamond shape when the tool is ground.

When you buy the Diamond Toolholder, you receive the four objects contained in Photo 3 and a set of instructions. In addition to the toolholder itself, the manufacturer supplies an Allen wrench, a blank tool bit, and a fixture to hold the tool bit for grinding and sharpening. In order to put the toolholder to work, you must begin by grinding the tool bit to sharpen it. Then you must mount the toolholder in a solid tool post. Finally, you must adjust the position of the tool bit in the toolholder.

GRINDING THE TOOL BIT

In order to cut cleanly in the lathe, there are four angles the cutting surface of any tool bit must present to the workpiece. These are the front clearance, side clearance, back rake, and side rake angles. The grinding fixture provided with the Diamond Toolholder holds the tool bit in the right position to grind the desired back rake and side rake angles in one operation.

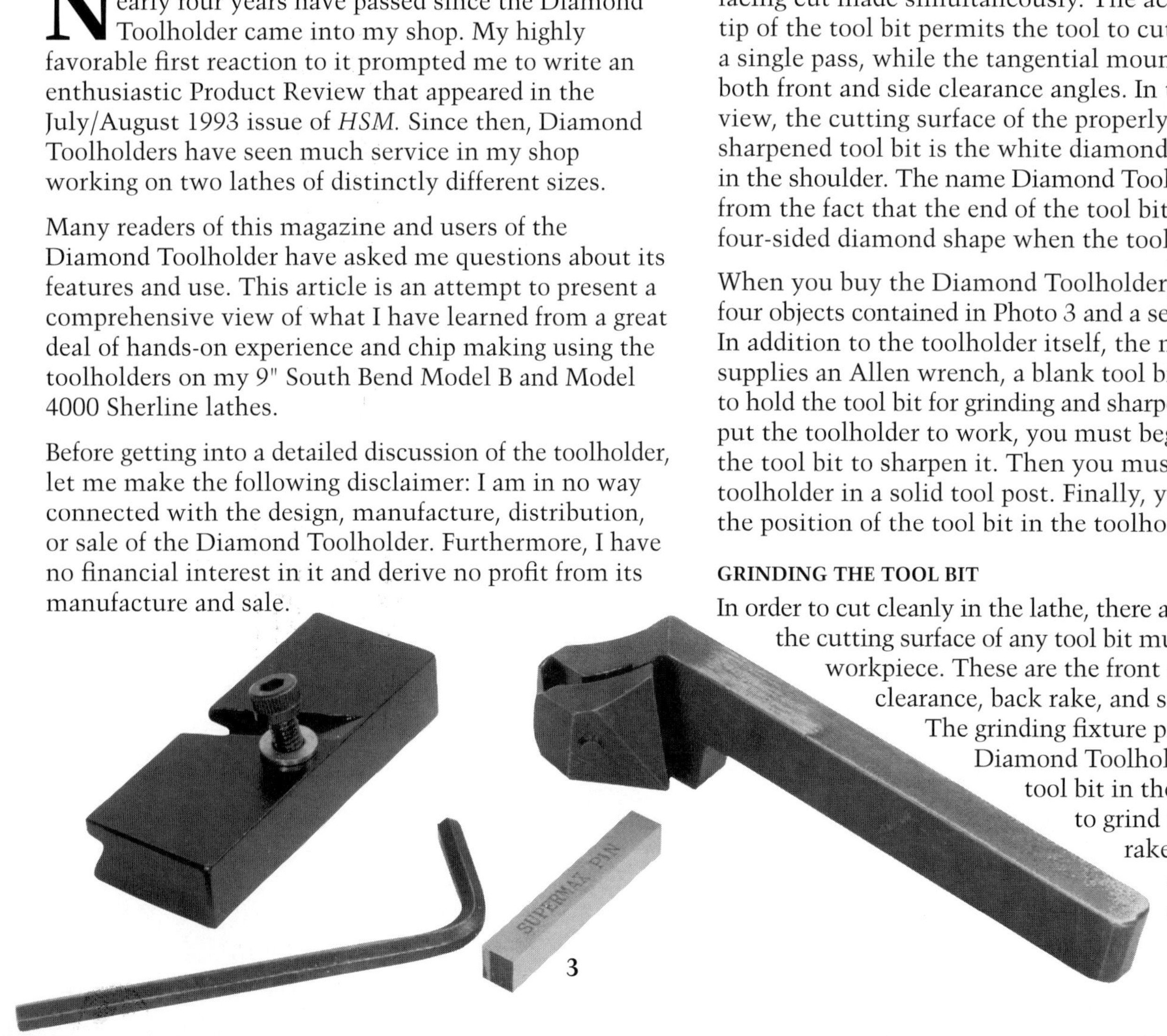

3

4

In Photos 4, 5 and 6, you see a 6" bench grinder being used to grind a tool blank. To obtain the correct shape on the end of the tool bit, you must first set the grinder table at the center height of the grinding wheel. Photo 4 illustrates how I made use of the bottom edge of a scale balanced on the grinding table to align its top surface with the center of rotation of the grinding wheel.

Notice in Photo 3 that the fixture that comes with the Diamond Toolholder has an angular groove in one flat surface. Place the tool bit in the groove with a little

5

6

sticking out in the position indicated in Photo 5. Tighten the socket head screw to clamp the tool bit in the groove. After turning the fixture with its long side toward the grinding wheel in the position shown in Photo 6, make slow passes from side to side until the end of the tool bit is fully ground to the shape shown. Grinding is a slow process that generates quite a bit of heat. As you grind the end of the tool bit, dip it in cold water from time to time to avoid the overheating that is injurious to the temper of the tool.

There are likely to be burrs along the edges of the tool bit after grinding. Remove the burrs by rubbing the tool bit by hand on the flat surface of a fine hand stone. This is the time to form a radius on the tip by rolling the tool bit between your fingers as you hone it on the hand stone. Mount the tool bit in the grooved end of the toolholder.

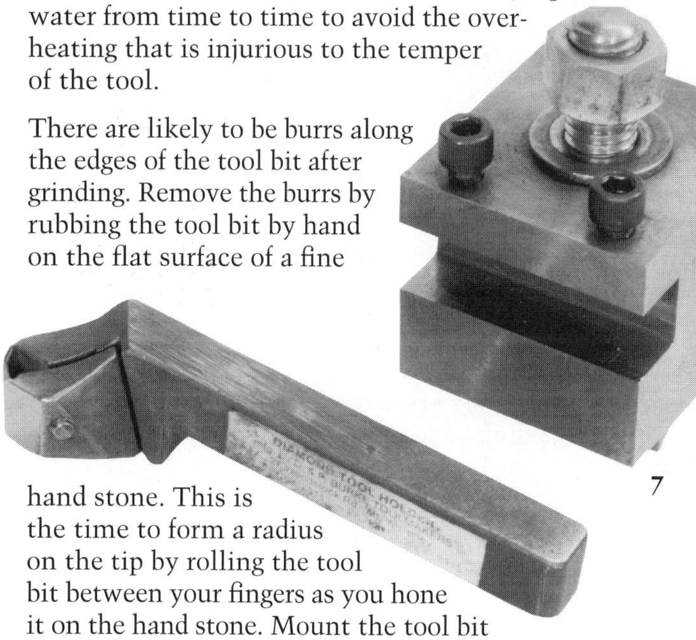

7

MOUNTING THE TOOLHOLDER

Provided the toolholder is mounted with its shank exactly horizontal, the grooves in the end of the Diamond Toolholder hold the tool bit in a position that automatically provides the front clearance and side clearance angles. In the November/December 1994 issue of *HSM,* I described how to make the open-sided tool post illustrated in Photo 7. The shank of the Diamond Toolholder fits into the groove in the side of the tool post, as you see in Photo 8. The bottom surface of the groove should be 3/8" below the center height of the lathe. There is some variation in height from the top of the compound rest to the center height in individual lathes; determine what is needed for your machine.

8

Many American lathes come equipped with a lantern-style tool post similar to the one shown in Photos 1 and 9. Normally a concave ring and rocker are provided with the lantern-style tool post for positioning a standard toolholder. It is a bad idea to use the ring and rocker for mounting the Diamond Toolholder, however, since the shank must be kept both horizontal and at the correct height. It is better practice to use a plain ring of a fixed thickness to get the correct height. By using a plain ring without the rocker, the shank of the toolholder is automatically secured in a horizontal position.

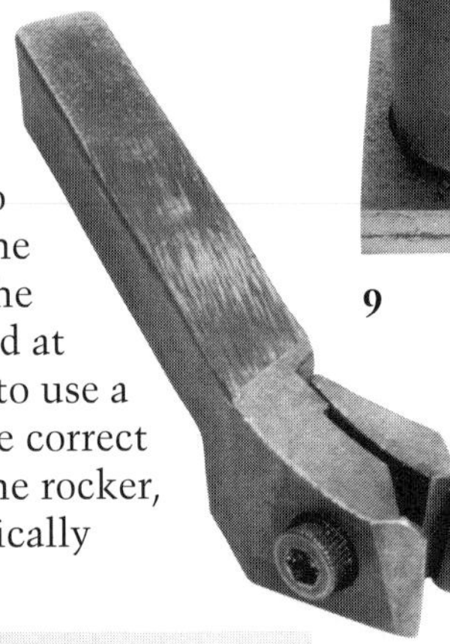

10

Refer to Photos 1 and 9 where you can see the plain ring I made to use with my lantern tool post. The ring was cut from a piece of 2" diameter steel rod. I mounted a piece of the rod in a three-jaw chuck to face off one end. Then I used the tailstock drill chuck to center drill the part, and drill and bore the central hole. The diameter of the hole is about .010" larger than the outside diameter of the lantern tool post to obtain a slightly loose fit. Next, I reversed the part in the chuck to face off the second end and bring the ring to the correct thickness. Once again, this dimension must be determined for each individual lathe.

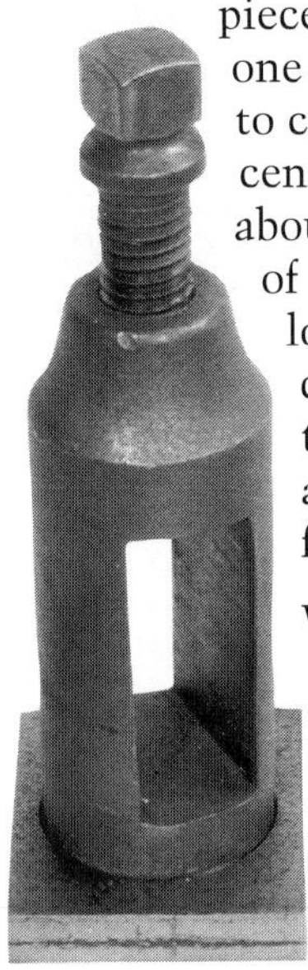

9

When the ring is in place, as in Photo 1, there should be a space of 3/8" between the top of the ring and the center of the lathe. In Photo 10, I was measuring the space to determine when I had obtained the desired thickness of the ring.

ADJUSTING THE TOOL BIT

The tool bit is in the correct position when the sharpened tip is exactly on the center height of the lathe. In Photo 1 you can see how I determined the setting of the tool bit by comparing its position to the tip of the lathe center. By sliding the tool up or down in the jaws on the end of the Diamond Toolholder, it was possible to make a precise adjustment of the tool bit.

The final proof of the setting of the tool bit comes when the lathe operator takes a facing cut across the end of a workpiece. When the tool bit is set at the right height, the facing cut is smooth right to the center. If the tool is set too low, a cylindrical pip remains at the center of the work. If the tool is set too high, a conical pip remains at the center. In either case, a slight readjustment is all that is required to set things right.

DIFFERENT MODELS OF DIAMOND TOOLHOLDERS

There are two Diamond Toolholders in Photo 11. The one at the rear of the picture is Model A9.5, and the one in the foreground is Model L9.5. Both are the right size for lathes in the 9 to 12" range. If you look closely, you will notice that the Model L9.5 toolholder in the foreground has a thinner shank than the other one. The shank of the Model A9.5 is 1/2" from side to side, while the shank of the L9.5 is 3/8" from side to side. The open-sided tool post will accommodate either model.

You may have noticed that Photos 1 and 2 each show a different lantern tool post holding a different Diamond Toolholder. The central slot in the lantern tool post shown in Photo 1 is a loose 1/2" wide, and the one in Photo 2 is a loose 3/8" wide. If you plan to use a lantern-style tool post and fit it with a plain ring, measure the width of the slot in your tool post to obtain the model Diamond toolholder that will fit it correctly.

11

The Model M6 Diamond toolholder is smaller than the other models described in this article. In Photo 12, you can see a Model M6 mounted in the standard tool post of my Sherline lathe. The standard tool post supplied with the Sherline is just right for it. The procedures for grinding and adjusting the tool bit are the same as for the A9.5 and L9.5 models.

The cross slide of the Sherline lathe has two parallel T-slots for mounting the tool post. You will notice in Photo 12 that the lower end of the tool bit comes below the top of the cross slide. This is no problem, as long as you put the tool post in the T-slot on the left side. If you must use the T-slot on the right side, it is a simple matter to cut a short piece off the bottom of the tool bit.

When the headstock of the Sherline lathe is raised on the elevator block, as in Photo 13, the elevated tool post must be used. As you can see, it can be used directly with the Model M6 Diamond Toolholder and provides all the needed clearance in all mounting positions.

12

13

The How and Why of Tangential Cutting

by Desmond Burke

This discussion had been drafted with the intention of simply demonstrating to the reader the difference between a conventional lathe tool and the Diamond Tangential System. However, as I have received so many queries regarding the use of the Diamond toolholder, and requests for more detail and a clarification for choosing the correct tool for each machine, I will make an effort to elaborate. First, however, I will offer an explanation of the term "tangential tool."

In this system, the tool bit is held vertically with the cutting point contacting the workpiece on center, thus putting the tool bit at a tangent to the work as opposed to the conventional tool which sits at right angles (Figure 1).

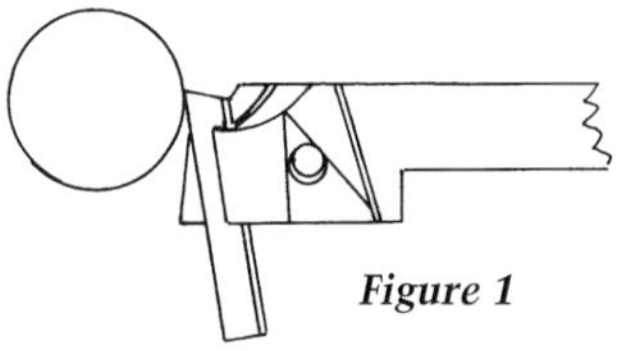

Figure 1

The cutting medium is high speed steel (HSS), which has been around since the turn of the century and, despite the advances of technology, has remained popular for many applications, such as low volume and jobbing work. Also, it is economical both in cost and life (usage) which helps it retain its place in so many forms, not only for turning but also for milling, drilling, etc. Solid carbide or stellite may be used in this system, but HSS is quite adequate for general purpose work.

The tool bits for the Diamond system are the standard square HSS available from any local supplier (only inch sizes are used). They are referred to by some manufacturers as precision ground, but usually the average types available are not always square, nor are the sides of equal dimensions. Nevertheless, clamping by friction proved not only adequate, but to us imperfect mortals, a blessing when we erred; under extremes, the tool bit may move down in the holder rather than hogging in or, just as undesirable, having the workpiece ride over the tool bit.

The attachment of the tool bit to a holder was the first design step. In the past, tangential cutting was achieved by specially designed serrated tool bits or a holder with a screw beneath the tool bit for adjustment and retention; the latter tended to make the whole assembly bulky.

CUTTING ANGLES

Cutting angles are an important aspect of the Diamond system. The aim was to achieve an effective tool to perform most of the basic turning functions, simply by grinding the end face of the tool bit.

The two angles that have the greatest variation for the cutting of various materials are the back and side rake. After various tests on a range of the most common workshop materials, a set of angles was adopted, based broadly on the American standard for single point tools. The front and side clearance for cutting are built into the clamping angles of the holder. The point radius is achieved by a light hone along the length of the tool bit and, therefore, will be maintained for the life of the bit.

SHARPENING

Sharpening has always been a problem for successful turning, and the amateur can take heart in the fact that many experienced machinists (using HSS) never have the time to spend or inclination to achieve correct cutting angles.

A grinding fixture seemed to be the most logical approach, and after the usual trials and errors, it became possible to grind a given set of general purpose cutting angles on a standard bench grinder. The only variation suggested is sharpening for brass-bronze (Figure 2). This will give a tool zero back rake. But remember –

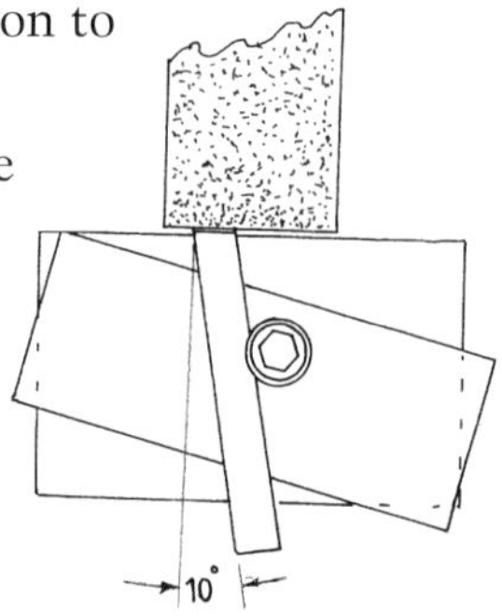

Figure 2

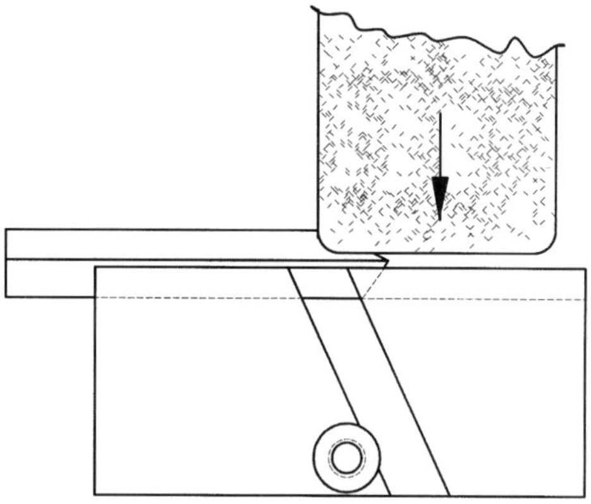

Plunge grind along the bit from the point for approximately 1/4"--rough out only.

Jack up with the screw to obtain 55° or 60°.

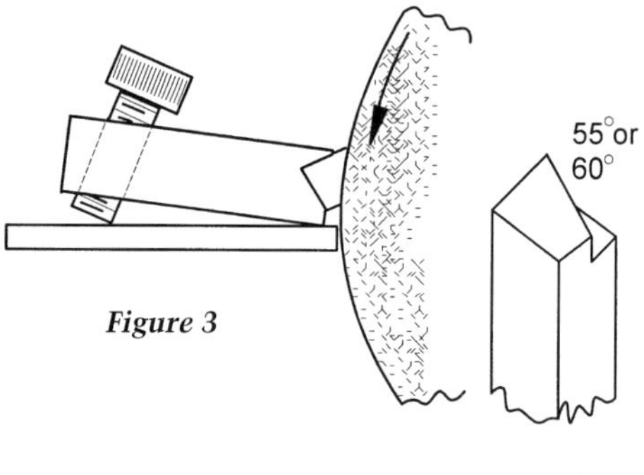

Figure 3

both of these metals use mechanical feed. Because these metals are soft, there is very little feel resistance with the saddle wheel, so a hog-in is quite common. One last small point: If the operator is inclined to hone the tool edge after grinding because the top end face is now hollow ground, a few rubs of a stone will give a keener edge.

TOOL LIFE

It is difficult to come up with any accurate figures on longevity, but because the fixture for sharpening gives so much control, it is necessary to remove only enough metal to achieve a sharp edge to the bit. This in itself adds considerably to the tool life.

THREAD CUTTING

At first, thread cutting was not foreseen as a part of the system, but a very simple and practical solution was found using the existing grinding fixture. First, the end face is ground as for general purpose cutting. Then the bit is placed in a large vee running along the length of the grinding fixture. The bit is ground along the length for approximately 1/4" from the cutting point (Figure 3).

Now comes the question of cutting angles; should zero (neutral) or positive rake be used on the top face of the tool bit? Either will give a correct thread, but the sequence of sharpening the tool is at times overlooked. The crucial point to remember is: whatever angle the tool is to have on its top face should be done first, and not altered after the sides have been ground to the required thread angle 55° or 60°.

Since this is an important aspect of the Diamond Toolholder, please bear with me while I illustrate this point. Figure 4 is an extract from the 35th edition 1939 of the South Bend booklet entitled "How to Run a Lathe." In this instance, a bit is being prepared for thread cutting using an inclined holder, and the tool is to have zero back rake. While the text is not explicit on the sequence, it becomes clear from the wording and the sketches that the top face is the first stage of grinding. If ground in the reverse (i.e., the 60° angle), then the neutral rake the thread produced will be considerably less than 60°.

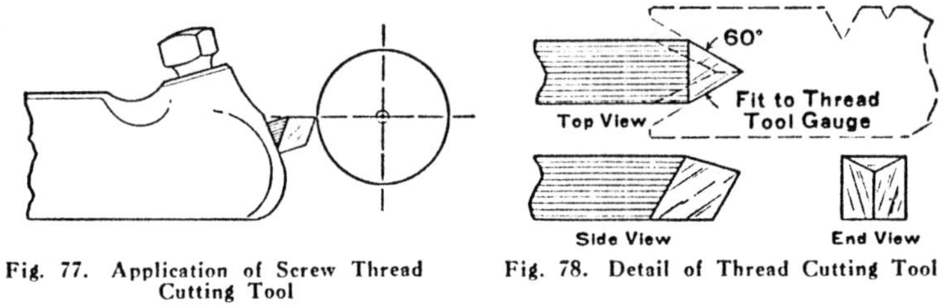

Fig. 77. Application of Screw Thread Cutting Tool

Fig. 78. Detail of Thread Cutting Tool

Figure 4

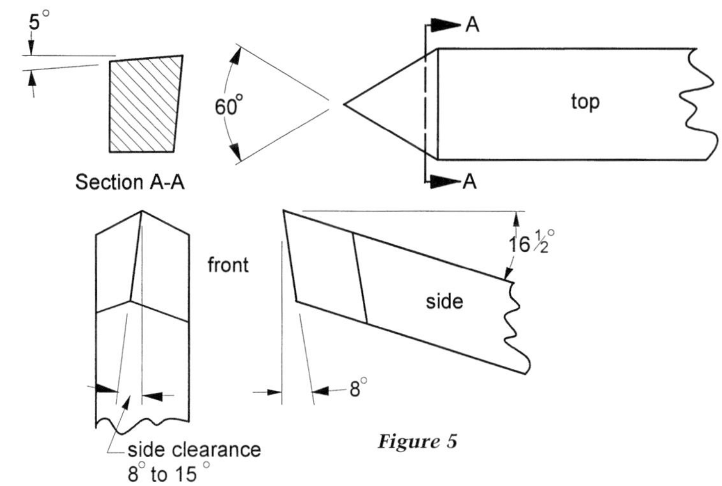

Figure 5

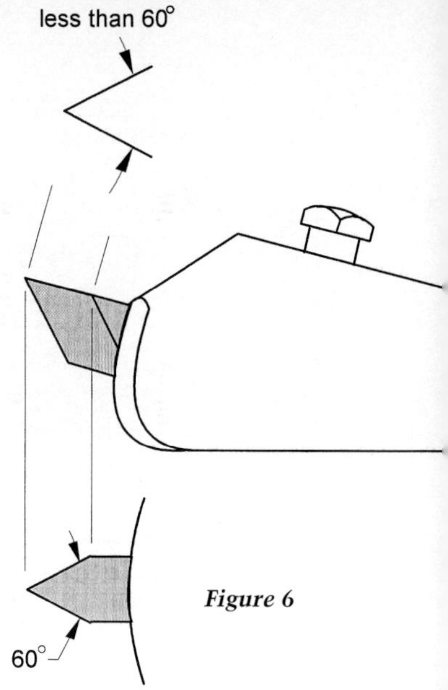

Figure 6

Here is a different approach. Figure 5 is reproduced from the 1941 edition of *Audel's Machinist*. This tool will also cut a 60° thread. The same tool bit located horizontally on center will give the same result (admittedly the front clearances will be excessive).

The instructions in Figures 6 and 7 have been taken from a textbook for technical education. The aim is to produce a 60° threading tool. The thread form will, however, be nearer 55° than 60°; the view in profile shows the gage positioned for a flat top tool not inclined to be put in another way. The angles are not the same where the gage is positioned as the angles of the cutting edge.

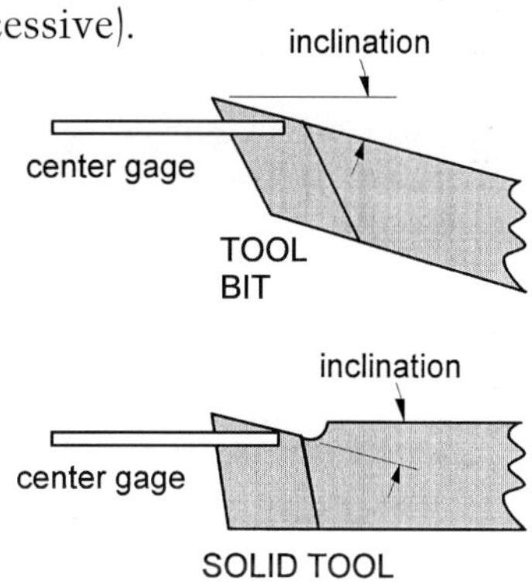

Figure 7

The Diamond system in Figure 8 shows the method of positioning the thread gage that will give a 60° thread. However, as a matter of interest, where the gage is positioned in the view marked X, the angle is nearer 70°.

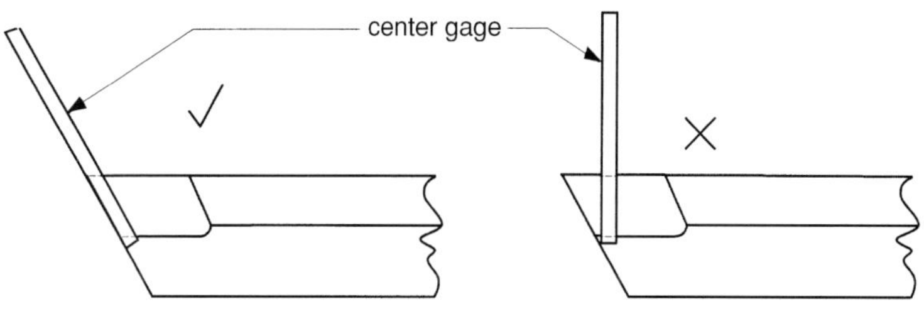

Figure 8

The Diamond Grinding Fixture produces general purpose cutting angles that reduce a 90° square tool bit at the cutting edge to approximately 80°. This is more than sufficient clearance to allow roughing cuts, finishing cuts, end facing and shoulder facing, all with the one tool bit. The other end of the tool bit can be

QUICK REFERENCE GUIDE
for RPM with HSS for Various Materials

TABLE 1		TABLE 2	
DIAMETER	RPM	MATERIAL	FACTOR
1/2"	800	aluminum	10
		brass	3
1"	400	bronze phos.	2
		copper soft	4
1-1/2"	300	copper hard	1
		cast iron soft	1.5
2"	200	cast iron hard	0.5
		free cutting steel	2
		mild steel	1
		medium hard steel	0.8
		hard steel	0.5

To find the correct RPM for various materials, multiply the rpm found in Table 1 by the factor found in Table 2;

i.e., free cutting steel 1-1/2" diameter = 300 × 2 = 600 rpm.

To calculate the correct RPM for diameters other than those in Table 1, a simple ratio may be established by applying the rule, "RPM is inversely proportional to diameter";

i.e., 1" : 400 rpm (Table 1)
hence 4" : 100 rpm

Figure 9

ground for thread cutting, all sharpened quickly and easily in the one simple grinding fixture, plus the angles required are maintained when resharpened.

CUTTING WITH THE DIAMOND TOOLHOLDER

As the top face of the bit is hollow ground, the swarf usually has a tendency to come away in a tight helical coil. Right-hand to left-hand tools are available only because of the volume requirements, but a good way of overcoming the need for a L/R-hand tool is to sharpen a cutoff blade as a left-hand tool, using it in a normal cutoff holder. This will meet most of the home shop machinist's needs. The usual methods of cutting apply: roughing and heavy cuts, medium feed; finishing cuts, light feed; and of course, each using the correct cutting speed.

Speeds are hard to judge and speed charts can be a nuisance. I have included yet another chart (Figure 9), but after having studied this for a short time, you will be able to file it along with a lot of other information that can't be found when you want it. Just remember that 400 rpm = 1" diameter and much of the rest will come with a little practice – especially mild steel and free machining steel.

How do you go about choosing the correct size of tool for a particular lathe? This has caused a great deal of trouble to some HSMs who do not know the terminology and machinist jargon. So, hopefully this can be clearly laid out.

Tool height is the measurement from the base of the tool to the cutting point. Lathe tool height is the measurement from where the base of the tool sits to the lathe center. So, normally, three things can occur. Let's assume the lathe tool height is 5/8". First, the lathe tool height and the cutting tool height can be the same, or the cutting tool can be less – say, 1/2" – which means 1/8" of packing is required, and last, the tool height is 3/4", which means you have a tool that is too high.

The Diamond Toolholder System tool heights:

M6	6 to 9mm	.236" to .354"
A8	8 to 11mm	5/16" to 7/16"
A9.5	9.5 to 12.5mm	3/8" to 1/2"
B16	16 to 19mm	5/8" to 3/4"

As can be seen from the above figures, each tool has a minimum and maximum tool height setting.

Example: If the lathe tool height is 9/16", a B16 is too high. A9.5 is too low, but add a 1/8" packing piece; this will bring an A9.5 to the midrange of the tool height.

Normally, however, a lathe with a four-way tool post has a lathe tool height of either 5/16", 3/8" or 5/8". Some Asian 12" swing lathes are being produced with a lathe tool height of 12mm. This then means that a fairly broad range of holders is available for the average HSM.

The last point, which I hope may be helpful, is that each tool has been designed to have a minimum setting of 1/4" from the cutting point to the top of the clamping area; as shown in Figure 10, this allows clearance for the front of the holder to the workpiece.

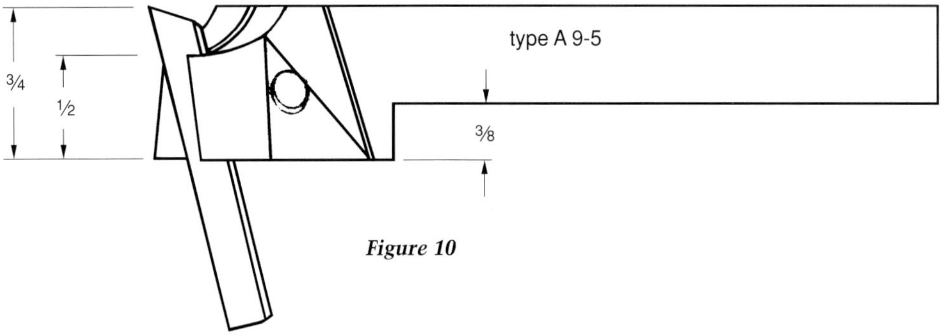

Figure 10

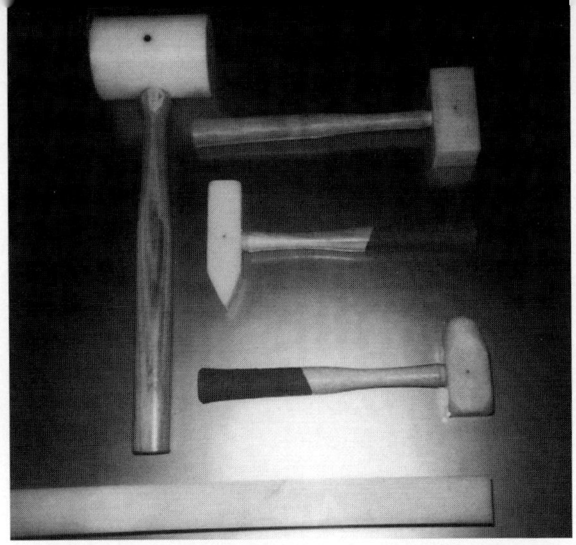

Engineering Plastics

by Thomas W. Dowling III

I would like to share a simple, useful project with the other readers of *HSM.* Most home shop enthusiasts divide projects into two classes: metal and wood. I have found engineering plastics to bridge the gap between the two traditional materials in properties and purpose. As I am also a woodworker, I found that wooden mallets are unsuitable for striking steel-capped chisel handles, while metal hammers quickly ruin handles made entirely of wood. Commercial "no-mar" hammers are available, of course, but I wanted to try to make one. I wound up making four.

I used nylon plastic for all four mallets, and fitted them with wood handles. The tools necessary are a drill press, a lathe (even a wood lathe will do if it has a faceplate or a chuck), a saw (band, circular – with a plastics blade or even one of those inexpensive "hard tooth" carpenter's hand saws), a plumber's reamer and some medium coarse files. A milling machine could also be used, but is not essential.

The first nylon mallet was made from a cutoff block 2' × 1-3/4" cut to 4" long. In two stages, I drilled the handle hole to 9/16" using an S&D drill in the drill press. I next reamed one side somewhat conical using the plumber's reamer; this became the bottom, the tapered opening fitting the shoulder on the handle quite well. All that was left before assembly was to file a slight bevel on the edges and face to avoid chipping and give a finished appearance. The handle, which I bought at a flea market, had a round top instead of square or oval, so I avoided any milling or filing of the eye, but I needed to drill a small cross-hole and pin the head to avoid turning or loosening. I chose not to use a fiberglass handle nor glue the wood handle in. This way, if the faces become worn I can press out the cross-pin and handle, then true up the face as needed and reassemble.

The next two were made for sheet metal work. I made them from similar sized nylon cutoff stock. The handles were 5/8" at the top so I selected a drill that size and reamed as before. One of them I centered in a four-jaw chuck and turned a series of bevels approximating a hemisphere. Then I lathe-filed the remaining ridges away leaving the nylon shaped much like a ball peen hammer. The other I sawed into a cross peen shaped and dressed up by hand filing.

The last is a big bopper, made from 3-5/8" round stock, 5-1/4" long; the head weighs about 2-1/2 lbs. I made the 18" handle from the butt of a sledge hammer handle that had broken off near the head. I use an old Craftsman cross center and a Delta live cup center to adapt my Atlas metal lathe for wood turning when I need to. I place a toolholder upside down in the tool post and use the flat bottom as a tool rest. I turned the part that goes inside the head to 1" diameter, plus a bit, as measured by old-fashioned spring calipers. With wood you don't need thousandths! Drilling was the only difficult operation. Nylon tends to be grabby, and my drill press is not that powerful. I used just about every "eighth" S&D drill to get the job done in steps. This violates the rules of hole enlarging, but the nylon is not likely to damage the drill bits, and loss of precision in the center was not a critical factor, either. I reamed one side conical as with the smaller mallets. I beveled the faces by chucking the round nylon block in a three-jaw chuck and lathe filing at a fairly slow speed. After assembly, I drilled a 5/16" cross hole and used a roll-pin. I cut off the excess handle close and filed it to neatly match the curve of the head. The tool cost about $8 to build and is comparable or superior to commercial products costing over four times that.

Cadillac plastics is my favorite source for nylon and other plastics. They have numerous branches, most of which will sell cutoffs and remainders at a reasonable price to small quantity users, ideal for the home shop. I have also bought, for other applications, acrylic, polycarbonate, PVC, ABS, PTFE, phenolic laminate, acetal, and lubricant filled plastics.

Different plastics vary in properties at least as widely as do metals or woods. Some operations on some plastics will be better accomplished treating the material as one would a soft metal such as aluminum and *machining* it. Most plastics cannot be worked with wood drilling bits, for example. Other materials and operations are more suited to a *woodworking* approach. A table saw with a triple-chip ground carbide blade suitable for hard woods is fine for cutting sheets, although special plastics blades are available. Sometimes both approaches work well. In nylon, you can bevel edges with a file (or milling machine!) or get a much smoother bevel using a block plane, although the plastic dulls the carpenter's tool much more quickly than woods will. I suggest that home machinists can extend their scope by more fully exploring the third class of materials – plastics, not just as an occasional item in metal projects.

Sharpening Tungsten Carbide Drills

by Walt Netzel

You can sharpen a tungsten carbide masonry drill on an oil stone, and with it you can drill hard items such as broken taps and easy-outs.

Masonry drills start at 1/8" and increase in 1/16" increments. They're available in most hardware stores and cost about $2. But they're not sharp enough to cut hard metal as they come from the package.

The preferred tool for sharpening this drill is a diamond wheel, but it costs about $100 and can be damaged easily. Another choice is a green silicon carbide wheel; it costs only about $30 but it wears rapidly. (If you use either of these wheels, don't cool the drill in water – it may crack.) For the occasional sharpening job, a silicon carbide oil stone is adequate.

An easy way to sharpen a drill on an oil stone is to use a jig. One can be made from a short length of 2 × 4, a large dowel, and a scrap of plywood. It will hold the drill at the proper angles. With these drills, the clearance angle is 24° rather than the usual 12°, and the lip angle is 64° instead of the usual 59°.

To make a jig, cut a 2 × 4 18" long and 3-1/2" wide. Tilt the saw blade 64° from horizontal, and rip a long wedge off one edge, leaving the 2 × 4 3-1/2" wide with one beveled edge. With the saw at the same angle, saw a kerf, or slot, parallel to the beveled edge 1" in from it and 12" long. Drill a wood screw clearance hole in the 1" projection, about 1" from the end, and just deep enough to go through the saw kerf. Finally, cut a 4" long roller from the dowel.

An 8 × 12" piece of plywood is used to hold the 2 × 4 at the proper angles while drilling several holes through the saw kerf.

It is temporarily nailed to the beveled edge. The width of the 2 × 4 should be held at 26° from horizontal; the length at 24°. To achieve these angles, draw a diagonal line on the plywood 24° from the bottom edge, starting at the front corner.

Nail the bottom of the beveled edge to the line; space it 3-1/8" back from the front corner of the plywood. If the plywood doesn't stand vertical with the 2 × 4 nailed on, pad the lower corner of the 2 × 4 until it does.

With the 2 × 4 held at the proper angles by the plywood, drill a series of holes straight down through the saw kerf. Drill a separate hole to fit the body of each different size drill you intend to sharpen. Start 1/2" behind the wood screw hole and leave about 1/2" between them. A drill press will insure that the holes are perpendicular. Remove the plywood after drilling.

To use the jig, insert a drill in the proper hole and turn it so the cutting edge is perpendicular to the long axis. Clamp it in place with a wood screw through the predrilled hole. Put the roller under the back end of the jig. With the drill resting on an oil stone, level the jig by adjusting the depth of the drill.

Make several passes over the stone, then turn the jig over and look at the bit. If the fresh-grind mark isn't centered on the flat face of the bit, you can adjust the lip angle by rotating the bit slightly and change the cutting angle by varying the depth of the drill in the jig. If you don't exactly match the original angles of the drill, you will have to remove more metal than is necessary. When the lip is about half sharp, loosen the wood screw and rotate the bit 180° and work on the other lip. It's easier to keep the lips equal in length if you grind them down alternately in small increments.

Be careful not to wear a rut in your oil stone. Use a circular motion and try to distribute the wear over the entire surface of the stone. Tungsten carbide will wear your stone faster than steel, but it doesn't require a lot of pressure to cut rapidly.

When sharpened, tungsten carbide bits will cut most hard metals. Try one on anything too hard for titanium nitride-coated drills. It will cut case-hardened surfaces, saws, drills, glass, even files. A sharp bit will also drill concrete faster. Turn it fast, keep it sharp, and use lots of pressure.

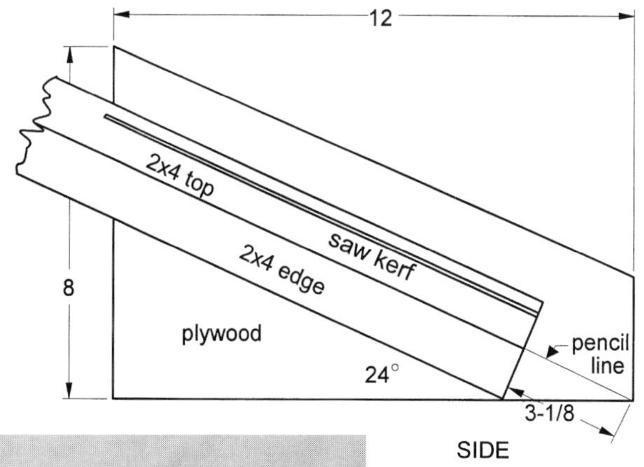

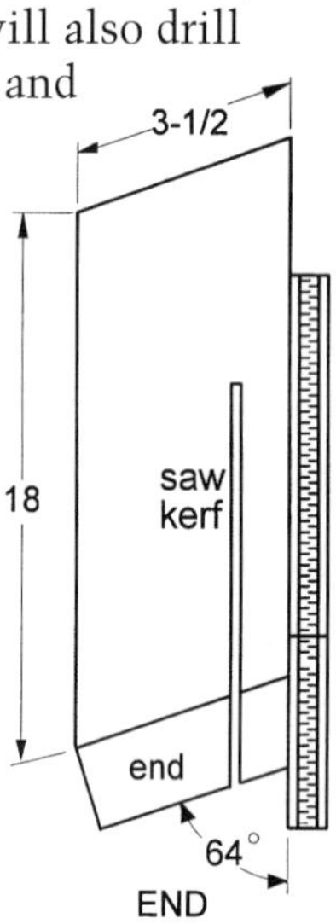

The jig in use.

The old Model C South Bend now converted to a Model B. This is a later photo, taken after I removed the leveling nuts and replaced them with regular bolts and shims. The bolts will be replaced shortly when I finish making new and larger leveling nuts. The original ones were too small. The photo also shows the new lead screw with the milled groove that powers the carriage. Below the lead screw to the right of the carriage, one can see an unpainted area. This is where part of the bed casting was removed to fit the new Model B apron. Later model lathes did not have this rounded area.

The Rebirth of a Model C South Bend Lathe

by Douglas Graham

Photos by Author

From day one, I had always wanted a small South Bend bench lathe. Between one thing or another, its purchase and possession had always eluded me.

While attending a trustees' meeting of the American Precision Museum in January 1994, I spoke to one of the other trustees about acquiring a 9" South Bend. He said, "I have one in my barn you can have." I had always thought an offer like this happens only in daydreams!

Due to an unusually heavy snow that year, I was not able to drive down to the barn where it was stored until April. However, in the meantime, the gentleman was kind enough to snowshoe to the barn from his house and get the serial number – 89,431. I immediately called South Bend Lathe and obtained a parts list and lubrication instructions. I was also informed that the lathe left the factory on April 19, 1939, headed to McKensie Machinery Company in Boston, MA, who then sold it to the Sanitary Service Public Works, also in Boston. From there, where it went was a blank. Working backward, the person from whom I got it had it in storage for 25 years, and the person from whom he got it made tobacco pipes with it for many years. So much for history.

When I got the lathe in my shop, I completely stripped it down into a lot of small pieces to see what was necessary to do to make it run again. I made up a list of missing parts: two gear covers, one end-gear guard, and some change gears. Thanks to Joe Bergamo of Plaza Machinery in Bethel, VT, and Barry Aronson in Massapequa Park, NY (both have been long-time advertisers in the *HSM* "Machinist's Marketplace"), I obtained what I needed. Several small parts were easily obtained from the South Bend people who are extremely helpful.

Finally, I re-assembled the little lathe, and gave the various parts a coat of South Bend gray – the color they had once been. I then carefully leveled the lathe, using a system somewhat similar to that described in Tubal Cain's book *Workholding in the Lathe* on pages 101 to 109.

Using new three- and four-jaw chucks, I made test cuts on various pieces of bar stock. All cuts indicated a .004" to .006" taper in the cut. Various adjustments using shims on both front and rear nuts and bolts of the tailstock end of the lathe did nothing to relieve the taper.

I removed the chuck from the spindle and inserted a 3/4" diameter piece of bar stock in the headstock spindle hole. Then I set up a dial gage made by Draper and Son (English manufacturer) to indicate on the spindle. By my lifting on the bar stock in various clock positions, the dial indicator, which reads in .0001 parts of an inch, indicated movement at the 7 o'clock and 3 o'clock positions. These tests indicated that the bearing hole in the headstock was definitely out of round – an interesting problem to solve.

At this point I did not want to give up. Remember, when this lathe was built in 1939, the bearing surfaces for the lathe spindle were bored straight through the headstock casting. There were no bearing inserts to change like those that came with the later model South Bend lathes. As I analyzed the problem, four choices emerged: 1) find another not-worn headstock casting; 2) make a whole new casting; 5) make a special bearing insert; or 4) bore out the worn bearing area and put a sleeve on the lathe spindle.

I called Rudy Kouhoupt, since he had acquired a very similar South Bend with the same headstock as I had. The main difference was his lathe was vintage 1942. Many of you have seen his lathe in numerous *HSM* articles. It also has a two-step sheave on the jackshaft. Rudy's advice was excellent: "Recheck everything – especially the chuck, to make sure it not bell-mouthed. A bell-mouthed chuck will give poor readings. Don't fix anything that is not broken."

The next step was to check with a close friend and mentor – Edward McCormack – who owns and operates a machine shop in West Rutland, VT. Eddie has operated a very successful machine shop for nearly 50 years, and there are few problems he and his sons have not encountered. Ed reviewed my tests and confirmed what I had discovered earlier – worn areas at the 3 o'clock and 7 o'clock positions. He decided that the best procedure would be option 4 – bore out the worn bearing area – then grind the large bearing surface diameter of the lathe spindle down just enough to accept a thin sleeve.

I then stripped and removed the headstock casting from the lathe. Prior to removing the casting, I carefully marked its location on the lathe bed, thoroughly cleaned the casting, and carefully set up and indicated on my horizontal Milwaukee milling machine. Prior to machining the headstock, I made a dummy bearing shim and fitted it to go in the large bearing slot with t he bearing cap screw removed. When satisfied with the proper fit, I put in and tightened the cap screw. This would keep the milling tool from damaging the edges of the gap where the regular adjusting shim fit.

When Eddie's son Gary (who is an excellent machinist) was satisfied with the casting alignment, he took a series of very small cuts, which started to open up the original and worn bore. The "new" bore (for the large bearing diameter of the spindle) had to be enlarged by .017"; just enough material was removed to return the bearing hole to round.

We turned our attention to the spindle. Attempts to reduce its diameter by turning in a lathe were not successful. This problem was quickly cured by grinding down the spindle bearing surface with a tool post grinder. Gary left a 3/32" shoulder on the end of the spindle against which to seat the new spindle sleeve. This same 3/32" shoulder also provides the seating surface for the chucks and faceplates that are screwed on the threaded end. We selected a piece of 4142 chrome moly to make the new sleeve for the large bearing surface of the spindle. In this case, the wall thickness for my new sleeve was approximately 0.070". The bore of the new sleeve was slightly smaller than the reduced diameter of the lathe spindle, and its OD was slightly larger than the new bore in the headstock casting. The new sleeve had a finished length of 2-1/4". Gary warmed it just enough to press it on the reduced diameter of the spindle, and butted it up tightly against the 3/32" shoulder left on the spindle. When the spindle and its new bearing surface had cooled to room temperature – not very long – it was set up between centers, and ground with a fine wheel. Then Gary polished it to provide an extremely smooth surface as well as a "just" fit into the new bore in the headstock. The job took Gary about four to five hours to complete.

I brought home the reworked parts – the headstock casting with its new bore and the spindle with its new bearing surface – and reassembled all the parts as before. During the reassembling, I made two new spindle oiling wicks with wool and springs to replace the ones long since missing. The original oil drain groove in the bottom of the large bearing hole was slightly deepened such as it was when new. I also inserted new adjustable shim packings in both front and back bearing slots. I then added South Bend spindle oil (most other spindle oils are also acceptable) to each oil reservoir, and gently snugged each bearing cap screw into position. I had to be very careful with these cap screws since it took little tightening to bind the new spindle.

In case the reader is wondering, absolutely no work was done on the rear bearing or spindle surfaces (that bearing next to the change gears). As Rudy said earlier – "If it isn't broke, don't fix it."

With the oil in the reservoirs, I rotated the spindle in its new bearing, and adjusted the two bearing cap screws to just allow the spindle to turn without binding. I screwed a four-jaw chuck on the spindle and centered a piece of 7" long round bar stock about 1-3/8" in diameter in the chuck jaws. Then I set the round bar back into the chuck jaws about 1-3/8".

I took a series of light cuts on the bar and miked the bar after each cut for any taper. In my case, it was necessary to add shims (or adjust the leveling nut) under the right rear tailstock foot. After the fourth cut and adding a few thicknesses of shims, I ended up with a bar that had a flat cut with no taper. The diameter of the bar across the length of the cut varied between 0 and .0001". I took two more cuts and got the same result. For me this was success, and certainly worth all the trouble I went through. It was once said, "Old fiddles play good tunes." This was certainly true in my case.

Now that the lathe would cut properly, the next job was to convert it from a Model C to a Model B. I wanted the power

This photo shows the large spindle bearing that was bored out .017", and the lathe spindle that was sleeved to fit the new bore.

cross-feed as well as the power longitudinal feed. The Model C had no power cross-feed and depended on the lead screw threads to supply the power longitudinal movement for the carriage, both for turning and cutting threads.

After another visit to Joe's Plaza Machinery, I had an excellent lead screw with a milled slot running the full length of the threads. There were no wear marks on the threads or anywhere else; I also obtained a complete apron in excellent condition and a cross-feed screw with the milled gear on its shaft. The screw was also in top-notch condition. From South Bend Lathe I got a new cross-feed nut and a new gear chart plate for the Model B (I had previously gotten one for the Model C). The old lead screw was greased and carefully stored for possible future use.

On the front of the Model C bed and at the bottom is a small semi-round bead, which is part of the frame casting. Enough of this bead was carefully removed to allow the new apron to fit in place. (Later South Bends for the A and B models did not have this bead on the bottom of the frame casting.)

I carefully checked the mesh of the cross-feed screw gear against the mating gear in the apron; a tiny amount of fitting was required. I carefully adjusted the new cross-feed screw nut height-wise to insure that the new screw with its drive gear would not bind.

As built, this lathe was designed for six speeds (three direct and three using the back gears). The speeds range from 41 rpm to 658 rpm with a countershaft speed of 300 rpm. At some point, I plan to install a second step on the countershaft to fill in the speeds between 658 rpm and 41 rpm. These speeds are fine for my metalworking projects. In the event I need higher speeds, I can fall back on my 12" Atlas, which has 16. The previous owner who used this lathe to make tobacco pipes may

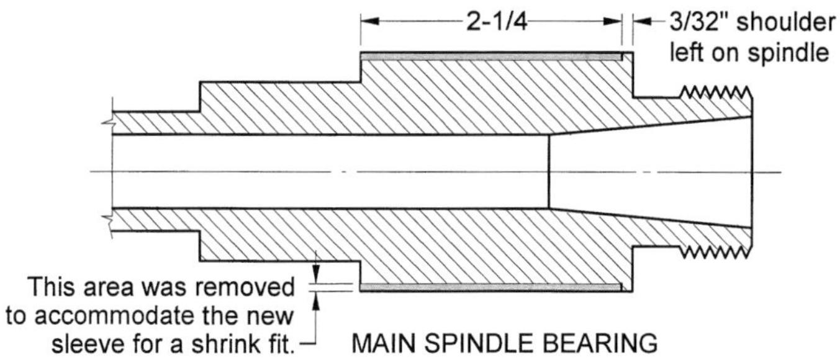

The 4142 chrome moly sleeve for the new large bearing surface has a length of 2-1/4" and a wall thickness of approximately .070". It was warmed enough to slip over the ground-down area of the original bearing surface and fitted up very tightly to the 3/32" shoulder left on the spindle. The new bearing surface was then ground and highly polished to fit the new bore in the headstock.

It is very difficult to give exact dimensions since each lathe and its bore wear are determined case-by-case. The machinist doing the work will quickly realize this. As I explained in the text, Gary McCormack removed only enough of the old bore to clean up the hole. He explained that in my lifetime, I would probably never wear out the 4142 chrome moly sleeve, and if I did, another could easily be installed.

Since this lathe is a bench type resting on two 1/2" thick steel plates (8 × 10-1/2") bolted to a 2 x 6-foot wooden bench with a 2" thick plywood top, and bolted to the shop floor, its accuracy is checked about every six months. Depending on shop temperatures, the wooden bench will change slightly over a period of time.

Shown here is the apron that was fitted to the saddle with a lever for cross and longitudinal feeds and a friction clutch knob.

have destroyed the large spindle bearing surface by running at the high speeds needed for good turning. A good book to have on hand for lathe adjustments is *Keep Your Lathe in Trim,* Bulletin H-4 by South Bend Lathe Works and reprinted by Lindsay Publications.

I am sure many readers have other solutions they could suggest, but this is the way I repaired my problem. There must be hundreds of these abandoned, older lathes around that could be put back into use with a similar solution. The old lathes, when repaired properly, will do almost anything a new one will. A repair like this would make a good technical school project. It does not, however, take into account *any* other problems the lathe might have, such as a badly worn, warped, or damaged bed.

If someone elects to tackle a job like this, please make sure you completely understand what needs to be done and how it should be done. Most good, reliable machine shops are capable of this sort of repair, and fully understand the work. While the job is fairly straightforward, care, accuracy, and patience are the key issues.

While ordering some parts from South Bend Lathe, I explained to the service representative what I had done, and asked if SBL could do an overhaul similar to this. He said they would have to rebuild the entire unit due to liability problems, and that cost could possibly run between $1,500 to $2,000, depending on what had to be done. He also explained that new import lathes were considerably less expensive than repair costs. (The service representative also told me *not* to run the lathe spindle much over 1,100 rpm due to the type of bearing it has.)

Another excellent reference book for the reader is *Machine Tool Reconditioning and Applications of Hand Scraping* by Edward F. Connelly. It was published in 1955 by Machine Tool Publications in St. Paul, MN. The book has several illustrations and photos of small South Bend lathes being reworked.

The reader is also referred to an excellent article "Aligning a Lathe" by Steve Wellcome, which appeared on Page 34 of *Projects in Metal,* August 1995. This article may help to relieve some of the frustrations inherent in lathe alignment, and will give the reader a hands-on understanding of the process.

LATHE ACCESSORIES

1 ***The compound slide is removed. At the left is Part 2, which fits into the base piece at the right. The base ring (Part 1) is also visible resting on the compound slide.***

A Unimat Headstock Adapter

by Conrad Huard

This is by no means an original idea. Anyone owning a Unimat lathe manual has seen how to machine the adapter to mount the Unimat lathe headstock onto a larger lathe. I am hoping some *HSM* readers would like to see how one home shop machinist made this adapter. My lathe is a 12" swing model LD1237 sold by Grizzly Imports – one often seen advertised in the magazine.

Start by measuring the exact lathe center height, measuring from the top surface of the cross slide to the lathe center. Also measure the height of the Unimat headstock from its base to center line. Notice that the adapter shown in the photos consists of two different pieces. You could machine it from one piece of stock if you don't mind wasting that much material.

The base ring (Part 1) shown is 4.875" OD, and such a big piece of round stock would create a lot of waste. This base piece need not be round. It can be square or rectangular. A firm base is required to support the Unimat headstock rigidly on the cross slide.

If you use round stock for the base piece, do not worry now about turning the OD to size between centers or in a three-jaw chuck supported by the tailstock center.

I started by chucking the base piece in a three-jaw chuck, and facing both sides. When facing the second side, make sure to face it to the thickness required, so when the Unimat headstock is mounted on it, it will be on the lathe center line. Drill and bore the ID hole to size. Now put the base piece aside and turn Part 2 from 2.50" round stock or whatever you have on hand. The small diameter .985" is a snug fit into the cross slide hole. The larger diameter is also a snug fit that goes into the ID hole of the base piece. As shown, both pieces are drilled and tapped for a setscrew. I used $1/2 \times 13$"

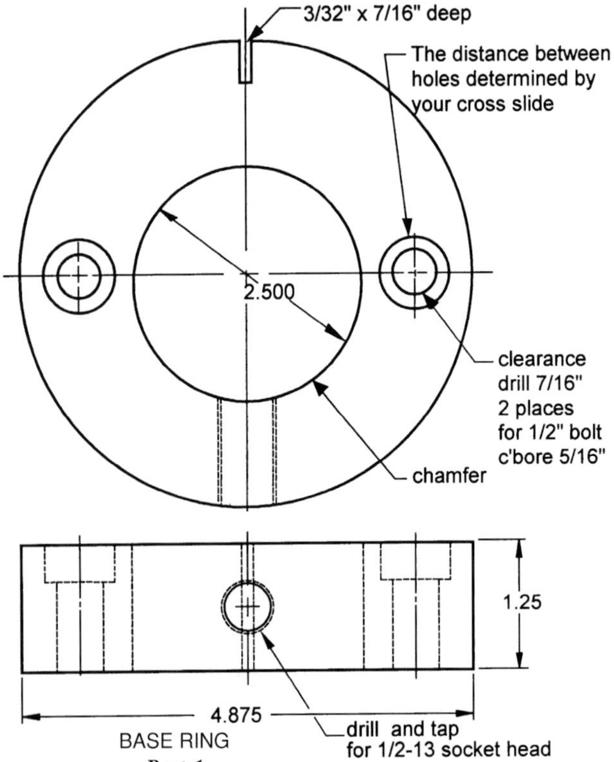

BASE RING
Part 1

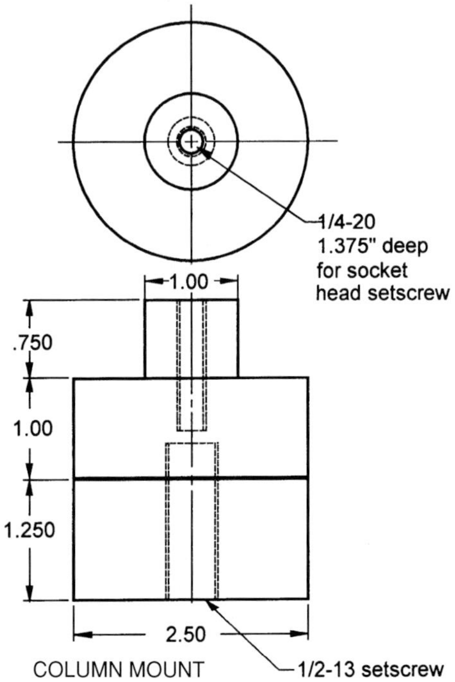

COLUMN MOUNT
Part 2

setscrews. The height of Part 2 need only be enough to hold the Unimat column rigidly.

For the most part, the dimensions shown in the drawings are not etched in stone. They are simply guides. Your own lathe cross slide and center height will be the determining factors.

Assemble Part 2 into Part 1, and tighten the base setscrew so the assembly will not turn. Mount the small end in a three-jaw chuck, and support the other end with the tailstock center. It would be a good idea to protect the end mounted in the chuck with a scrap piece of brass or aluminum to prevent marring. Turn the base to the required OD. This completes the machining.

2 ***The adapter is shown mounted on the cross slide. The Unimat headstock is now in the center line with the lathe.***

The adapter makes a very versatile tool. The Unimat heasdtock can be mounted on the base ring alone, and when it is, both headstocks will be in the center line. That being the case, it can be used for drilling, milling, or grinding from 0° to 180°. When the Unimat column is used, you are then able to drill, mill or grind above the work at any compound angle.

3 ***The Unimat headstock is shown mounted on the cross slide with a drill or milling column.***

The photos don't show it, but the adapter is finished with a product called *Amer-lene Gray*. This product is primarily used to produce a gray military phosphate-type finish on rifles, pistols and shotguns, and meets military specs for corrosion.

It is quick and easy to use. Full directions come printed on the container. The 8 oz. bottle makes 2-1/2 gallons of working solution. All that is needed is a stainless steel dish or container to hold enough liquid to cover the parts to be treated. Heat the liquid to 165°F and maintain this temperature while the part is immersed for three to five minutes. Rinse immediately under cool running water for 30 seconds. Dry with a clean, absorbent cloth, and spray immediately with *Amer-lene Seal*, which is a companion product to the *Amer-lene Gray*. Parts can then be handled right away.

I recommend cleaning the parts with *Acetone*, and then bead-blasting them prior to the *Amer-lene Gray* treatment. If you don't want the finish adhering to a given part, coat it with machinist's layout dye, such as Dykem machinist dye.

Amer-lene Gray (No. 3601) and *Amer-lene Seal* (No. 3702) can be obtained from Brownell's, Inc., or by contacting the manufacturer: Mount Labs, Inc., P.O. Box 1179, Golden Rod, FL 32733, phone (305) 3317659.

Mount Labs only sells in case lots. The minimum order is one case of each. *Amer-lene Gray* ($12.95) and *Amer-lene Seal* ($10.95) both come packaged in 8 oz. bottles, 12 per case.

4 ***The Unimat headstock is shown mounted on the adapter.***

A Quick Tumbler Gear Reversal Mechanism

by Ted Myers

Needing a wrench to reverse tumbler gears on the small lathe seemed to be a slow and awkward way of doing things. I changed my threaded bolt stop pins to a spring-loaded plunger that was quicker and self-contained. The idea was okay but too sloppy. Photo 1 shows a third generation reversal mechanism, which is an improvement on the first two. It is not only quick and self-contained but is adjustable for wear. A strip of 4 mil poly, which mikes .002", can be put between the gears to indicate the correct clearance when reversing. This mechanism uses a threaded rod to move the tumbler plate through the short segment of arc needed for reversal.

The drawings should be self explanatory, and only a few comments will be in order. Obviously, the dimensions will need to be adapted to any other small lathe. The ones shown are only to provide a starting point reference. The yoke is free to swing on the tumbler plate while the drive nut is locked down tight. The rod is extra long to provide clearance for the handle over the headstock index plate. It should be shortened if there is no plate. The two jammed 10-24 nuts on the threaded rod need to be in place before drilling and pinning. I lightened the cast iron disk by drilling and countersinking six 1/4" holes in a circle through the disk in the blank spaces. An aluminum disk would even be better.

1

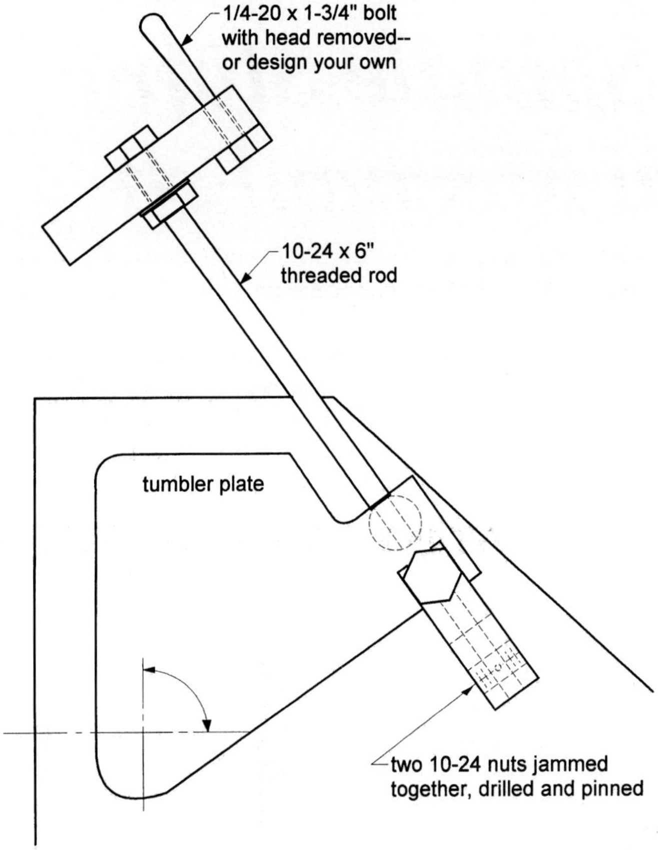

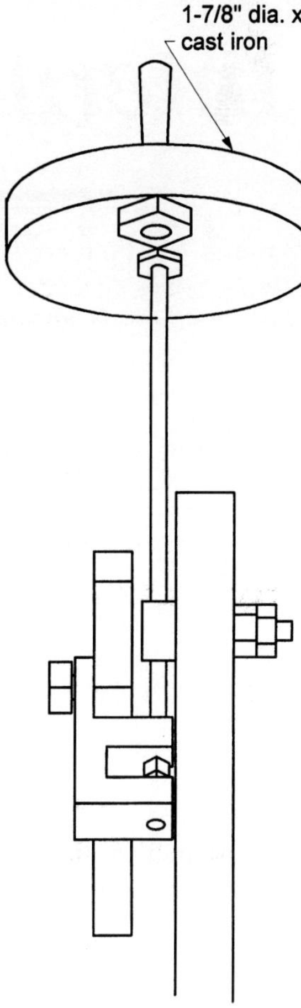

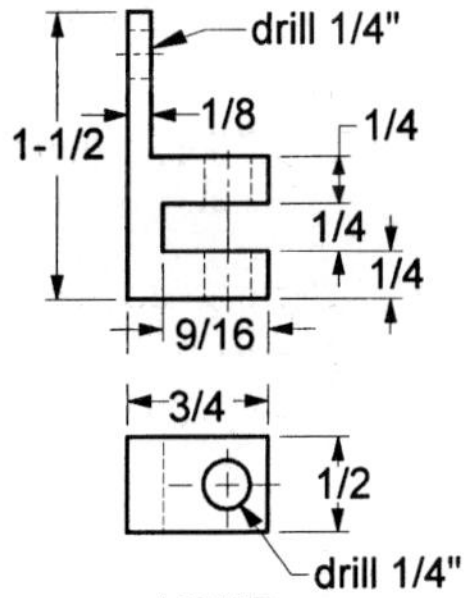

YOKE

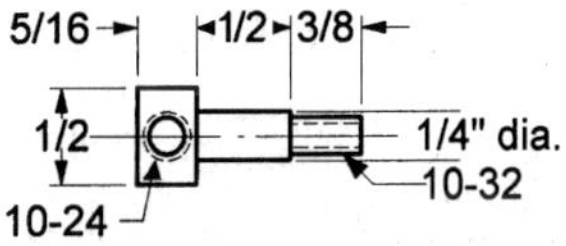

DRIVE NUT

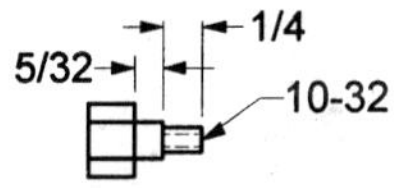

YOKE BOLT
1/4" bolt modified

Internal Spindle Stop

by Billy J. Blackmon

I had the occasion to make 760 shafts of CRS, 7/16 × 4-3/16" long, with two grooves spaced 7/16" apart on one end. Obviously, some system was needed. I am sure one can find many uses for a spindle stop of the type I devised for the job.

My lathe is a Jet 1236Y and the spindle bore is 1-3/8". It is actually just a few thousandths over that. In order to adjust the length of the 3/8" rod used as the stop for the lathe, the body has to be the length shown in Part 1. I obtained a length of 1-1/2" crs and reduced it to fit the spindle bore, and found the aluminum tube for Part 2 in the scrap pile. (Some days I get lucky.) This piece of tubing was 1-1/4" ID and 1-3/8" OD.

I then turned Part 1 to 1.25" so the tube (Part 2) would slide over it. The 1.25" diameter was sort of worked up the front end and the 15° turned so the two would meet. This taper needed to be polished fairly bright so it wouldn't wedge up the tube on removal.

I turned the rear portion to a diameter of 1" to accept the expander sleeve (Part 3), which was made from 1-1/2" crs. I cut the threaded portion to 7/8-14 to take the nut (Part 4). The end of the body was reduced to .760" (something less than 7/8" – I'm sure .750" would also do). This part was tapped 1/4-20 for the setscrew that would lock the stop bar in place. I drilled the opposite side 1/4" to take a length of rod so the nut could be tightened. Then the whole thing was locked in the spindle.

Part 2, as mentioned earlier, is aluminum tube. Make sure this tube is long enough to extend out over the 1" part of the body so the sleeve can push it forward and cause the expansion. The tube was held very carefully in the three-jaw chuck and trued up. I cut the taper into one end, leaving 1/32" flat, and then cut the eight slots with a band saw.

At first, the tube was left hollow and the resulting attempt was disastrous. The first two cuts went satisfactorily, resulting in four slots. The next cuts, however, bent the little fingers and the results promised to be a ruined tube. The second attempt was more successful. I cut the next ones, therefore, one at a time – a total of four cuts. These cuts were accomplished by holding the tube up at an angle of 45° to 50° and being very careful. Looking back with perfect 20/20 hindsight, a wooden dowel snugly fitted inside would probably have eliminated all difficulty.

Part 3 is also 1-1/2" crs. I turned the end next to the expander to 1.380" so it fit inside the spindle. This kept the whole thing from wobbling and the stop bar approximately centered. Tightening the nut pushed the sleeve forward driving the tube up on the taper; then everything was locked in the spindle.

The nut (Part 4) made the whole thing work. It is of 1-1/2" crs sliced off the end of the bar, faced on both sides, bored and then tapped. The flats were done on the milling machine by cutting the first flat to the desired depth, turning it over, and milling the same amount off the other side. If a mill is not available, cutting the flats can be done in the lathe chuck by turning off the desired amount, reversing, and then turning the other side.

The stop bar is 3/8" of the variety one purchases at the local hardware store, and isn't too expensive. A disk can be threaded on for work so long that it might miss the stop.

I've found this stop to be a very useful addition to my lathe. If for some reason the back gears are needed, the gear cover door cannot be opened because of the protrusion of the stop. At times that can be a frustration, so I altered the gear cover door to accommodate the stop.

If a collet chuck is used, the handwheel, naturally, is on the outside of the door. Everything is fine for turning to the required diameter. If it is to be threaded, you then have to remove the drawbar and disturb the work as well as the time involved to do all this. After much time studying the problem and the location of the lever, I decided it was possible to cure the problem.

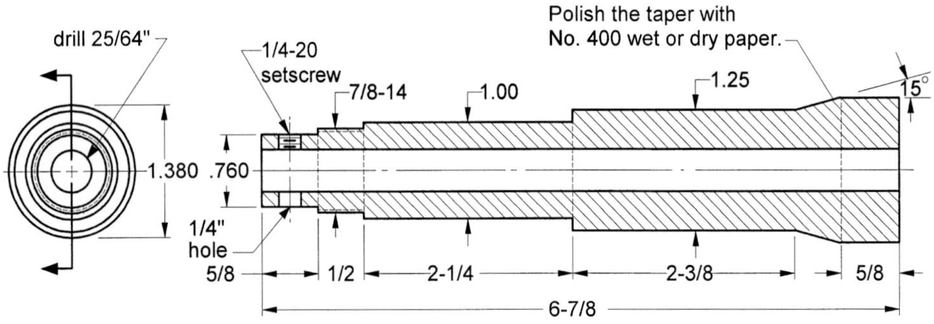

(1) BODY

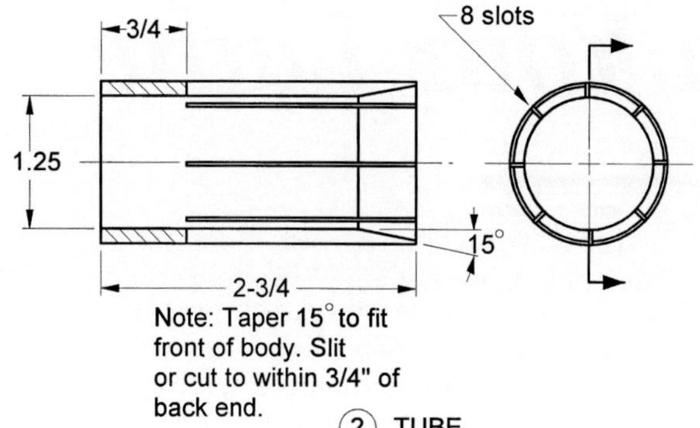

(2) TUBE
aluminum
1-3/8" OD

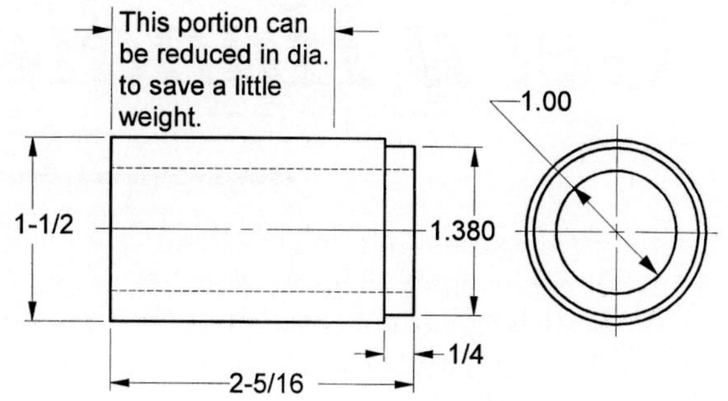

(3) EXPANDER SLEEVE

First of all, I turned off the power source (good thinking!) and took out the switch. The latter activity is a bit of a job, but not all that bad.

I then determined all the cuts and marked them in with a felt tip pen. Across the line at the rear of the door, I bored and tapped a hinge so all the holes were located properly. The main hinges will determine where the new gear cover door will go.

I then mounted the door on a Milport 9 × 42 on one end of the table so it would hang off the front of the table. The door was clamped to an angle plate mounted on the table. I put a couple of pieces of 4 × 4 wood between the C-clamps and the door. Then the ram was extended out and the head and ram rotated on the column in order to reach the work. I cut the back first, using a 3" slitting saw. On cutting the back and top, I backed off the saw to just the thickness of the door casting. The next cut was made on the top.

At this point, I turned the piece to align the saw with the ends of the cuts just made. I made a couple of dry runs down the table to make sure the bottom of the saw was lined up with the end of the two previous cuts, then I was ready to make the final cut.

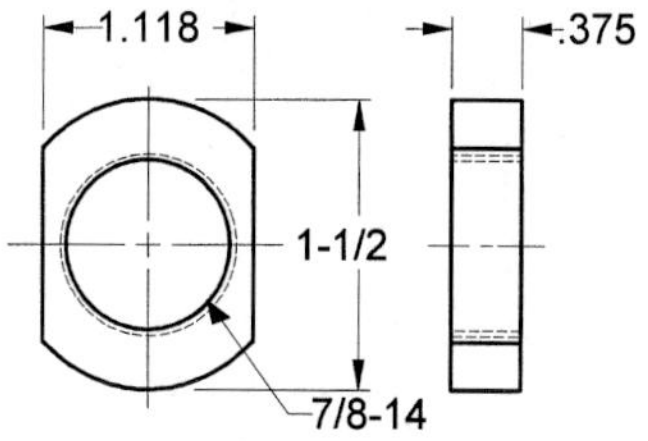

(4) NUT

If you're doing this, you won't be able to cut all the way through at the ends, but what is left is minimal and can easily be broken away and then filed smooth. After cleaning up the last cut, the door can be put back on the lathe. Put on your new door and admire the fruits of your labor for a few minutes as you contemplate the chore of reinstalling the switch.

If a milling machine is not available to handle a piece as large as this, the cuts can be made with a hacksaw. I first considered this since I was not sure the door could be mounted. The material is aluminum and cuts quite fast.

After you've used this little door a few times, you'll be glad you made the changes.

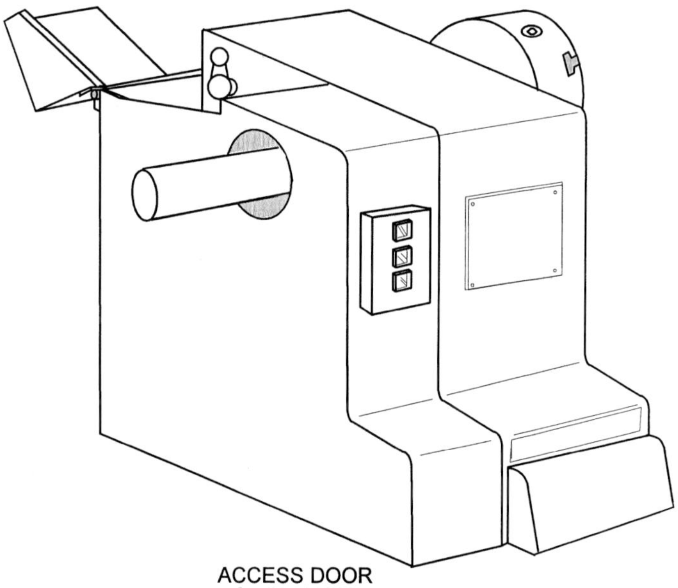

ACCESS DOOR
for Jet 1236
in opened position

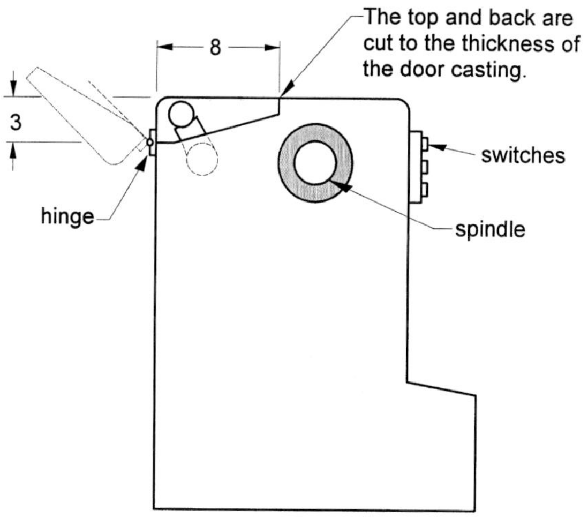

Jet 1236 Gear Cover Door

Get a Handle On Your Lathe

by John B. Gascoyne

A rather simple extension to a lathe lead screw will permit the addition of a handle, thus providing a means of accomplishing lathe carriage travel by hand. Travel by hand! Add a handle to the tailstock end of the lead screw on your Atlas Model 618 (or Craftsman) 6" swing lathe, and obtain fine, controlled, lateral carriage travel without using geared power feed.

In the late 1960s and early 1970s, I was still doing a lot of model railroad-related construction, some requiring very small parts. I found the rack carriage feed on my lathe much too coarse for the short, and somewhat precision, distances required for most of my small work. Also, the various powered carriage feeds, using the several gear setups available, were unsuitable. What to do? The solution to my need was to add a length of 3/8" diameter rod to the end of the lead screw on the tailstock end of the lathe. Over the years, I have found this little piece of 3/8" rod, which permitted installation of a handle on the lead screw, to be a bonus for all manner of projects, including some with rather lengthy diameter reductions.

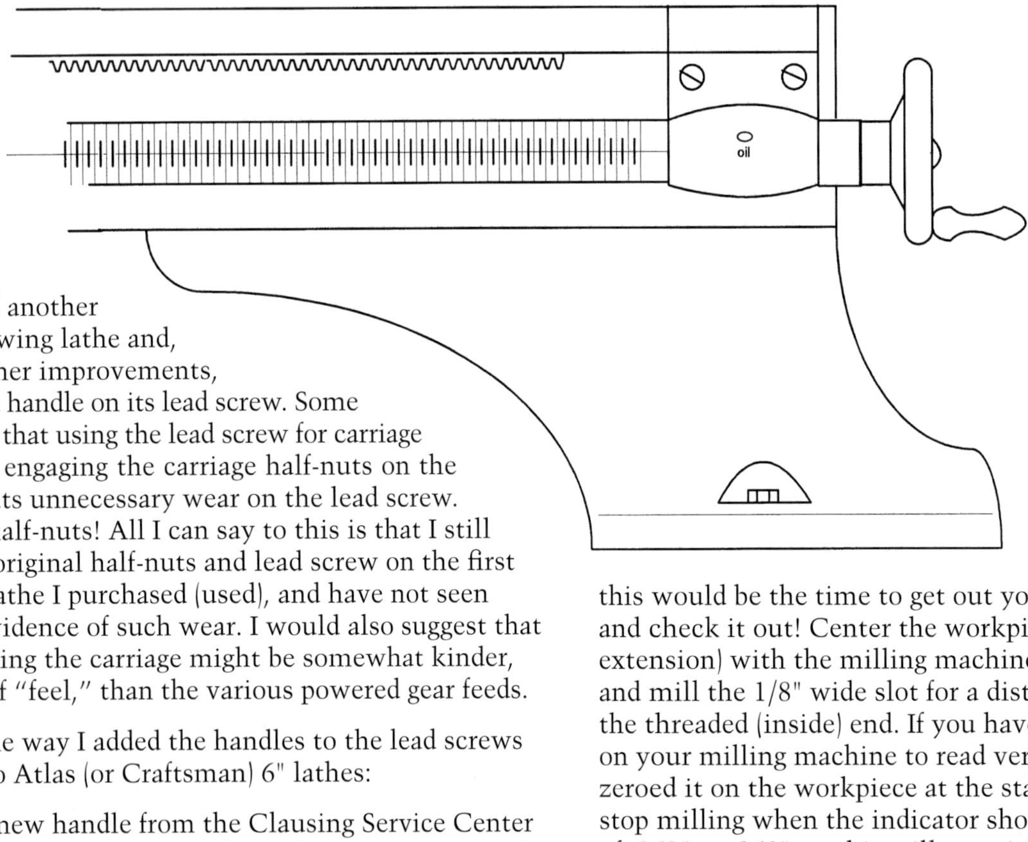

I obtained another Atlas 6" swing lathe and, among other improvements, installed a handle on its lead screw. Some might say that using the lead screw for carriage travel (by engaging the carriage half-nuts on the apron), puts unnecessary wear on the lead screw. And the half-nuts! All I can say to this is that I still have the original half-nuts and lead screw on the first Atlas 6" lathe I purchased (used), and have not seen blatant evidence of such wear. I would also suggest that hand-feeding the carriage might be somewhat kinder, because of "feel," than the various powered gear feeds.

Here is the way I added the handles to the lead screws of my two Atlas (or Craftsman) 6" lathes:

Obtain a new handle from the Clausing Service Center (811 Eisenhower Drive, South, Goshen, Indiana 46526). This is the same handle as used on the end of the tailstock on the 6" lathe (my catalog No. 3950-20, old catalog No. M6-23). This handle has a 3/8" bore, with one 1/8" wide slot the length of the bore, for a .124" square piece of keystock about 7/8" long.

Select a piece of .374" to .375" diameter cold-rolled or unhardened drill rod and install it in the lathe chuck. Face off the end. Center drill (No. 2) the faced end about 1/8" deep, then drill a No. 25 hole in this end for a distance of approximately 1/2". Leaving the workpiece in the lathe chuck, get the micro tap guide out, install it in the tailstock chuck with 1/2" capacity, and adjust it to the end of your tap wrench. Thread the hole with a 10-24 tap. This may require a little tightening down of the jaws of the lathe chuck to prevent rotation during the threading process. The spindle can be locked in back gear.

Now, cut off the workpiece so it is at least 1-1/2" long. Remove the workpiece from the lathe chuck, turn it around, and re-chuck it. Face the cut-off end, and reduce the diameter to .1875" (3/16") for a distance of 1/2". This 1/2" long, 3/16" diameter stud will be going into the end of the lathe lead screw.

Next, the portion of the lead screw extension that will have a handle on it has to have a 1/8" wide × 7/8" long slot milled into it. If you are not certain the *fixed* jaw of your milling machine vise is zeroed in with whichever of the horizontal axis movements you will use to mill the 1/8" slot, this would be the time to get out your dial test indicator and check it out! Center the workpiece (the lead screw extension) with the milling machine spindle center, and mill the 1/8" wide slot for a distance of 7/8" from the threaded (inside) end. If you have a dial indicator on your milling machine to read vertical distance, and zeroed it on the workpiece at the start of the milling, stop milling when the indicator shows a depth of cut of .062" to .063", as this will permit approximately one-half of the key to go in the workpiece slot and one-half into the bore of the handle slot. The lead screw extension is now finished.

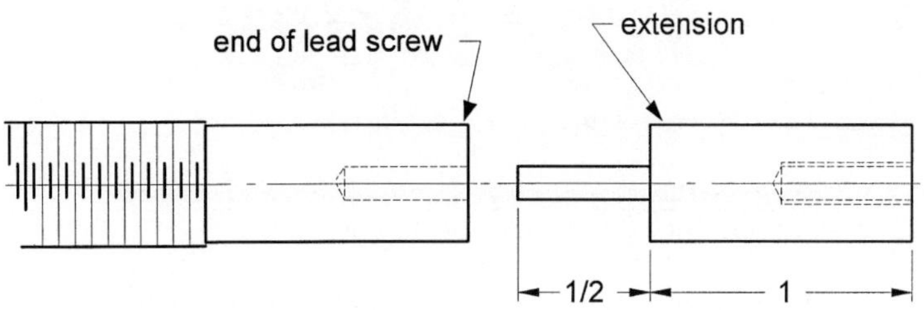

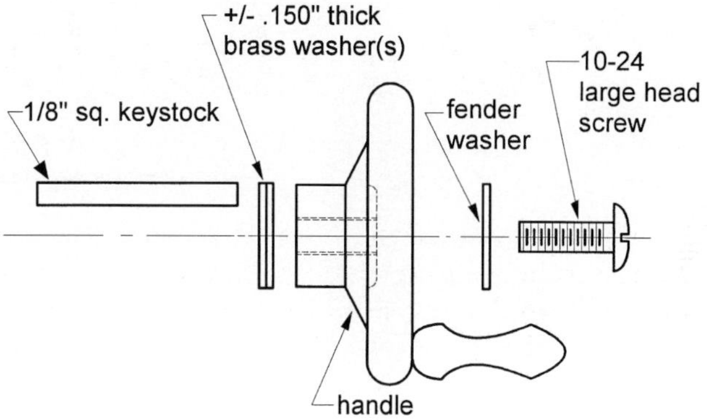

With the extension piece completed, all that remains is to get a matching hole centered in the end of the lathe lead screw and to join the two together.

Remove the four screws holding the lead screw bearings – two on the front bearing and two on the rear – and the lead screw. Remove these bearings and the thrust collar between them from the lead screw. Also remove the gear bracket from the head end of the lead screw. Insert the tailstock end of the lead screw into the hollow lathe spindle through the rear of the spindle, and tighten the chuck jaws down on the smooth tailstock end of the lead screw. Face the end of the lead screw just enough to impart a flat surface (which will later mate with the shoulder on the end of the extension 3/16" stud). By now, you have noted there is some disquieting amount of lead screw sticking out of the end of the lathe spindle. If you have never made provision for this situation in the past, now is a good time to make up a little means of supporting the outboard end of a long rod, etc. (Also note: It has been assumed that the lead screw, for removal from the lathe, was withdrawn through the carriage half-nuts, toward the front, and after removal of the rear bearing. If this proves cumbersome, the lathe tailstock and carriage can both be removed, allowing ready removal of the lead screw and related parts.)

Spot center drill the just-faced end of the lead screw. Drill a 3/16" hole in the end of the lead screw to a depth of at least 1/2", allowing sufficient depth so the 3/16" diameter end of the extension piece will go all the way in with its shoulder tight against the end of the lead screw (on trial fitting). Remove the lead screw from the lathe. With suitable solvent, clean off any oil that may be on the end or in the hole in the lead screw.

Do the same on the extension piece. Flux the 3/16" stud and the shoulder area of the extension piece, and insert the stud into the lead screw hole to distribute the flux in the hole and on its end. Silver solder the two together. A propane gas torch or a MAAP gas torch will do the job nicely. This method of joining the two leaves the lead screw bearing end clean of any marked condition.

You are urged, when soldering the extension piece to the lead screw, to wear a full face mask. This admonition is based on learning the unhappy, unfortunate way, particularly in the case of soldering jobs where the hot solder is sandwiched. While this is less likely in the case of silver soldering, where the pieces are first mated and then the solder travels in by a capillary-like action, take no chances!

After it is apparent that the silver solder has flowed in nicely (hopefully with no excess outside anywhere), and the piece has cooled down sufficiently so there is no possibility of it separating, cooling can be speeded by dipping and moving the end into a can of used motor oil. Clean up the area and inspect it carefully. The idea is to have two clean parts joined into one, both the same diameter, and the bearing end of the lead screw to be in the same good condition. Before remounting the lead screw, with its added length, to the lathe, trial mount the new handle so as to determine that no problems need correction. Also, I suggest this may be a good time to get a pan, a suitable brush, and some solvent to clean the lead screw. Over a period of time, a lot of residue can collect in the screw grooves, even though it has been wiped from time to time. Lightly re-oil the lead screw with light way/slide oil, or the oil of your choice.

Reinstall the lead screw and its related parts. Proper positioning of the lead screw on the lathe (in the case of the Atlas and Craftsman) is essentially adjustment of the front and rear lead screw bearings. This positioning is most helpfully achieved by moving the carriage as close to each bearing as possible, closing the half-nuts, and then tightening the screws holding the bearing to the lathe.

Install whatever packing washers that may be needed between the tailstock end lead screw bearing and the new handle. Be frugal! These washers do not normally take any thrust. They are there simply because some extra space was allowed on the extension piece for any needed later adjustments required by the lead screw or its bearings. Install the new handle. Rotate the handle and check the lead screw for smooth rotation. If no glitches are present that need attention, engage the half-nuts on the carriage apron. If the carriage is properly adjusted to its ways, the carriage should travel in forward and reverse direction with satisfying smoothness.

Tap Dogs

by Glenn A. Pettit

Photos by Author

I've often wanted to tap holes by putting the tap in the three-jaw chuck of my Atlas 6" lathe, using back gear. The problem of the tap slipping in the chuck makes it almost impossible. I didn't want the strain on my only three-jaw chuck as it's still fairly accurate after ten years of TLC.

When I looked at the space between the headstock spindle and the chuck jaws, I saw that maybe a "tap dog" could drive the tap by having a square hole to fit the tap end and a projection to catch one of the jaws. The three jaws center the tap, and the dog provides necessary torque to the tap.

Your chuck dimensions might be different, but the dog dimensions to fit my setup are: 7/8" diameter and 9/32" thick. The drive pin is 3/16" OD 2 9/32" long.

Figure 1 hole is for a 3/8" tap. Figure 2 handles 5/16" and 1/4" taps. Figure 3 handles No. 12 and 10 machine size taps. Figures 2 and 3 each have two 6-32 setscrews to secure the smaller size taps. Figure 1 uses no setscrews.

When measuring the distance between the headstock and jaws, be sure to allow for the scroll teeth projecting from the rear of the jaws.

The square hole is filed from a central drilled hole, and the projection was turned offset in the four-jaw chuck. Silver soldering a pin in a drilled hole might be a viable alternative.

Taps less than No. 10 are probably too short to have a usable projection beyond the chuck jaws and possibly subject to breakage from too much torque.

A drilling pad mounted in the tailstock makes the tapped hole perpendicular to the rear surface.

I hope this makes shop work a little easier. Good tapping!

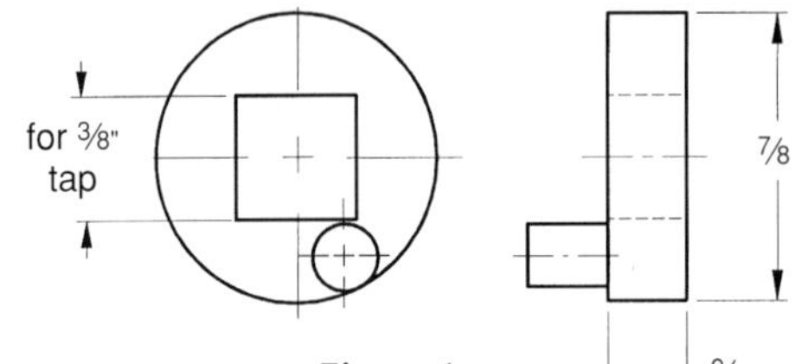

Figure 1

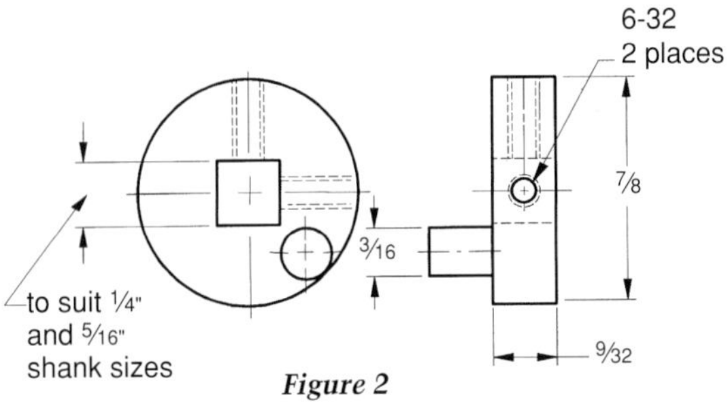

Figure 2

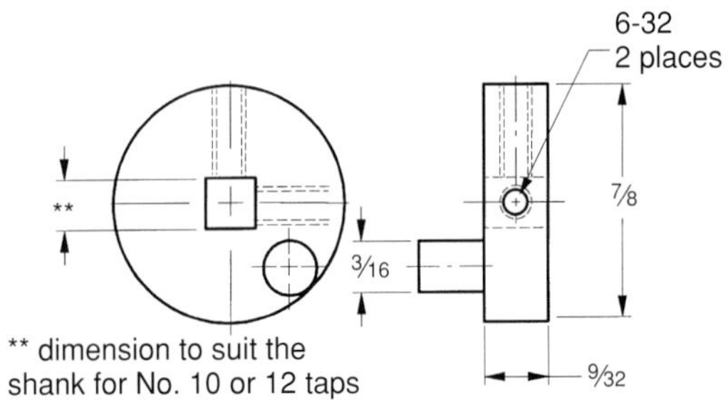

Figure 3

Atlas Tailstock Modification

by Bruce Jones

Photos by Author

Regardless of the type of lathe owned by the home machinist, there is always some feature that is less than desirable.

The most annoying feature of Atlas-type lathes like the one I own is the tendency of the tailstock clamping mechanism to seize when clamped securely. Sometimes it can be worked free in such situations and sometimes it can't. When the clamps just won't give, the whole darned tailstock has to be disassembled to free it. To say the least, it is a very frustrating and time-consuming experience.

After it had happened to me on more occasions than I would like to remember, I finally broke down and fixed the problem once and for all.

Photo 1 shows the completed parts for the modification. Other than the new parts, which must be fabricated (shown in Figures 2 and 3), the only other work to be done is to counterbore the inside faces of the two clamps .390" diameter × .172" deep.

1

Figure 2 shows the spring, which works quite well. Hardened steel washers (Figure 3) must be used to prevent the spring from digging into the softer material of the clamps.

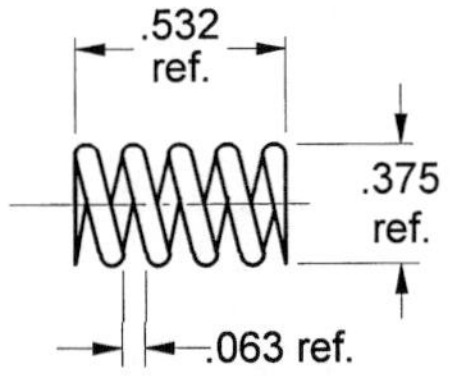

TAILSTOCK CLAMP SPRING
1/16" music wire, wound to slip on 1/4" dia. bolt heat treat

Figure 2

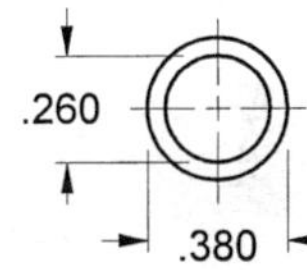

TAILSTOCK SPRING WASHER
.380" dia. x 1/16" thick drill rod
heat treat
2 required

Figure 3

After all the work is done, simply assemble the bottom clamp with the square head bolt in place. Drop in one of the washers so it rests on the bottom of the counterbore, then drop the spring over the bolt so it comes to rest on the washer. Next, drop the other washer in place over the bolt, and insert the whole assembly into the bottom of the clamp bore. Hold it in the bore with something handy, then drop the top clamp half into place and screw on the clamp nut.

After making this simple modification, I find that no matter how tight the clamp is tightened it always frees up. After loosening the pressure on the clamp nut, I simply move the tailstock handwheel a little in either direction and I can hear the spring snap the clamp apart. What a joy to be free of that problem.

A Pseudo Drawbar for the Tailstock

by Phil Nyman

Photos by Author

When using a 5/8" drill chuck in my Maximat V10 P tailstock, I have found it difficult to manually seat the No. 2 Morse taper tight enough to prevent rotation. A light hammer tap is sufficient but exerts undesirable force on the tailstock thrust ball bearing and on the precision chuck. My solution uses a threaded ring attached to the chuck, a split clamp mounted on the tailstock ram, and a tightening ring. In use, the knurled chuck collar is held in one hand while the tightening ring is rotated with the other hand to tightly seat both the tailstock taper and the chuck arbor taper. For chuck removal, the tightening ring is loosened sufficiently to clear the threaded ring, and the tailstock ejector is used in the normal manner to pop out the drill arbor. Photos 1 and 2 show the pseudo drawbar seated and released respectively.

1

2

Most tailstock attachments, such as centers and small chucks, can be used with the tightening ring and split clamp in place on the ram. However, two of my live centers required tang end extensions of approximately 7/16" to allow complete compatibility. One extension used a modified brass pipe fitting, and the other consisted of a brass cylinder held with a screw in the tang end of the taper. Please note that this design fits a 30mm tailstock ram and a 5/8" Albrecht drill chuck. Usage with other rams and/or chucks will require appropriate modifications.

The tightening ring (Part 1) is detailed in Figure 3. I started with 2.25" diameter by 1-1/16" long stock mounted in a three-jaw chuck. Face and chamfer the end, then turn end-for-end and rechuck. Face and chamfer the remaining end, and finish bore all inside dimensions including the 24 tpi threads.

The single depth of this thread is 0.0316" for a V-thread. Next, turn an aluminum arbor for a tight slip fit into the 1.750" dimension of the ring. Finish turn the ring OD, and cut the flutes. To cut the flutes, use a custom sharpened tool bit mounted in an axial direction in the toolholder. Marked tape around the chuck provides the twelve index locations, while manual carriage movement toward the headstock does the cutting. *The Home Shop Machinist* has described this process in other articles. The fluting serves only to provide a grip, so use any alternate pattern you desire.

A few notes will be helpful when making the split clamp (Part 2) shown in Figure 4. It is desirable to make the faces parallel to each other and perpendicular to the 1.181" bore. This allows an accurate pulling surface for Part 1 with any mounting configuration of the split clamp. Face one end and finish bore the 1.181" hole. Remove the part from the chuck and set aside for later finishing using a self-locking stub mandrel. This mandrel will also be used in finish machining Part 3 as described later.

Next, begin machining the chuck pull ring (Part 3) detailed in Figure 5. Face one end and finish bore the 1.181" hole. Remove Part 3 from the chuck. Now, machine the self-locking stub mandrel for a diameter of 1.181". (See "Temporary Self-locking Stub Mandrel" by Philip Duclos, *HSM*, July/August 1984, page 28, or *The Shop Wisdom of Philip Duclos*, page 54.) My mandrel was made from aluminum and used a 0.100" diameter steel locking pin. Make the mandrel short enough to enable facing the chuck pull ring (less than 1/4"), and with an outside diameter small enough to clear the threading tool (approximately 1.5").

Mount the split clamp onto the mandrel, face the remaining end, and final turn the 1.740" diameter.

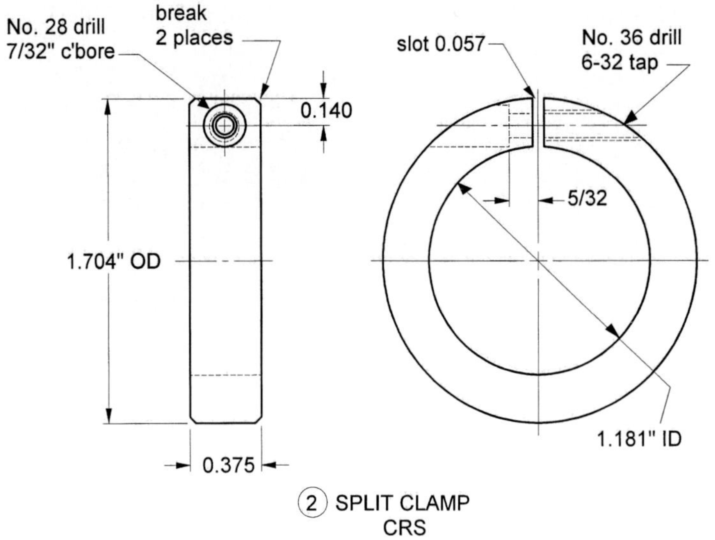

Figure 4

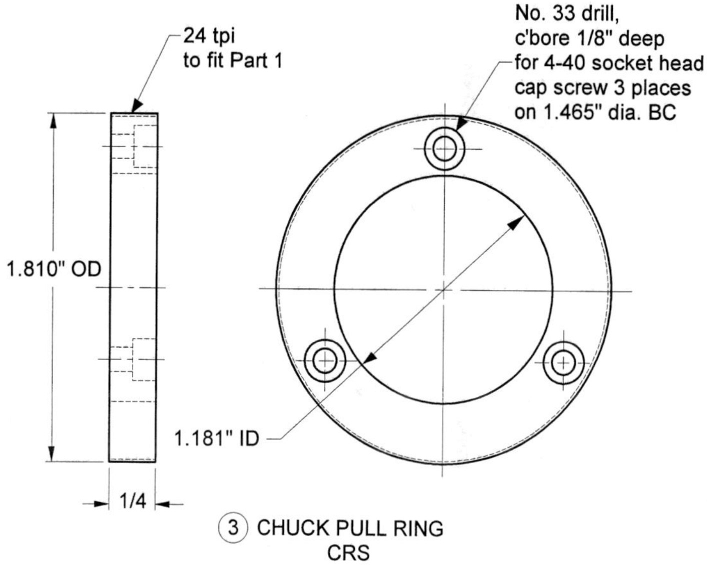

Figure 5

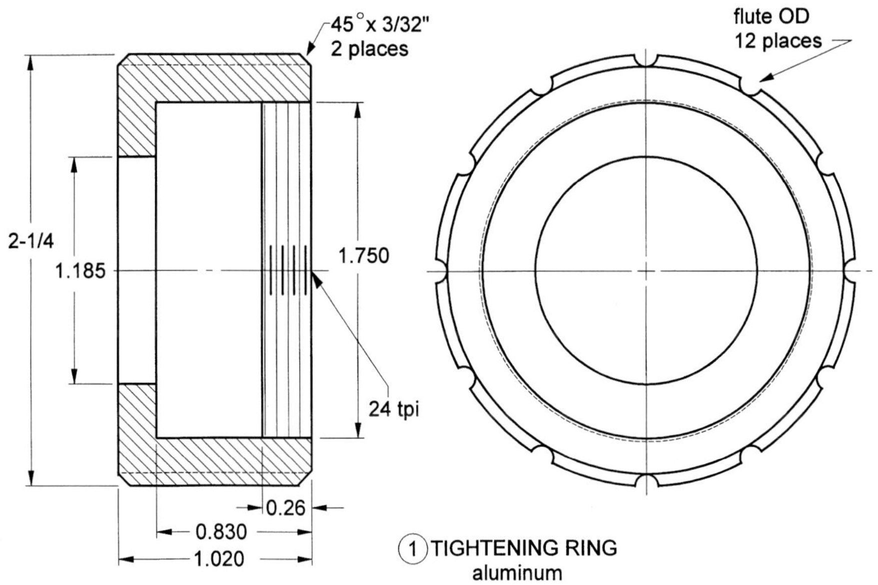

Figure 3

BILL OF MATERIALS

PART	DESCRIPTION	QTY	MATERIAL	SIZE
1	Tightening ring	1	aluminum	2-1/4" dia. × 1-1/16"
2	Split clamp	1	CRS	1-3/4" dia. × 7/16"
3	Chuck pull ring	1	CRS	1-7/8" dia. × 5/16"
4	Clamp screw	1	stainless	6-32 × 1/2" socket head cap screw
5	Knurled collar	1		from 5/8" *Albrecht* chuck
	Assembly screws for Parts 3 and 5	3	steel	4-40 × 3/8" socket head cap screw

Use the milling machine to locate and machine the clamp screw hole prior to cutting the 0.057" slot. Due to stress in the CRS, my clamp bore size shrunk about 0.005" after the slot was cut! Therefore, to mount the clamp, a 6-32 screw is temporarily inserted into the threaded side of the clamp and tightened against a washer, held in the slot, until the clamp is forced open to a convenient size for mounting.

The chuck pull ring can now be mounted onto the mandrel for final facing and turning of the 1.810" diameter. Cut the 24 tpi for a smooth fit with the tightening ring. Before removing the pull ring, scribe the three screw hole locations on the (non-critical position) 1.465" diameter bolt circle.

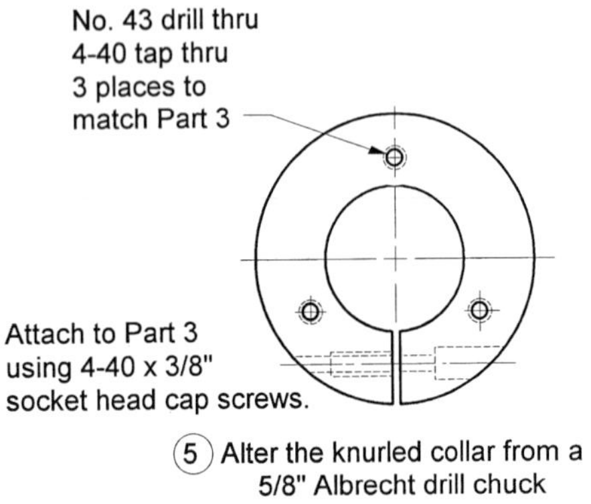

Figure 6

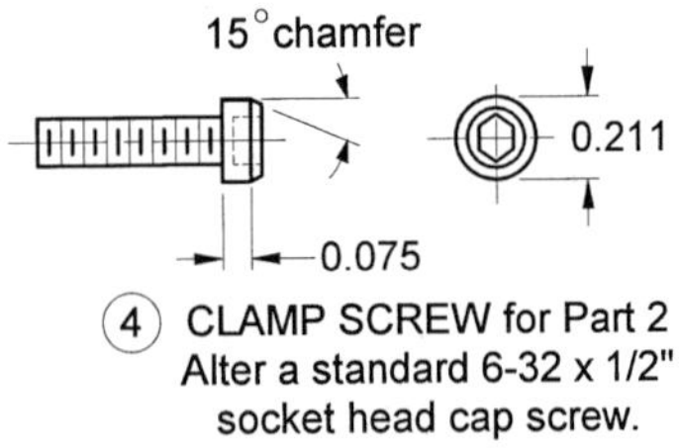

Figure 7

Remove the knurled collar from your 5/8" Albrecht drill chuck. You can see that the collar bore = chuck pull ring bore = tailstock ram diameter. Therefore, slide these parts (3 and 5) onto the tailstock ram, orient for proper hole location per Figure 6, and clamp Parts 3 and 5 together. Carefully slide the clamped assembly off the ram, and then drill and tap for three holes per Figures 5 and 6. Using 4-40 × 3/8" screws, assemble Parts 3 and 5 together before removing the clamps. Remount the knurled collar assembly onto your drill chuck (Photo 8).

Machine the clamp screw as detailed in Figure 7, and install it in the split clamp. After final assembly and checkout, apply oil at the clamp pulling face and the threads to allow greater usage of the theoretical mechanical advantage provided by the fine threads. Neglecting friction, this advantage is approximately 169 times. With oil, I would estimate the actual advantage to be between 10 and 20 times.

8

MILLING & DRILLING MACHINE ACCESSORIES

Table Stop

by James Madison

Drawings by Author

When setting up machine operations, a table stop is needed in many situations. The best design is one that provides an easily installed, yet rigid and durable, construction. The design featured here is best suited for milling machine table; however, it may be found to be useful on drill presses, grinding setups, lathes, and so on. The basic features to be considered are:

1) A flat, stable base;
2) a vertical member with clamping holes for a stop arm; and
3) an adjustable stop arm with finger.

If possible to combine 1) the flat base and 2) the vertical member in a one-piece L-shaped construction, do so. The ultimate shape of the base is an L-shaped, solid member that is clampable to the machine (table) with one bolt.

The bottom surface of the base must be flat within .001" or less. This flatness will assure no stop movement and no finger deflection through tightening. The advantage of one-piece base construction is having less parts with joints that may loosen or shift during machining operations. Aluminum is a suitable material, preferably 6061-T6 for its rigidity. The determining factor is having access to a piece of 6061 of the 2" thickness (and length and width needed) to allow one-piece, L-shape construction (*Figures 1, 2,* and *3*).

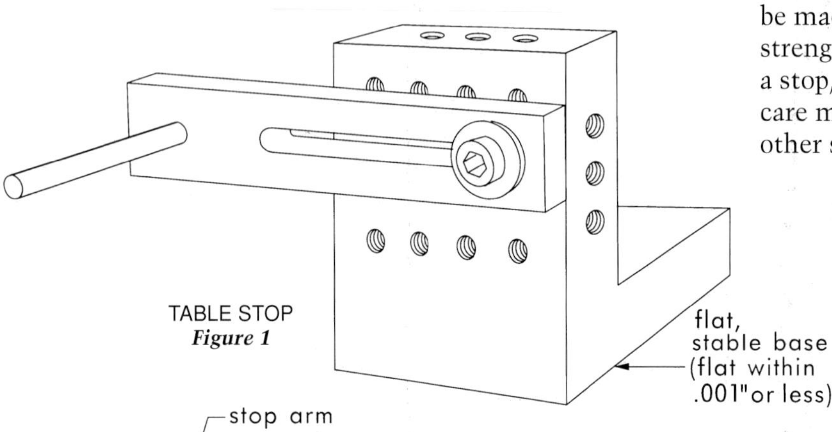

TABLE STOP
Figure 1

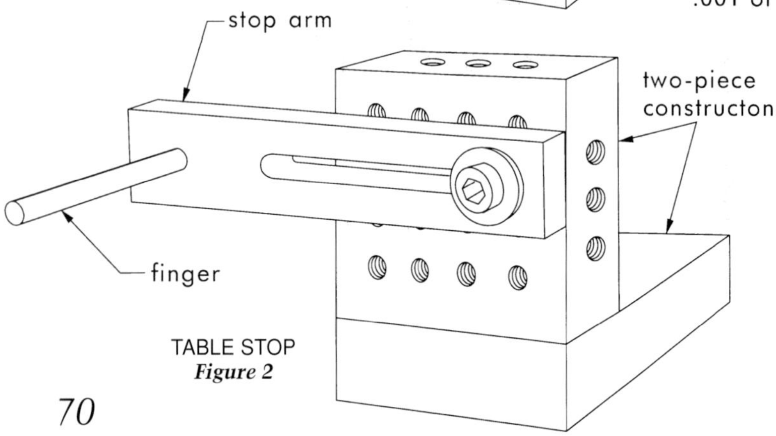

TABLE STOP
Figure 2

The vertical member of the base has many 1/4-20 tapped holes for attaching the stop arm. The greater number of holes, the more versatile the stop-finger will become.

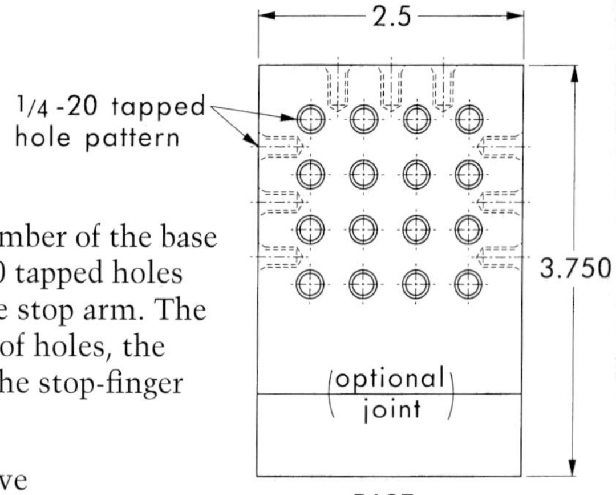

BASE
Figure 3

If you do not have access to material large enough to cover one-piece construction, it will be necessary to use the alternate method – two-piece construction. This does, of course, entail more work and more hardware to complete. The two parts must be bolted and pinned with dowel pins (*Figure 4*).

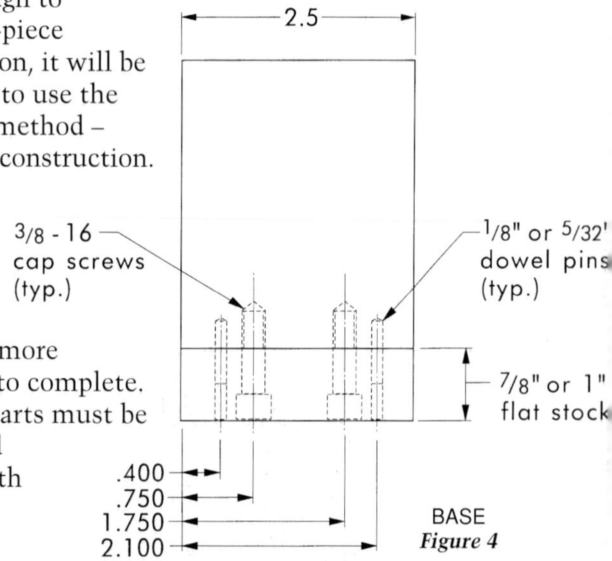

BASE
Figure 4

STOP ARM AND FINGER

The stop arm and finger are to be constructed of steel (cold or hot rolled are suitable). The stop arm should be made out of 1/2 × 3/4" or 1" bar stock to provide strength without deflection. When sliding parts against a stop, a solid, rigid stop will not move and, therefore, care must be taken when constructing these and any other setup accessories. The stop arm is slotted to

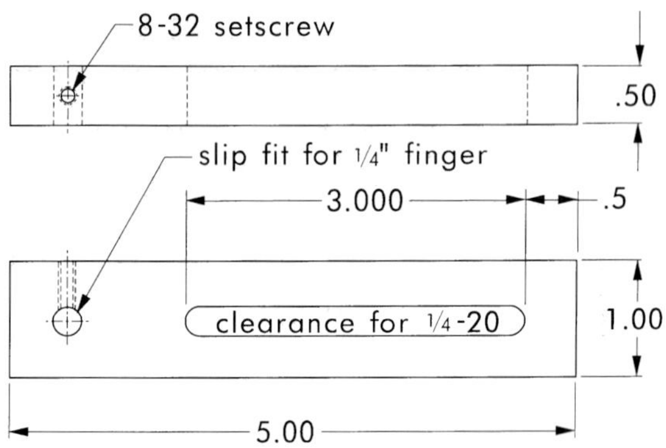

STOP ARM

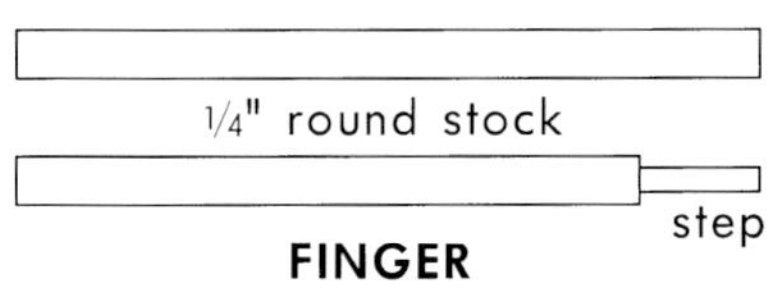

FINGER

hot or cold rolled steel

Figure 5

accommodate a 1/4-20 screw to attach it to the base (*Figure 5*).

The main purpose of the slotted arm is to make use of the tapped hole patterns and sliding movement of the base to position the finger in the needed stop location. The .250" diameter reamed hole is a nominal size because of the easily available size. 1/4" stock can be bent, turned, tapered and milled to fit any part requirement (*Figure 6*).

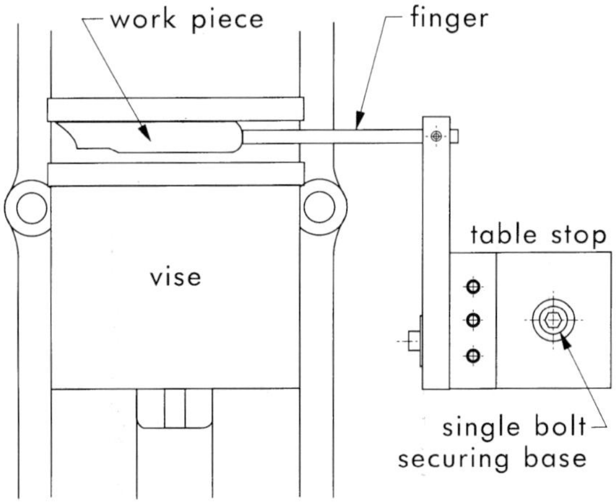

LOCATING PART IN MACHINE VISE
Figure 6

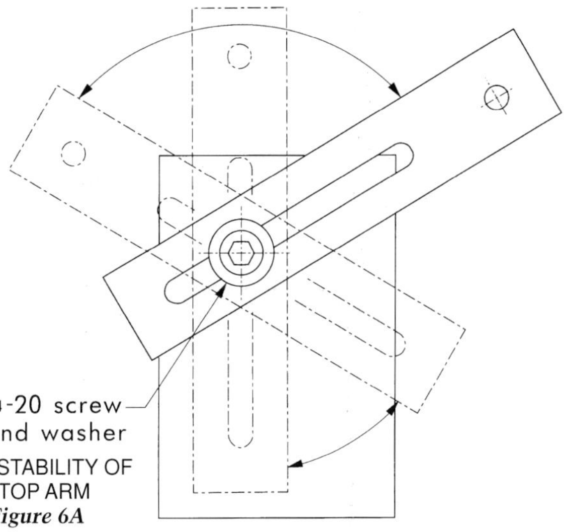

ADJUSTABILITY OF STOP ARM
Figure 6A

The stop finger can be fashioned to fit into virtually any machine setups. After deciding where the stop is needed in a particular setup, pick out the finger necessary to do the job. If a custom-shaped finger is needed, 1/4" stock can be easily worked with, especially cold or hot rolled (*Figure 7*).

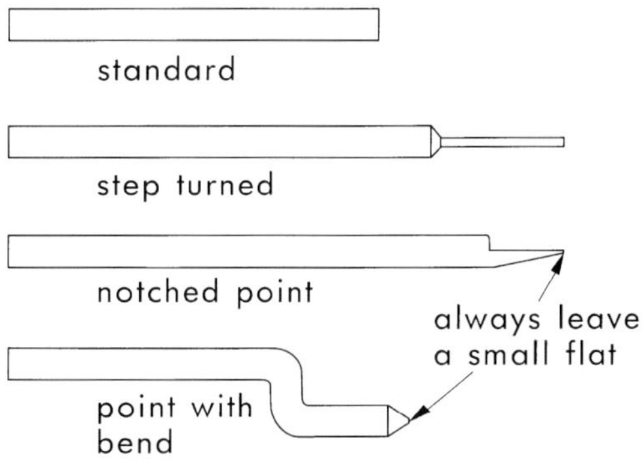

DIFFERENT FINGER CONFIGURATIONS
Figure 7

ALTERNATE APPLICATIONS

One of the real beauties of the table stop (besides its ease of installation and adjustability) is its use with many machine tools besides the milling machine. It can be clamped on the cutoff saw (*Figure 8*).

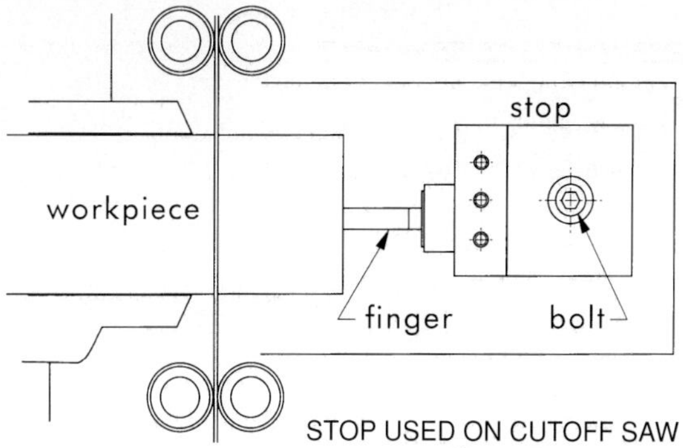

STOP USED ON CUTOFF SAW
Figure 8

It is easily placed on the chuck of a surface grinder. One steel strap can press over the aluminum base or, if available, a single threaded hole can be accessed (*Figure 9*).

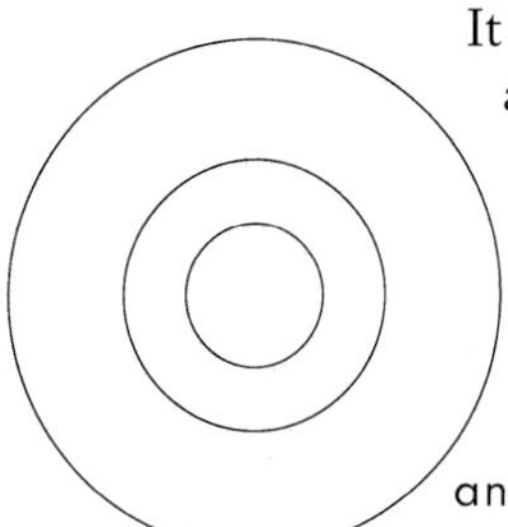

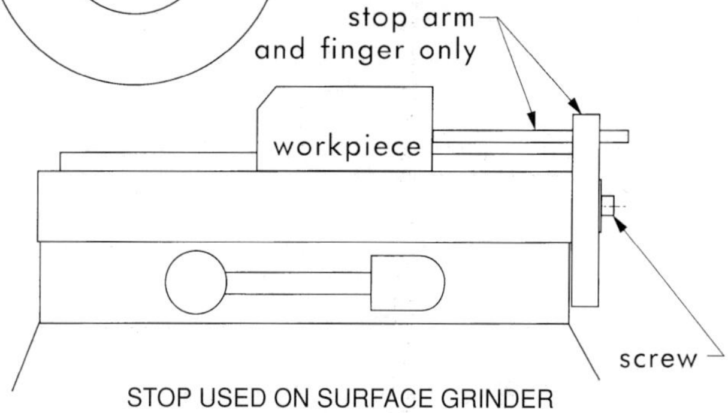

STOP USED ON SURFACE GRINDER
Figure 9

The stop may be mounted on the moveable carriage of a lathe for the purpose of: a) part location during loading; b) an additional carriage stop for an alternate tool; and c) a cross-slide stop for a certain operation (*Figure 10*).

The stop bar and finger can be utilized without the base for many purposes, also.

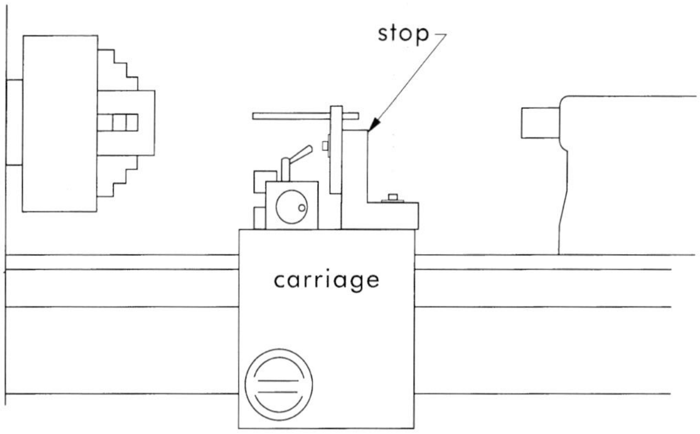

STOP USED ON LATHE
Figure 10

Remote Switch for the Mill

by C. O. Voss

Anyone with a lathe and mill can make a remote switch for your mill that will be much more convenient. The only items needed not shown in the photos are four bushings used to take the pressure from the aluminum head housing. These bushings should transfer the support of the levers to the top. Replace the original screws with longer ones, counting the threads used to hold the head together. Use the same number of threads on the longer screws.

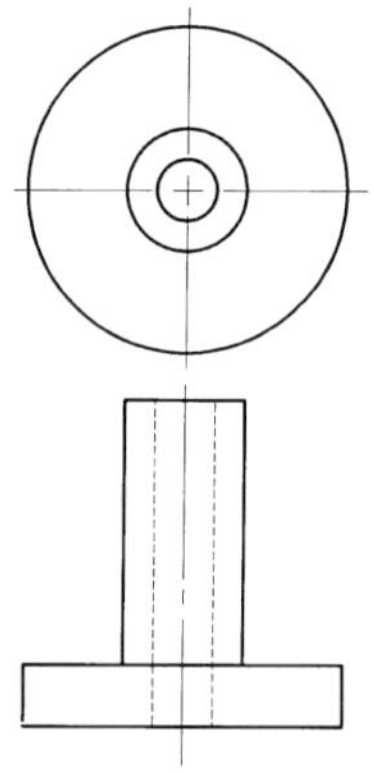

BUSHING
(to suit)

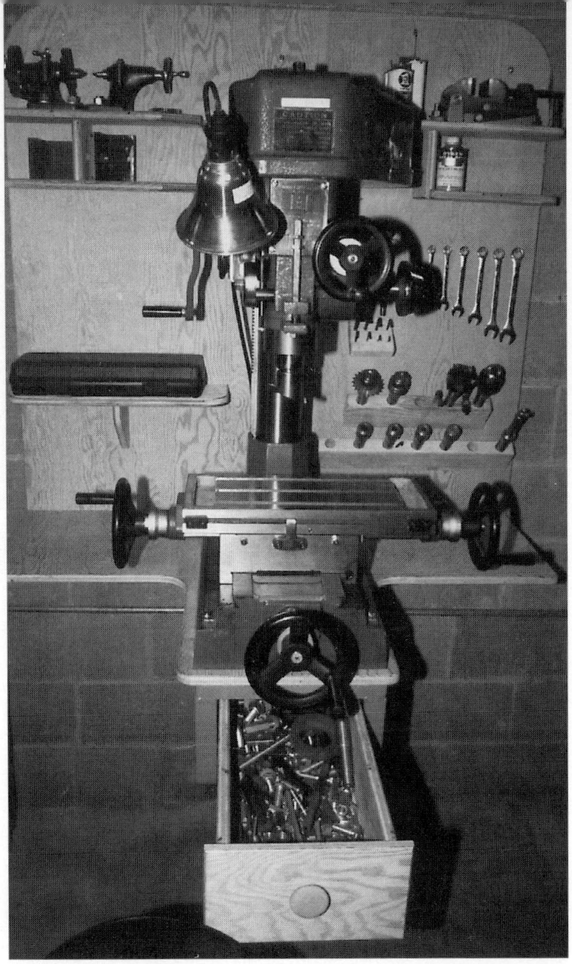

1

3

A Mill/drill Support

by Paul Holm

Photos by Author

To gain a little more space in my shop, I decided to mount my Jet-15 mill/drill on the cement block wall (Photo 1). The design is given here for those of you who may have a bench-type mill/drill, and would like to buy more of your own space. The dimensions, of course, are for the Jet-15, so if you have something different, you must adapt to your own needs.

Figure 2 shows the basic layout. The support is made from 1-1/2 × 1-1/2 × 1/4" angle. Photo 3 shows the support before the wood components were added.

Drill 11 evenly spaced holes 5/16" in diameter in the back side along the 44" angle to mount the unit to the wall using lead anchors and 1/4-20 capscrews 1" long. There should be two holes in each of the 6-1/2" angles, and one hole in each of the bottom angle supports.

To hold the mill/drill in place, I drilled four 7/16" holes: the front two are 11" apart, and the rear two are 9-1/4" apart to suit the Jet-15. From front to rear, these holes are 17" apart.

Seven 3/16" holes are drilled on each side – each countersunk 3/8" for 3/4-10 flathead screws to hold the side wood panels. These panels give you the extra strength and support you need to hold up the weight of the mill/drill.

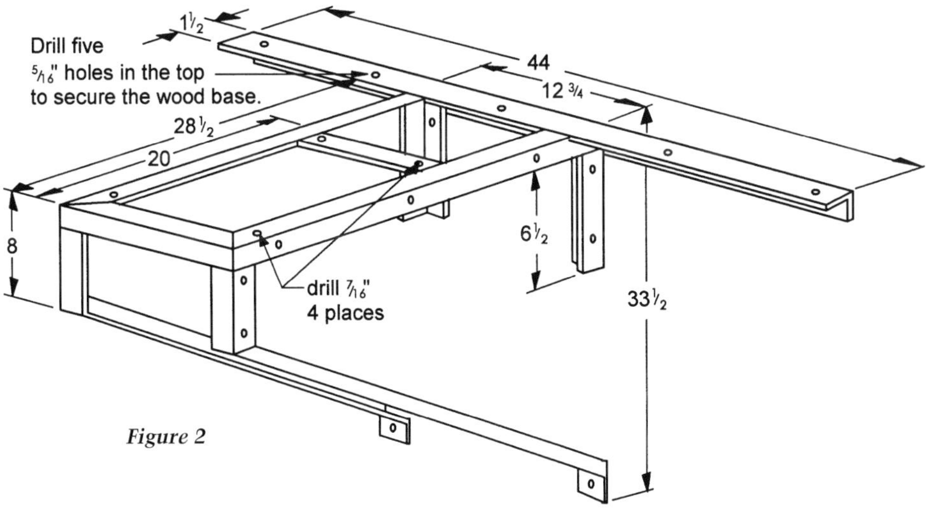

Figure 2

6

7

The mill support table top (Figure 4) is made from 3/4" AC plywood. Drill all holes to suit the frame in Figure 2. Note in Photo 1 how handy the drawer is. It holds all those items that are required but impossible to keep neat and organized.

I built a backboard for tooling (Figure 5) that is especially handy. Make it from 1/2" AC plywood. Secure it to the wall with 1/4" lead anchors. Make brackets to suit your tooling.

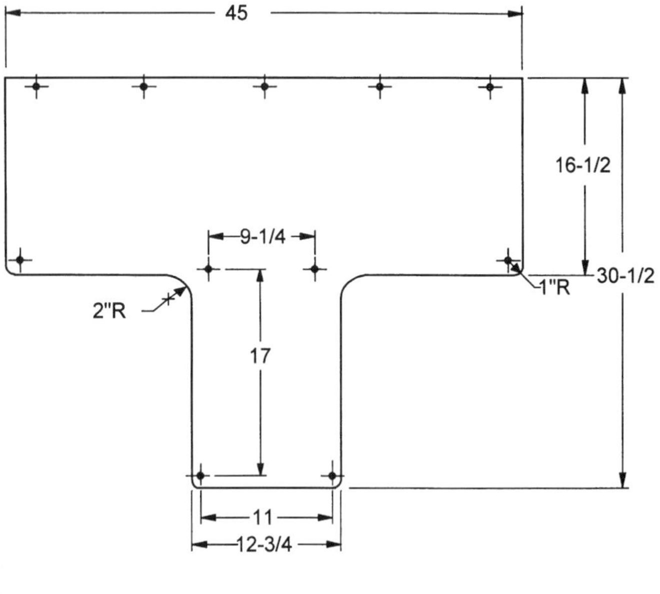

MILL SUPPORT TOP

Figure 4

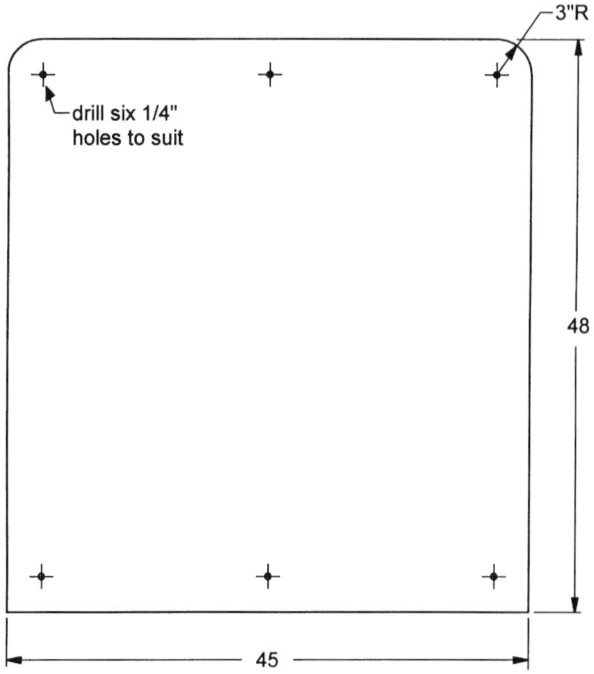

BACKBOARD FOR TOOLING

Figure 5

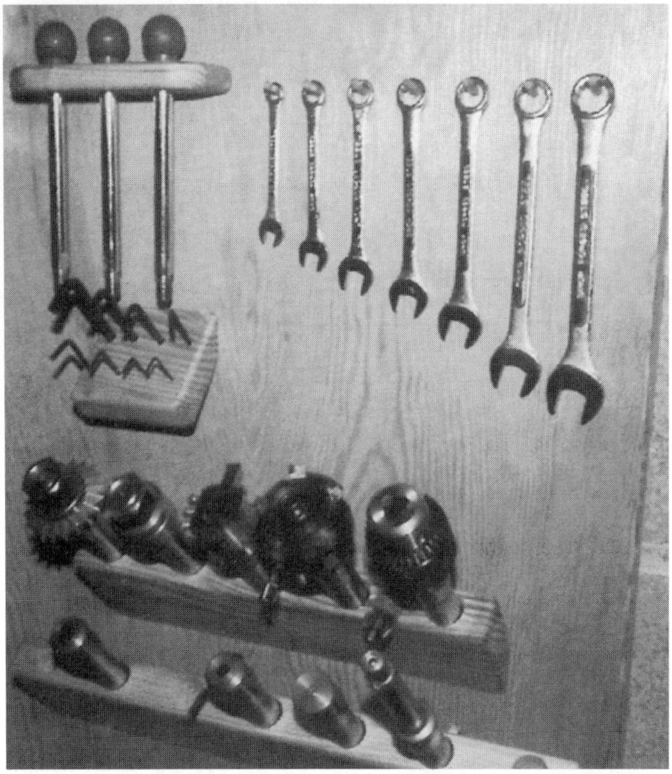

8

9

Be sure to leave clearance for the head to travel up and down, and swing. Photos 6 and 7 show my solution. I sealed all the wood surfaces so oil could be cleaned up easily.

Photos 8 and 9 are detail shots of my tooling on the backboard. Later, after those two photos were taken, I added some extra accessories to the backboard: a T-bolt bracket (Photo 10), and a holder for my clamps and end mills (Photo 11). The whole project has been a very efficient solution to my space and convenience problem. Perhaps it's a good one for you, too.

10

11

Chatterless Countersinks

by R. S. Hedin

Photos by Author

A chatterless countersink in action in aluminum with a nice curled chip.

Commercial countersinking tools generally scrape and chatter their way into the hole making a faceted and rough countersink (usually when the part is complete). A single-lip unit is guaranteed not to chatter, and with a positive rake or hook is free-cutting. This "sink" cuts well in all materials from steel to wood. It also works well for hand deburring of holes.

The tools shown are made of water-hardened drill rod in .50" and .37" diameters. Drawings and photos illustrate the steps in fabricating a "sink." The blank is first turned to a cone. This surface must be smooth and polished, as it will determine to a degree the finish of the countersunk hole.

An unhardened, completed blank.

The cone is milled flat to its center and back past the corner, as shown. The blank in this shape, hardened, would make a cutting tool. However, the next operation – hollowing out or putting a rake on the cutting edge – makes the tool cut freely. The blank is held in the vise with the cutting edge and the flat face parallel to the mill spindle. Substitute a round rod in the end mill holder to line up the blank in this position. Using a .18" or .25" diameter end mill, set its center line about .06" back of the cutting edge. With the end mill above the blank as shown in the photo, move the blank into the mill path in small increments and feed the end mill down the length of the sink cutting edge.

Proceed slowly until the radius of the groove meets the edge of the sink creating a sharp edge. Deburr lightly but do not touch the cutting edge. Heat-treat the blank, keeping the flame away

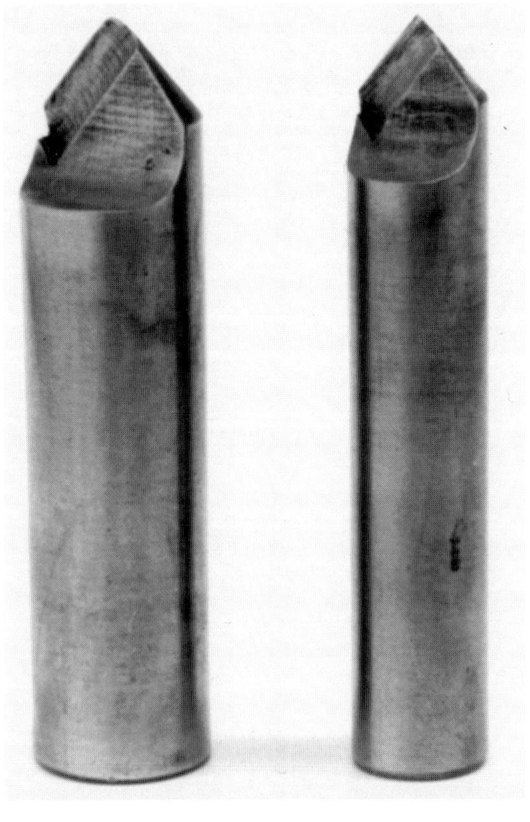

A pair of finished countersinks – .37" and .50" diameters.

A countersink blank in the mill vise. Note that the flat goes past the cone angle.

The hook or rake angle being milled into the blank.

from the cutting edge. Quench the blank and draw to a light yellow. After hardening, the sink can be sharpened by laying a round stone in the groove to hit the cutting edge. Do not stone away the hook angle. Don't stone the cone side or the tool may not cut.

Use lubricant when countersinking steel or aluminum; other materials cut well dry. Keep the speed down as the tool is made of carbon steel and can lose its hardness if overheated. Other configurations are suggested in the drawings. A .25" diameter shank is useful for an electric drill or a long shank for deep and tight spots.

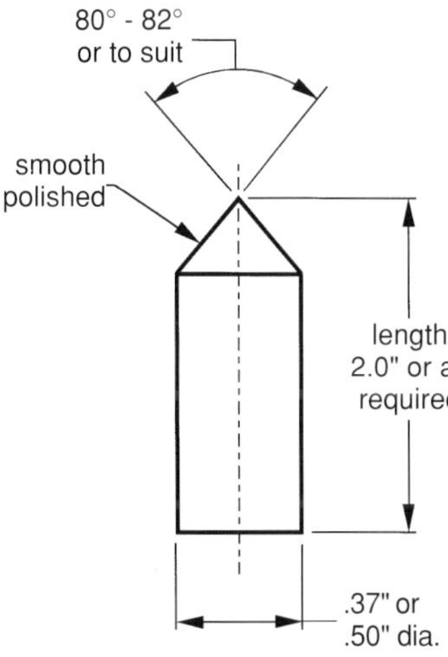

TURNED BLANK
water or oil hardening drill rod

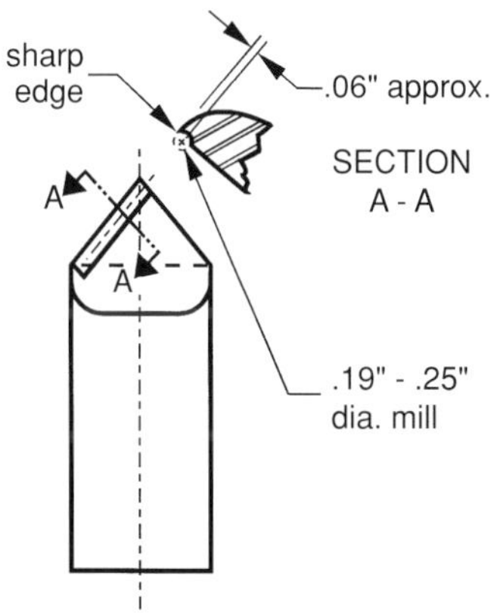

CUTTING EDGE
ON MILLED BLANK

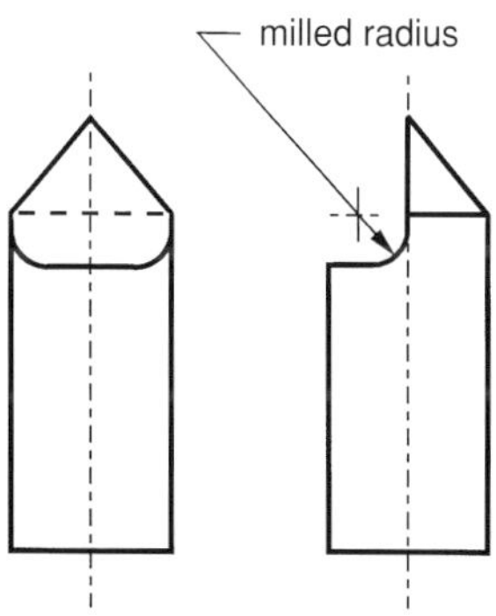

MILLED BLANK

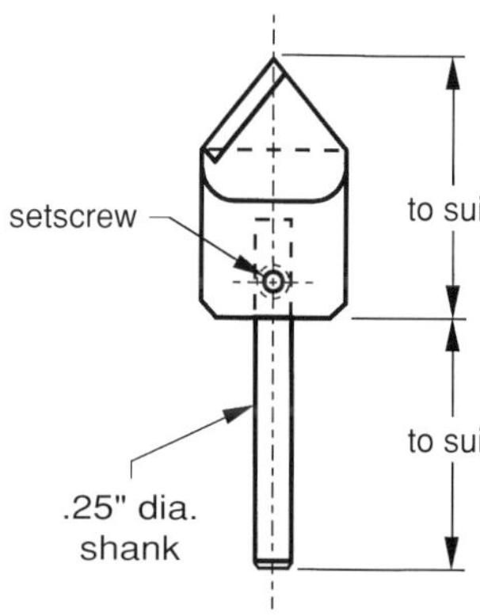

POSSIBLE EXTENDED
SHANK UNIT

Mill Spindle Indicator Holder

by Roy Rice
Photos by Author

So you have that new milling machine safely ensconced in your shop, have mounted it on a proper stand, or set it up in its appointed place. You have even sprung for an extra couple of hundred for a proper vise. Now it's time to take a few measurements to verify your new baby's accuracy. It sure would be nice to have a convenient way to attach that dial indicator to the spindle to see if the table is really level! You could call up your local supplier and order out an *Indicol* or similar device, but what better way to save a buck and put your new mill to practical use than to make your own? That's what I did, and found it to be a fairly simple project, but with a few interesting challenges thrown in. Here's how to do it.

The body of the holder is 3/8" aluminum flat stock, the linkage is CRS flat and bar stock. You will need a piece of aluminum at least 2-1/4 × 4", a short piece of 5/8" and/or 3/4" aluminum bar stock for the knobs, a scrap of nylon about 3/8" thick for the pads, some small nylon washers and a few inches of 5/16" and 3/16" crs bar stock plus 1/16" flat stock for the linkage. I have given the dimensions I used, but they aren't critical, so feel free to modify them to suit the contents of your scrap box. (If you take a careful look at the photo of my indicator holder, and compare it to the drawings, you will see I "cheated" a bit to make do with a body blank that was about 1/4" short.)

Begin by making the holder body. The basic dimensions are 2-1/4 × 4", but if you have a piece 2-1/2" wide, it will allow you to mill out the inside as a whole. First, square up the blank for the holder body and bring it to size. Lay out the body per the drawing. It is quicker to waste the inside by chain drilling and then cleaning up with the mill, but with aluminum, you could simply drill one hole to start the end mill, then mill out the inside. I set the vise square first, milled out the straight sides, and roughed out the sloped leg. I then set the vise over to 45° to finish the angled cuts.

If you have been working with the oversized blank, remove it from the mill and saw off the waste. Trim the sawn ends, then mill out the relief for the clamp linkage. Drill the clamp pad holes, drill and tap the clamp screw hole and the body is finished.

Make the clamp pads and the clamp screw pad from 3/8" thick nylon, or any reasonably

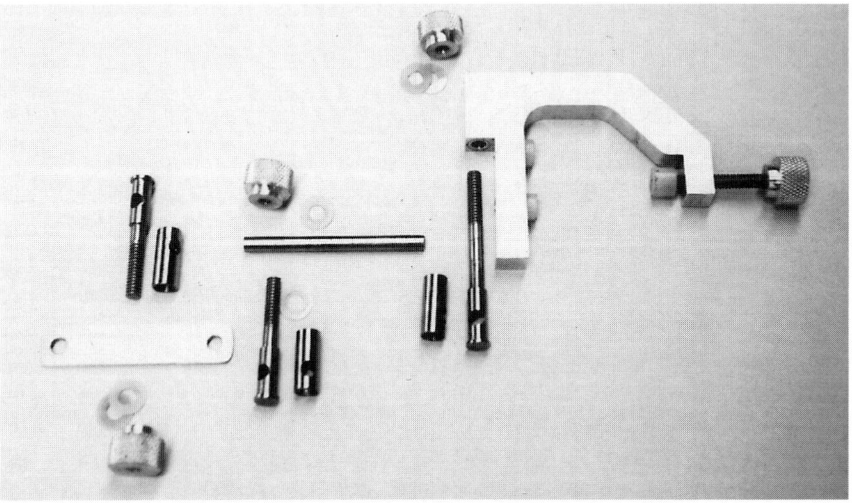

All the components for the indicator holder are shown.

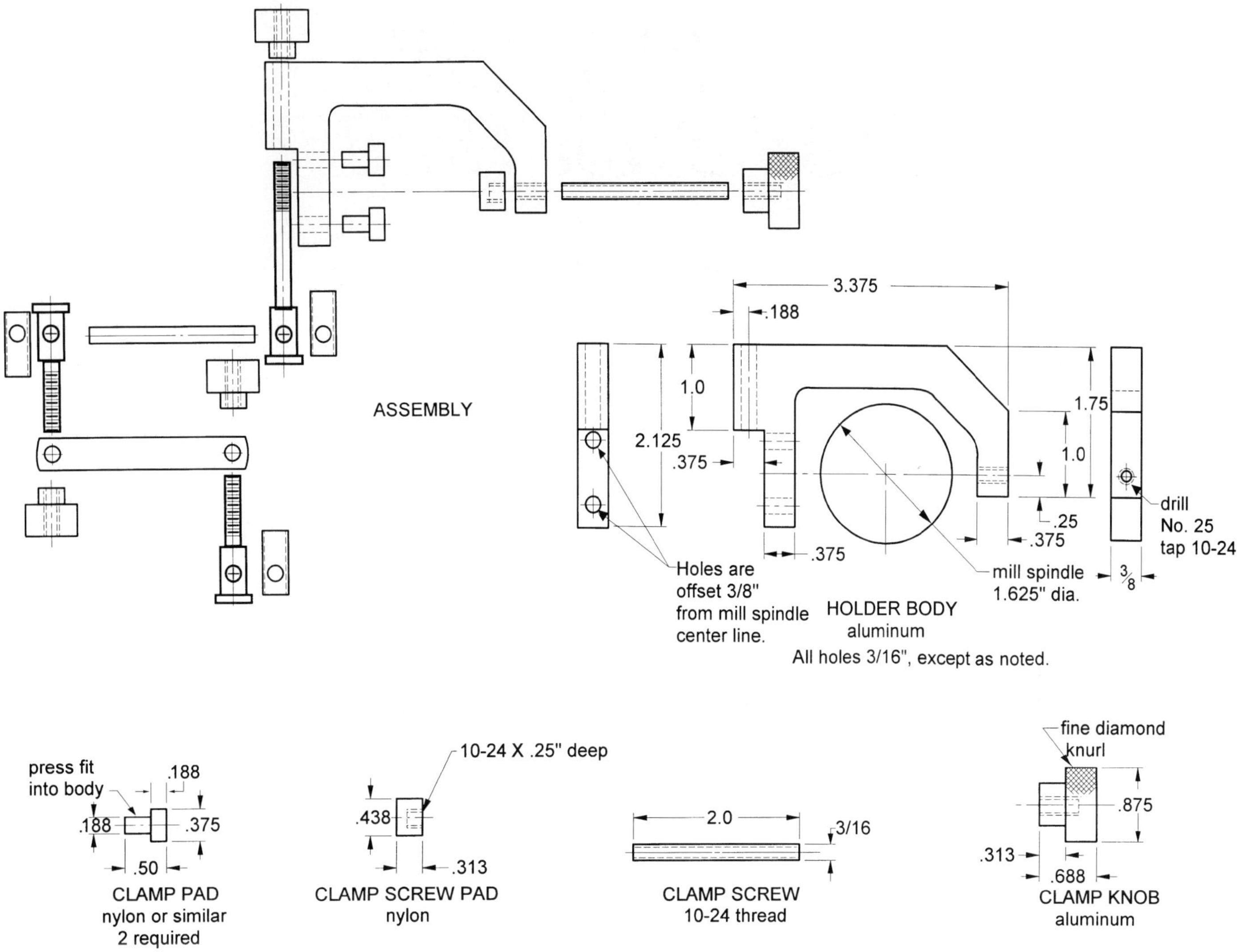

tough plastic you have in the scrap box. Make the pads a press fit in the back of the body, and tap the screw pad to fit the screw. The clamp screw is a piece of 10-24 all thread, or a threaded piece of 3/16" crs. Make a knob for it from 3/4" or 1" diameter aluminum bar stock finished with a medium diamond knurl and fix it to the threaded rod with a drop of *Loctite*. While you are set up for knurling, you might as well make up the three locking knobs. They should be smaller than the main clamp knob. I show them as 5/8" diameter, threaded 10-24 through.

The clamp linkage consists of a round bar at the top, linked to a flat bar below (see the Assembly Drawing) with nylon friction washers at each link pivot. Note that the holes through the locking pins are offset from the end shoulder by .031" (1/32") more than they are from the end of the locking sleeves (more on this later). This provides the clamping action when the knobs are brought up tight.

Things have been relatively simple up to here, but now we are going to make some parts that require a bit of precision to function well. Here's how:

Clamp a piece of 5/16" or 3/8" bar stock in the three-jaw chuck of your lathe with enough overhang to make the whole length of a locking pin (note that there is one long locking pin and two short ones), plus a little for starting a threading die. Center drill the outboard end with a small center drill, about a No. 0 or 1. We are going to turn a fairly small diameter, so we need the support. Turn the blank down to head diameter, then to the sleeve ID. Make this diameter as precise as you can, so the sleeve is a good slip fit on the pin. Finally turn the remainder down to the screw diameter and give the outboard end a slight taper to help start the die. (If you are really in to this, you could cut the threads, but a die is much quicker.) Make all three locking pins, then make the locking sleeves from the

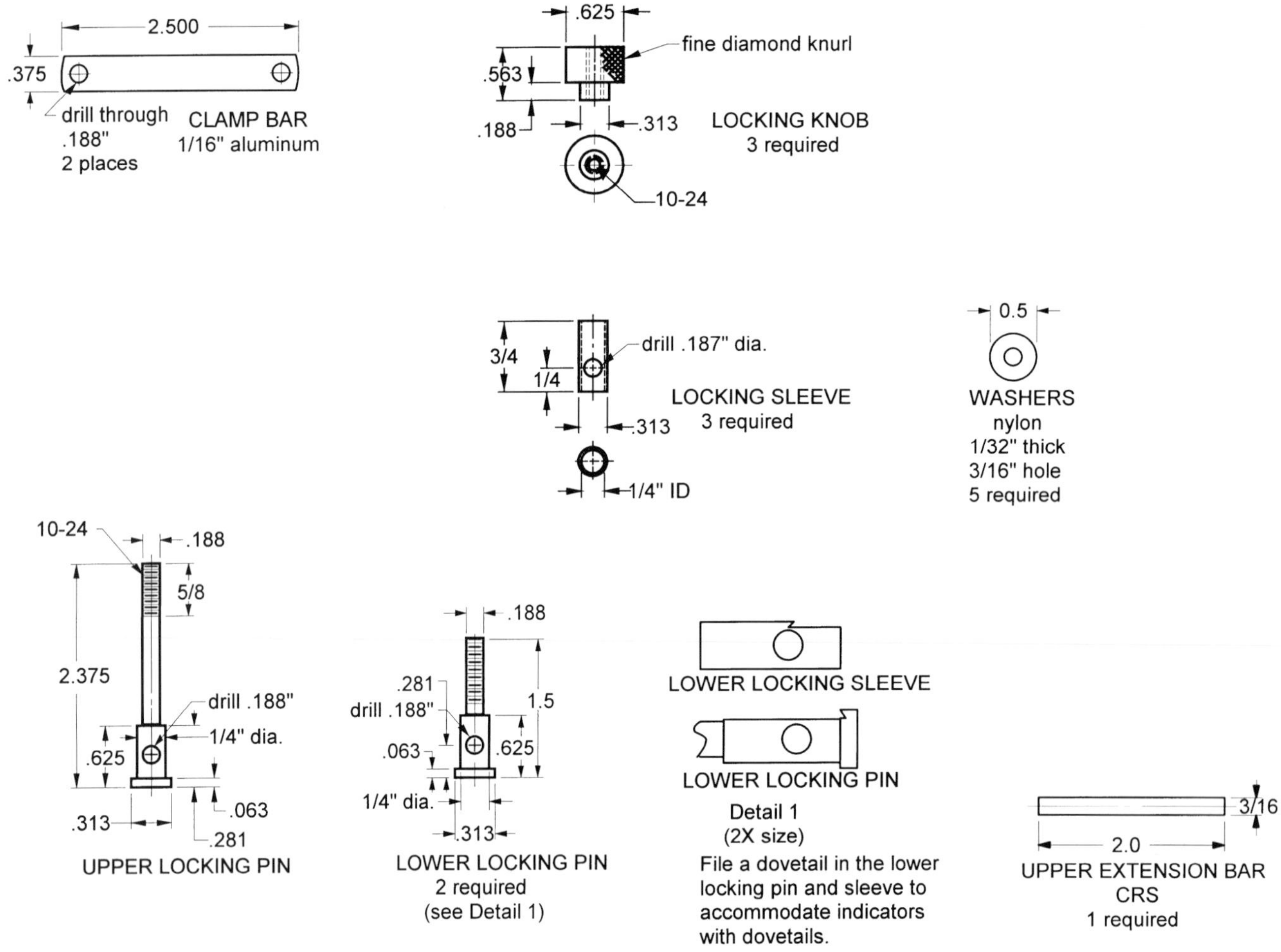

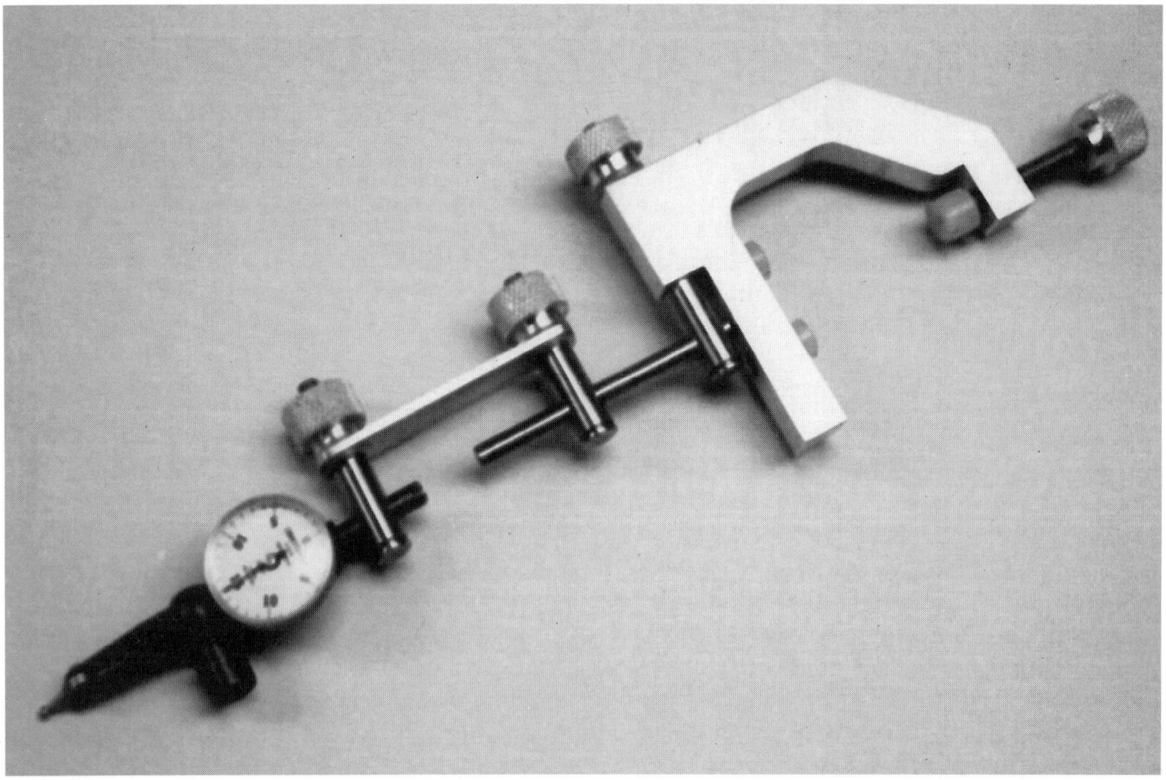

The author's indicator holder assembled and ready to use.

The author's indicator holder in use.

same stock. A bit of judiciously applied fine emery paper will bring these parts to a fine polish, and really enhance the appearance of your project.

First turn the outside diameter, drill undersize about 1/32", then drill or ream the final ID with plenty of lubricant. Part off to the proper length, and test assemble to make sure they are a good sliding fit on the pins. Good workmanship here pays off in a smoothly working indicator holder.

Now drill the holes in the locking pins and sleeves. I set this up in the mill rather than the drill press to take advantage of the positioning ability of the mill. Assemble the pin and sleeve with a short piece of copper wire wound around the head of the pin. This offsets the sleeve enough to provide the proper clamping force. Using thin cardboard to protect the finish of the sleeve, clamp the assembly in the mill vise, being careful to keep the clamp parallel with the mill table. Locate the center of the pin/sleeve assembly and center drill with a No. 0 center drill. Now drill through slightly undersize, and finally to size. If all has gone well, the hole should be straight through the center of the pin, and a close sliding fit on the extension bar.

If your indicator is the type with a dovetail back, carefully file a matching dovetail on the lower locking sleeve as shown in the drawing. Use a small triangular file, and put the extension bar through the hole to keep the assembly properly aligned.

Make the lower clamp bar from 1/16" flat stock to the drawing dimensions, assemble your new indicator holder, and you are ready to check out your new mill in style.

SHOP IMPROVEMENT PROJECTS

Variable Speed DC Motor for the Home Shop

by Charles Eyer

Photos by George Panos
Drawings by Author

This is not another article on how to build a controller for your drill motor or your router motor. It is an article on how to build a controller for direct current (DC) motors that were in the past available to only the industrial user because of the cost.

Due to the low cost of high power semiconductors available today and the developments in DC motors, the home machinist can build a constant torque DC motor controller.

The controller described in this article features: speed variation of over 20 to 1 (depending on the motor used); reversibility; electronic braking; and remote control.

A REVIEW OF THE DC MOTOR

Most readers will not have a very good background in direct current (DC) motors unless they have used them in industrial applications. The average home machinist is probably familiar with the alternating current (AC) series motors (used in drill motors) and the split phase motors used on the majority of equipment in the home shop. Probably a DC motor has not crossed his path before. Therefore, I believe a short review of the DC motor would be beneficial at this point.

Simply stated, a DC motor rotates as a result of two magnetic fields interacting with each other. These two interacting fields are produced by the current flowing through the windings of the field and the armature. The DC motor can be configured in several ways. The two most common configurations are the series and the shunt wound motor.

The shunt wound motor, shown in Figure 1, has characteristics that make it very desirable for the home shop. The first and one of the most important features is the variable speed capability. The DC shunt motor differs from the AC series (or universal motor) in this respect in that as the speed is reduced, the torque remains constant. You have probably experienced torque loss when reducing the speed of your drill motor or your router. The second characteristic of the DC shunt motor is that its rotation can be easily reversed. The third feature is that it can be braked electrically, simply by shunting a load across the armature after the power has been disconnected.

A variation of the DC shunt motor is the permanent magnet (PM) motor. The permanent magnet replaces the field winding on the DC motor. This motor can be treated as a DC shunt motor except that the power does not have to be applied to the field winding. The advantage of the permanent magnet for the field is that no field power is required. PM motors are hard working and should not be confused with PM motors that power toys.

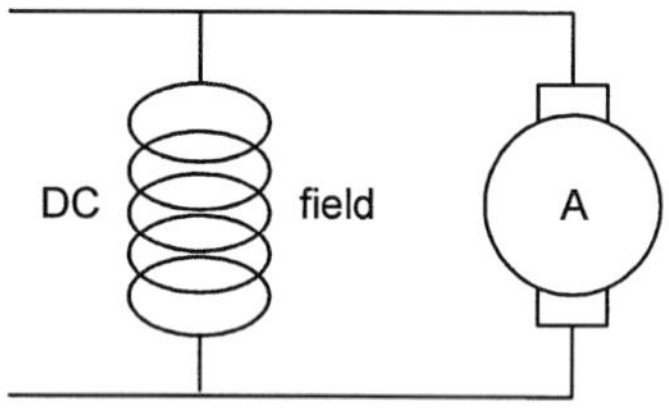

SHUNT DC MOTOR
Figure 1

The permanent magnet motor has really come into its own since the development of rare earth and other exotic magnetic materials. An analogy of these new magnetic materials can be made by comparing the amount of air stored in the portable tanks used for inflating automobile tires to the smaller air tanks that skin divers use that can be pressurized to 1,000 psi. The magnetic materials manufacturers have done the same thing with these new materials. A high magnetic field can now be concentrated in a much smaller space than ever before possible.

COST

In one of the past issues of *The Home Shop Machinist,* one reader stated that he had purchased a DC motor for his shop but was lamenting the cost of a commercial unit. He was right, the price in the Grainger catalogue for a 1 hp motor with controller is $400. I have built a controller and acquired a motor for under $20. In the remainder of this article, I will tell you how to acquire and to build a controller for minimal cost.

MOTOR ACQUISITION

Permanent magnet motors are showing up in a lot of strange places today. I acquired mine from the electric lawn mower that my neighbor put out on the street for trash pickup. The mower contained a 1.5 hp permanent magnet motor as shown in the various photographs in this article. The only problem with the motor was the bridge rectifier. This is used to convert the AC power to DC for the motor. This was easily replaced for under $2. The mower came equipped with the 1 ohm braking resistor. This resistor is used for braking the mower to a quick stop after the power is removed.

The mower I got was a Black and Decker (B&D) model. Not all electric mower manufacturers use a permanent magnet motor. Some of the cheaper units still use the universal motor. Do not bother to carry these units home. You also can make the rounds of the garage sales and yard sales to acquire one of these units cheap if you cannot find one at the curb as I did. I checked with the Black & Decker parts department and the price for a replacement motor is $125, so it is probably not economical to purchase them new.

Another source of the DC motors is from many industrial surplus houses. I have included a list of these at the end of this article. Some of these companies are advertisers in *HSM.* The motor specifications to look for are permanent magnet DC motors requiring 90 volts DC input. Some motors require 48 and 24 volts; do not buy these units. They require a transformer that will make the project more complicated than it needs to be.

MOTOR PREPARATION

If you are lucky enough to find a used B&D lawn mower, start by removing the motor from the mower chassis. Attached to the motor you will find a ceramic 1 ohm resistor; this is the braking resistor. Also you will find a plastic cube about 1″ square. This is the bridge rectifier used to convert the AC line power to DC (be sure to note the AC leads and the motor leads). Save these two items as they will be used to construct the control module. If you are constructing this controller for a motor other than the B&D, don't worry – the resistor and bridge rectifier can be purchased separately.

You will find that most of the surplus motors and some of the industrial motors are flange mounted. This tends to be a problem for the home machinist, since most of the normal machine tools use a standard horizontal base mounting.

I have constructed a horizontal mounting bracket, shown in Photo 2. This mounting procedure also provides a convenient place for the mounting of the controller circuit.

2

I will not attempt to describe a universal mount for the many different styles of motors you will find. This is an area for which you are probably more suited than I am, since most of you are machinists and this is your craft.

CONTROLLER CONSTRUCTION

The electronics of the controller is mounted in a standard 4 × 4 × 2" electrical junction box, available at any hardware or building supply center. The main power cord is secured to the box by a standard cable clamp that is normally used with this box in house wiring applications.

Since most motors are usually located in remote or inaccessible places on machinery, the controls need to be located remotely from the motor itself.

3

This point occurred to me when I installed the motor on my band saw. The motor is mounted in a cabinet under the saw practically impossible to get access to in less than five minutes, so a remote control for this application is a must. I am sure you have similar situations with your equipment.

The design of this controller allows remote control of the speed of up to several feet, if necessary. The mechanism that controls the speed is a potentiometer (POT). The POT controls the charging rate of a capacitor. When the capacitor reaches a predetermined charge (this charge level is determined by the diac, about 28 volts) it causes the silicon controlled rectifier to turn on and thus allow current to flow for the remainder of the cycle.

This POT can easily be remotely placed by mounting it in an outdoor-type PVC electrical utility box. It is connected to the main controller by a phono cable. Since there is very little current flowing in this part of the circuit, the wire can be of a small gage. I used standard RCA phono plugs on each end for easy disconnection. These cables can be easily purchased from your local Radio Shack or stereo store, complete with plugs on each end. The remote control unit is shown in Photo 3.

This is a PVC box with the POT inside.

The heart of this circuit is a silicon controlled rectifier (SCR), shown in Figure 4. This semiconductor can be purchased in many of the surplus catalogues very reasonably or can be ordered from one of the many electronic parts supply houses. (A complete bill of materials is shown with the circuit diagram.)

It is important that an SCR is used and not interchanged with a triac (an SCR back-to-back) which conducts on both cycles versus only one for the SCR. The SCR is manufactured in two styles; one is stud mounted (Figure 4), and the other is a flat pack-type used for mounting on a printed circuit board (PCB). Purchase the stud-mounted type because the flat pack is too hard to mount when not using a PCB.

The SCR electrical specification should have a peak inverse voltage rating of 400 volts and current rating of 10 amperes. These specifications are not that critical, so if you can find some surplus units that are close they will probably function all right. The reason for the 400-volt rating is that when some motors are switched on and off, they generate some high voltage spikes that can damage the SCR. (For you tekkies, this is known as the L-di/dt effect.)

The stud type SCR comes complete with the mounting kit. The kit consists of the following: two insulating washers; one regular metal washer; one lock washer; one solder lug; and one *Teflon* insulating bushing.

It is important that the SCR be electrically isolated from the chassis. The stud type (Figure 4) can be mounted by drilling a clearance hole the size of the stud insulating bushing in a metal chassis. Assemble the SCR on the chassis in the order shown in Figure 4. This ensures that the SCR is electrically isolated and, at the same time, conducts heat for proper cooling.

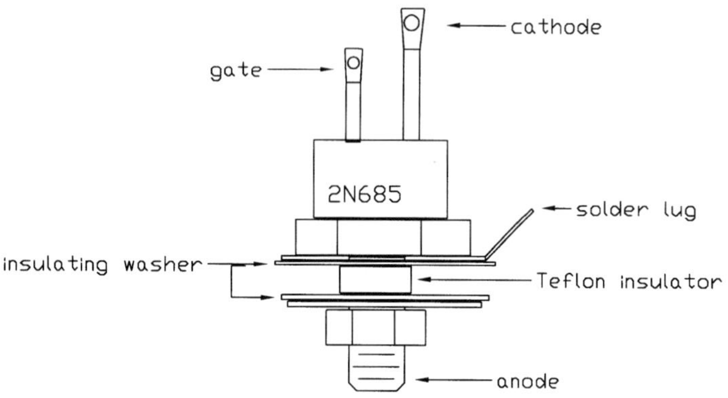

Figure 4

The current requirement could be adjusted lower or higher depending on the motor you use. Check the manufacturer's plate on the motor for the maximum current drawn for your motor. A higher rated SCR will function properly with a lower current requirement but not the other way around.

I do not recommend using a printed circuit board (PCB) for this application, because the current requirements in some parts of this circuit are too great for the PCB foil.

The circuit is constructed on a .125" piece of aluminum. The reason for this is that the SCR requires some heat sinking, so by mounting the SCR on this metal plate a heat sink is provided and, in addition, it provides a convenient place for mounting the remaining circuit components.

Start the construction by cutting the .125 " aluminum plate so it will fit in the electrical outlet box. This box must be at least 4" square and 4" deep. I happened to have some L-shaped extruded aluminum in my scrap box that allowed me to bolt the plate to the box easily, but you can fasten it to the box any way you desire. The important requirement is that it must be secured so it won't move during operation of the motor.

The next step is to mount the SCR to the aluminum plate. The placement is not critical. I put the SCR on the right side and the diode bridge on the left (Photo 5).

Start by drilling a clearance hole for the *Teflon* bushing. The bushing may have to be trimmed to the thickness of the chassis.

Assemble the SCR on the chassis in the order shown in Figure 4. Tighten the nut firmly; this establishes electrical connection with the solder lug.

The rectifier can now be mounted. This is a straight-forward procedure. The rectifier has a hole through the center that is used for mounting with a screw. Drill a clearance hole on the left side of the mounting plate. The rectifier can be fastened directly to the plate – no need to be concerned about insulation from the chassis.

These rectifiers come with two different types of leads for electrical connection. If you salvaged the one from the lawn mower, it has flying leads. The other type of connector is the spade lug. The type of lug is not critical. The only difference is in the way the connections are made to the rest of the circuit.

At this point I should mention the wire gage needed for hookup. Generally the wire size should be no smaller than the leads of the individual components, with some exceptions. For example, the wire size of the rectifier leads should be that gage or larger for leads that go to the motor and SCR. However, the current through the resistors, capacitors and pot is quite small, so you can use a smaller gage here.

I have indicated on the schematic the leads that need to be a heavier gage by showing these in bold.

In addition I should mention about soldering these components. Do not use a soldering iron any bigger than 40 watts and do not use a quick heat type. That single coil of copper wire used in these guns carries a large current and in some cases can induce a damaging current into delicate electronic components. However,

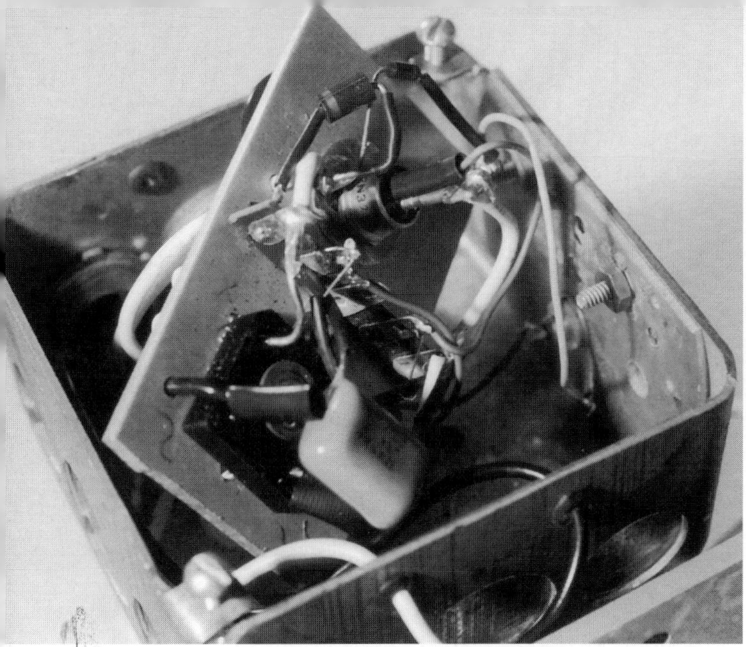

5

in this circuit the current would probably not harm anything, but it is best not to tempt fate.

The biggest cause of damage to electronic components is heat. Use a low melting solder made for electronic circuits and apply just enough heat to melt the solder. Never use plumber's solder or any solder containing acid.

WIRING PROCEDURE

Start the wiring with the bridge rectifier. When connecting the power cord be sure to connect the ground or green wire to the metal electrical box; this is a safety measure. The labeling on these components varies with manufacturer; some label the AC side and some label the DC side. The DC side will have a plus and minus symbol (+ −) and some will have a miniature sine wave symbol on two leads (∩U). Connect the AC power cord to the terminals marked AC or the leads opposite the plus and minus symbols.

I am assuming 110-volt 60-hertz residential wiring for the AC line. At this point, stop and connect the AC cord to a wall socket. Using a voltmeter, measure the voltage across the DC leads. You should have a reading somewhere around 100 volts; if not, recheck the wiring again. If you are sure everything is wired correctly and are still getting no voltage, the bridge is probably defective. Replace the bridge.

Electronic components don't always come from the manufacturer perfect. You cannot be sure of the handling they get before you received them, especially if you are using surplus components. This is especially true if you are using the bridge rectifier supplied with the mower.

When you have the proper voltage at the DC leads, temporarily connect the two leads of the motor to the two DC leads. Reconnect the plug to the wall socket. The motor should run at the top rated speed at this point.

Warning: Always disconnect the AC wall plug before making any connections or adjustments. There are lethal voltages here.

Before you continue on with remote control and braking, be sure the speed control is operational. Finish connecting the remaining components, making sure the polarity of the diodes is observed.

If you are going to remote the speed control temporarily, mount the POT on a piece of plastic or any non-metallic material that will provide some support during testing.

When you have completed the wiring of the controller, connect the motor and turn on the AC power. The motor should be running at some RPM. The speed is varied by adjusting the POT.

If the motor is not operational, check diode D1 for a high forward resistance. By this I mean the resistance in the conduction or forward direction. This can be accomplished with a voltmeter using the resistance measurement mode. Connect the positive lead of the volt meter to the anode or the positive side of the diode and the negative lead to the cathode. The resistance should be several ohms. If this resistance is high, replace it with a diode with a lower value.

I found several diodes fresh from the distributor that were out of range for this application. These diodes are a fairly cheap item, so don't panic if several diodes have to be discarded.

If this doesn't solve your problem, measure the voltage at the gate of the SCR; this should be several volts. If there is no voltage at this point, check the voltage at the capacitor end of the diac.

If you find several volts here, replace the diac. The diac has no polarity but is fairly sensitive to heat, so be careful with the soldering process.

The only remaining source of trouble is the SCR. If you are still having trouble, confirm the polarity of the SCR with the manufacturer's literature. There is no standard polarity for the SCR. In other words, the stud is not always the cathode. The small terminal on the top is always the gate, though.

This is not an overly complex circuit as electronic circuits goes. I believe the average person with some knowledge of voltmeter operation can easily complete this project. So I am going to assume the motor is working at this point and proceed to the next step.

Now not all applications will require electronic braking and direction reversal, so these items can be added, if necessary. I believe braking for a drill press or milling machine is probably a very desirable feature.

The reason I am separating these functions is that if remote control is desired, the cabling for every feature adds another pair. For example, speed control requires two wires, braking requires two large gage wires, and rotation reversal requires two wires. This brings the total to six wires. This makes for a complex connector on each end. So if you don't need all of these functions use only what you need.

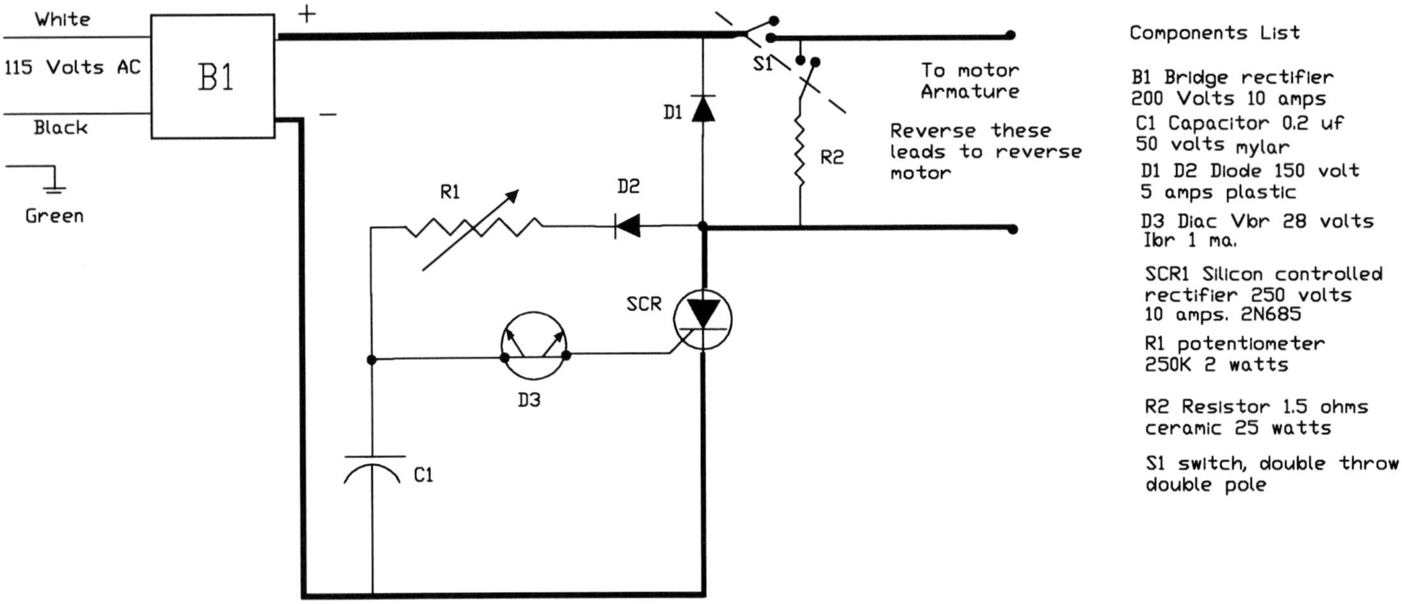

SCR SHUNT OR PM MOTOR SPEED CONTROL
Figure 6

BRAKING

Electronic braking is accomplished by shunting a low resistance across the motor after the power has been removed. There is still a considerable amount of energy remaining in the motor due to the flywheel effect of the armature and the coupled machine tool. This means the resistor will have to dissipate this energy without destroying itself. The resistor must be a ceramic type with about 10 watts rating, and have a resistance of about 1.0 ohm. This is not a critical value; use whatever value you can find in this resistance range (under 10 ohms).

To remove the power and shunt, the resistor requires a double-pole, double-throw switch with a 10-ampere, 110-volt rating. I prefer the push-button type. The main consideration is that the switch not be the momentary type. This is a safety consideration; you do not want the motor to start again after you release the braking switch.

If you are using remote control with your motor, the braking adds a little more complexity to the cable connecting the remote controls and the motor. The cable now must be a four conductor and have a greater current-carrying capacity than the speed control. The difference in connectors and wire is broken out in the components list.

This circuit is a fairly simple design and should give you little trouble in the construction. However, if you're interested in gaining a better background in electronic control of motors and industrial electronics, I have included a reading list of manufacturers' literature and technical school level textbooks.

The textbook on Industrial Electronics should be in every home machinist's library since these new motors and controls will be showing up on new machine designs soon. For example, Sherline – the company that makes the miniature lathe and milling machine – has an electronic controlled motor on its latest product. This is only the beginning. I can't imagine machine tool companies making complex gears and pulleys to control the speed of machines any longer.

DC MOTOR SOURCES

Surplus Center
P.O. Box 82204
Lincoln, NE 68501-9973
(402) 474-4055

C&H Sales Co.
P.O. Box 5356
Pasadena, CA 91107-9988
(800) 325-9465

Herbach & Rademan
P.O. Box 122
Bristol, PA 19007-0122

ELECTRONIC SUPPLIERS

Newark Electronics
4801 N. Ravenwood Ave.
Chicago, IL 60640-4496
(312) 784-5100

Digi-key Corp.
P.O. Box 677
Chief River Falls, MN 56701
800-344-4539

REFERENCES

Industrial Electronics
James T. Humphries
Leslie P. Sheets
Delmar Publishers
ISBN 0-273-3828-7

Thyristor Device Data
Motorola
P.O. Box 20912
Phoenix, AZ 85036

Electronic Power Control
Circuits, Devices & Techniques
Irving M. Gottlieb
TAB Books
Blue Ridge Summit, PA
ISBN 0-8306-2453-8

Supplement to the Home Shop Motor Controller

by Charles Eyer

I have received many requests from the readers of *HSM* for catalogue numbers for the parts used in the construction of the PM motor controller. They all inform me that the parts are difficult to acquire. I have to apologize; since I wrote the article in 1995, a lot of things has happened in the electronics world. The key component, the SCR (2N685) with the stud mount is no longer manufactured. So the solution is to use an SCR designed for use on printed circuit boards (PCB). This is not all that difficult. In the following discussion the catalogue numbers and mounting techniques for the PCB SCR are shown.

First, a little theory on the way the controller operates. I am surprised that so many readers were able to find the motors as described in the preceding article, which originally appeared in *HSM*. But many readers are using other surplus units. This may require some adjustment to the resistor capacitor circuit to make the motor operate over its entire range.

All motors when running also act as generators. They create a voltage in the opposite direction to the supplied voltage. This is called back electromotive force (EMF). Back EMF is beneficial in this case because it can be used to determine the load on the motor. When the motor is loaded, the RPM decreases and, thus, the back EMF decreases. Voltage that appears at diode D2 (see the preceding article) is the difference between the supplied voltage and the back EMF voltage. So when the motor has a decreasing RPM, the voltage at D2 is increasing, since there is less back EMF to buck the supply voltage. This in turn will charge the capacitor more quickly which will cause the SCR to conduct sooner and, thus, tend to keep a fairly constant torque over the RPM range of the motor.

Some readers have attempted to test the controller without the motor attached, using a voltmeter to observe the change in output voltage. This controller will not function without the motor attached. This is due to the above discussion on back EMF. The capacitor is charged from the back EMF, so without the motor attached, there is no voltage there to accomplish this.

Every motor will have a different back EMF. If you are having trouble with your particular motor, you may want to try changing the value of C1. Try, first, doubling the capacitance to .4 uf. This is quite easy to do. Connect another .2 uf capacitor and put it in parallel. This will double the capacitance to .4 uf. To halve the capacitance to .1 uf, put two capacitors in series. Remember, this is just the opposite of resistors.

Different SCR's will also have an effect on the value of C1. Since the original SCR is no longer available, some capacitor adjustment may also be required with the new SCR I am about to describe.

I have found a source of components. The company is very responsive and representatives don't mind dealing with hobbyist. The name of the company is:

All Electronics
P.O. Box 567
Van Nuys, CA 91408-0567
(800) 826-5432
Internet http://www.allcorp.com

You can order a free catalogue or order the following parts direct:

SCR Tecor S4020L	catalogue No. S4020L
TO-220 Heat Sink	catalogue No. HS-17 (This is used with the SCR)
DIAC No. ST2	catalogue No. ST2D

If you haven't ordered the remaining parts, here are the catalogue numbers for those:

Full wave rectifier bridge
10 amp 200Vcatalogue No. FWB-1021
Diodes D620 catalogue No. D620
Capacitor catalogue No. RMC-23
Potentiometer 250K catalogue No. PT250

This company also has a good selection of metal and plastic boxes that could be used for mounting the components.

MOUNTING THE SCR

The new SCR will look similar to the drawing. The three prongs protruding from the bottom are the cathode (1), anode (2), and gate (3). These will be connected to the corresponding connections in the preceding article.

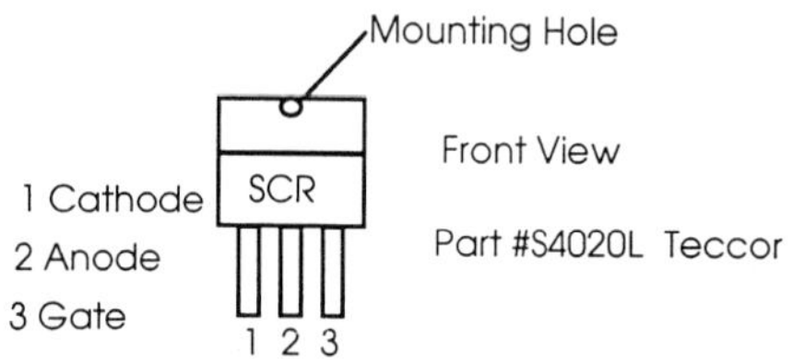

A word of caution. The above terminals are believed correct, but check with the package that accompanies the SCR; some manufactures have their own ideas about the terminal configuration.

This SCR should be mounted with a heat sink. The correct sink is shown in the parts list. The heat sink fits on the back of the SCR, and they are bolted together. They should probably be mounted on an L-shaped bracket so the air can circulate freely around.

Connection to the terminal is made by wrapping one or two turns of wire around each terminal, and apply a small amount of solder. Do not use any more heat than necessary for soldering. Be sure no terminals touch and, if you are using a metal box, be sure to isolate any bare terminals from the box. If you have some heat shrink tubing, it would be a good idea to place some on each of the terminals after the soldering.

TROUBLESHOOTING HINTS

Some readers have tried to test the circuit without connecting it to a motor. The circuit will not operate in this manner, for reasons discussed in the theory of operation.

If you want to check your wiring as you progress, start with the bridge rectifier. After you have wired the AC line to the bridge, connect the DC output to the motor. At this point, the motor should run at full speed. If not, check your wiring and/or components for failure or wiring mistakes. Another hint, be sure the motor is operational; don't assume anything.

Caution: all semiconductors are delicate devices; they will not tolerate shock, excess heat or abuse. For example, do not drop a semiconductor from the workbench to the floor. The shock from this distance is sometimes great enough to destroy the semiconductor.

There is also some misinformation circulating on the Internet that the Black & Decker lawn mower motor will not function in a horizontal position. This is false; I have used it in a horizontal position for two years without difficulty. Most motors will operate in either position without any problem. I have never seen restrictions on any motor I have used. Most motors today have sealed bearings, and lubrication is not an issue. Early motors that used complex lubrication of the simple bronze bearings did have this requirement.

A Modified Arbor Press Ram

by Alan Andrews

Drawings by Author

I modified the basic imported arbor presses presently available so they are "ratcheting" with minimum expense. See the drawings.

The top drawing shows the ram as manufactured. The lower drawing shows the modified ram. I removed the two lower teeth of the rack. I installed the ram with the modification closest to the work. If I find that the handle isn't in a good position when I go to use the arbor press, I just raise the ram up until it slips a tooth on the pinion and try again. No disassembly, no lost time.

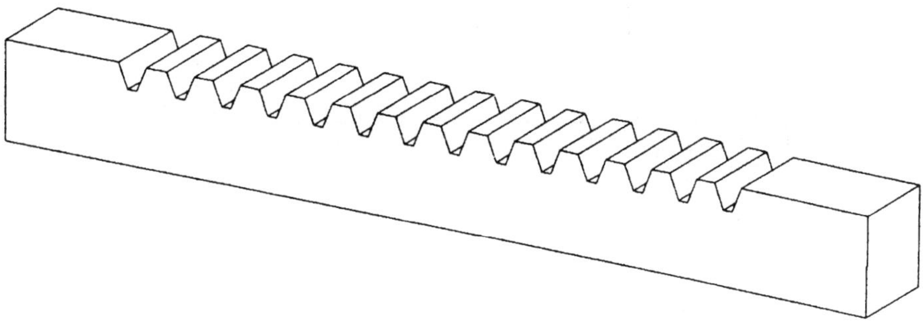

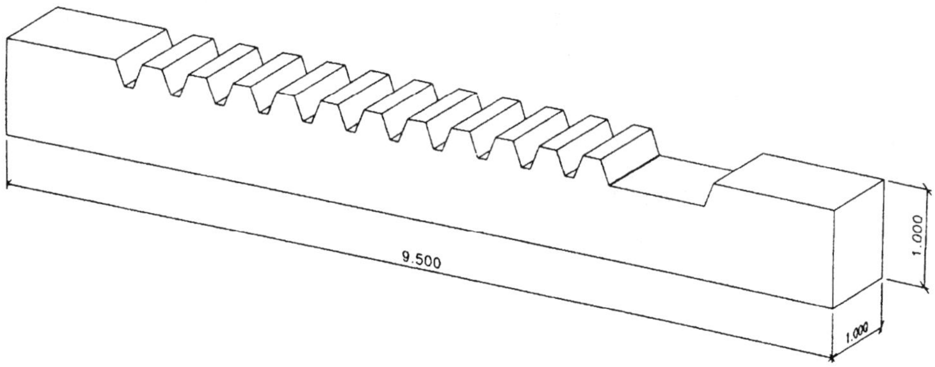

By unplugging, rotating a half-turn, and replugging these two wires, the motor can be reversed. This makes tool post grinding more convenient.

Low-cost Reversing Switch for Electric Motors

by R. L. Halbert

Photo by Author

A quick way to reverse the rotation of electric motors without the cost of installing expensive reversing switches can be done by practically anyone.

Manufacturers of small metalworking lathes, for instance, will sell reversing switches to fit their machines for $125, or even more. For the rare need to change shaft rotation for such operations as tool post grinding, the home shop machinist must "re-wire" his motor to get the desired results.

Anyone can achieve the same effect with a simple unplug-and-replug operation that takes no more time than flipping the more expensive accessory switch.

Most power tool motors – for instance, a 1/2 hp capacitor-wound motor – have two wires inside the connection box (often the black wire and the red wire) that must be switched to reverse the rotation of the motor. This fix uses two short pigtails to take the operation outside the connection box, and allows it to be done almost instantly, again and again, as often as necessary.

First, the two wires are disconnected from the inside terminals. Then a short two-wire cable of at least 16-gage wire is connected to these terminals with a two-prong male plug at the other end. The two wires from inside the motor that formerly connected to the terminals are then connected to another short two-wire cable with a female two-prong plug at the other end.

The cover is then replaced on the motor. The male plug is plugged into the female plug and the on-and-off switch is flipped to start the motor, just as always.

To reverse the motor while the switch is off, unplug the two cables and rotate the plugs 180° degrees – or half a turn – and plug back in. Now when the switch is flipped to "on," the motor will turn in the opposite direction.

Often the terminals inside the motor connection box will be fitted with spade connectors, and connection is easily made by crimping or soldering complementary spade connectors to the two pigtail cables. Should these reversing cables no longer be needed, they can merely be unplugged and the original wires reconnected as before.

1

A Machinist's Clock

by Bill Reichart

Photos by Author

Here is a project for you home shop machinists I'm sure will be well received. It is a clock; however, it takes more of a machinist to make it rather than a clockmaker.

Furthermore, a short study will show that this is not a project full of dimensions and details, because each individual will have to cut and fit for his own clock. The finished clock, hanging on a wall, is shown in Photo 1. It is basically described as a long screw, driven by a gear motor, and as the hand creeps along the screw the time is shown.

To construct the clock, you first must find a small 110v gear motor. A good source is Surplus Center, P.O. Box 82209, Lincoln, NE 68501-2209. You can get their catalog by calling (800) 488-3407. This company offers wide range of all kinds of electrical and mechanical devices that are suitable for home shop machinists. The motor I found was about 3" in diameter and 4" long, and the output shaft ran at 22.8 rpm.

Once you have decided on a motor, you build a mock-up of your project (Photo 2). I built my mock-up on a 2×6 about a yard long. The bearings and mounting on this mock-up are all wood and only roughly made. Now study Photo 2. The motor is indicated at A. Then to further slow it down, I incorporated a geared unit from an old 3/8" electric drill. This is B. At C, I further reduced the speed using a couple of gears.

The shaft D was a piece of 1" OD cold rolled steel with a hastily cut 1/2-20 thread. Next, at E is the mock-up hand. This was just a 1/2-20 nut with a thin sheet metal strip wrapped around it. This hand laid upon a piece of wood shown at F. This F indicates the face of the clock.

Next comes the big moment when you switch on the motor. As the screw turns, the hand creeps along. Please understand that you don't need any complicated mathematics to figure how fast the hand creeps. Instead you simply put a mark on F, and run the motor for 30 minutes or less and see just how far the hand moves. On my first try the hand moved almost 6" in one hour so I knew I had to slow

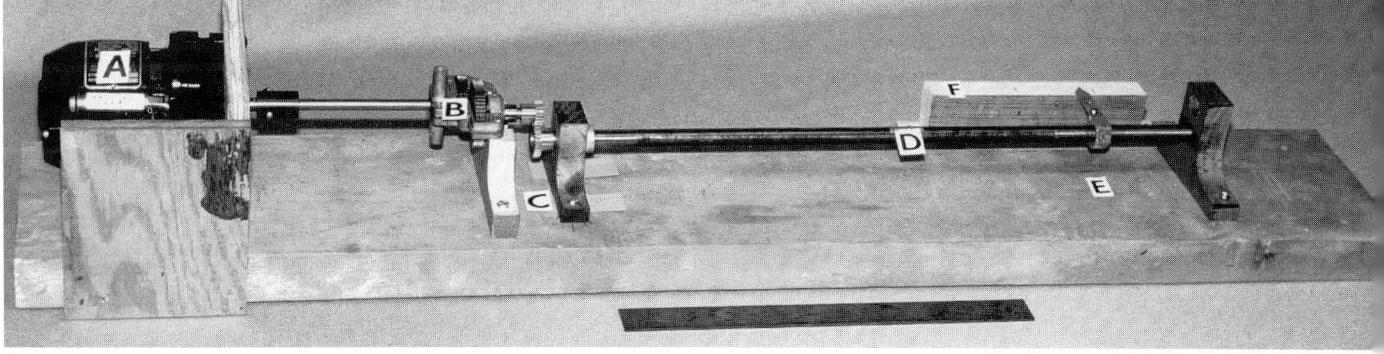

2

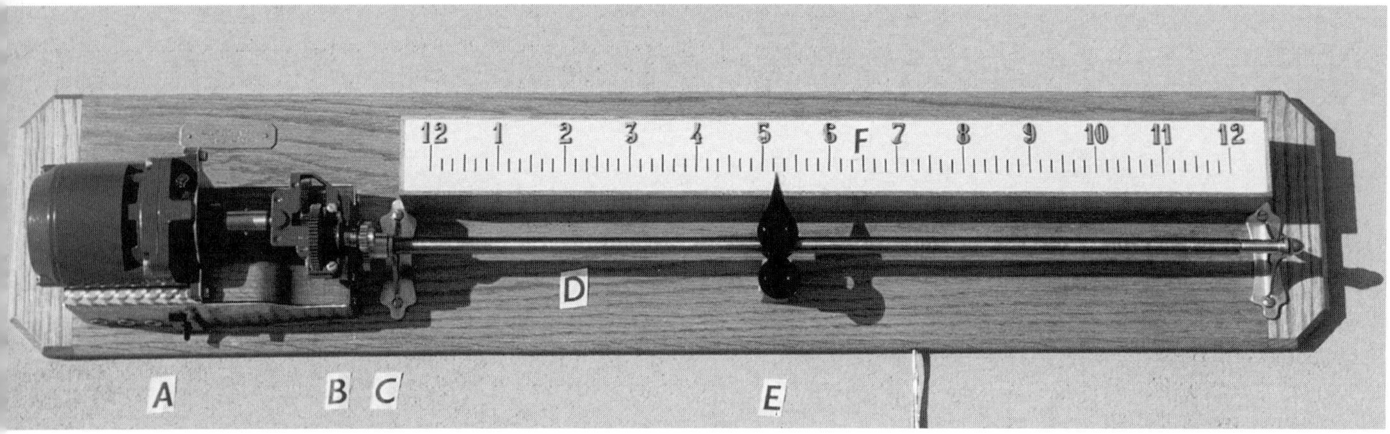

3

down the shaft D. In my case, I used different gears at C, and soon found the hand moved 5-9/16" in 3 hours, or 1-7/8" per hour, less than 1 rpm. This speed suited me because if the hand moves 2" per hour, in 12 hours I would need a 24" long screw.

In making your clock, please don't try to get my speed, but rather get a speed that suits the clock you want to build. If you are happy with a hand that moves only about 1" per hour, your clock would be much shorter than mine. Or if you prefer, once you have found a

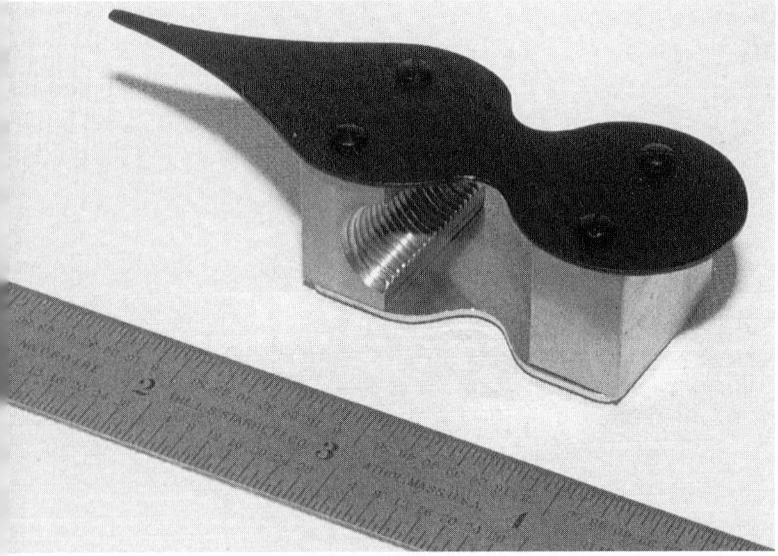

4

5

6

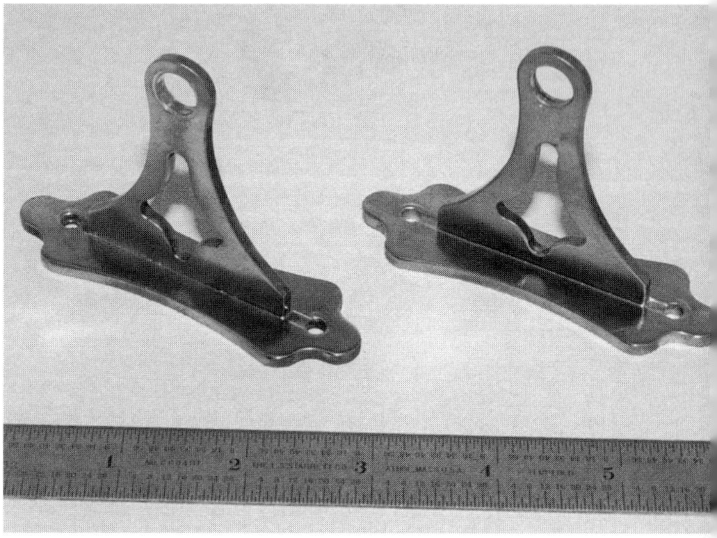

7

suitable speed, you can make a face for 12 hours such as mine, or you could divide the spaces and have 24 hours on the face. In my case, once I found the hand to move 1-7/8" per hour, I simply laid off a 12-hour space for the face.

Now look at Photo 3. This is my finished clock with the same letter designations as shown on the mock-up in Photo 2. As you can see, my finished clock is about one yard long.

To make the hand, please see Photo 4. My hand is about 3" long and is a piece of sheet metal filed to the shape and painted black. This is attached to two pieces of brass. A shorter piece of sheet metal also joins these two brass pieces at the back. At the point end on Photo 4, the brass piece is really a half nut. I made this half nut (Photo 5) by threading a piece of brass that was about 1" long with the 1/2-20 thread. Then I cut it in two. The piece of brass at the round (bottom) of the hand was simply a counterweight to hold the hand vertical. My hand was assembled using eight small machine screws.

Now because these two pieces of brass are spaced apart, it is possible to lift the hand and position it along the screw, thus setting the time. In making your mock-up, it is important that the long screw rotates so as to cause the hand to want to bear against the face. If the screw rotation is reversed, the hand hangs somewhat vertical but the point lays away from the face. So be sure the rotation causes the point of the hand to turn into the face.

There is a lot of leeway in making the final clock parts. Photo 6 shows my setup to chase the 1-20 thread in the lathe. A follower rest is necessary. Mine looks rather clumsy, being homemade, but it does incorporate two ball bearings. My long screw is stainless steel, but cold rolled would be fine. I just don't want rust down through the years, so I use either stainless steel or brass for the parts. When making this long shaft, you must turn the right end down below the thread line for 1-1/2". This is done so that, if the clock hand goes all the way to the right end, it simply rests on a smooth shaft and does not feed.

Photo 7 shows the brass brackets I made to support the shaft. These are made of 1/8" brass plate. Each bracket is two pieces silver soldered together. I tried to make mine look a little "clocky" and somewhat ornamental.

My clock face is about 26" long and about 2×2" across. The white face was a piece of 1/8" masonite with white already on it. But it could be painted white or whatever color you desire. The numbers are pressure sensitive plastic from a craft store. I laid off the divisions with a black pen. On my clock, each division is 10 minutes.

In final assembly I mounted all parts on an oak board. The motor was painted green. I made a fancy sheet metal enclosure around the motor to hide electricals. In this housing, I put a switch and a pilot light. My enclosure was made of stainless steel, and I put a damascening finish on it. This finish is done in a drill press using a small wood dowel with a felt tip. An abrasive is applied to the tip, and as pressure is applied, a small shiny spot is made on the metal. By somewhat carefully indexing, you can put these spots all over the housing.

Finally you can say, "What good is a clock that only shows 12 hours of time, and can't run constantly?" Well, the good is that you had a lot of fun making it, and if finished up nicely, it is a trophy to your machine shop skills. I put a brass nameplate and date on my clock, and I'm sure my great-great-grand-children will love having it. I use it if the gang comes over for happy hour and a game of cards. Just flip it on and it becomes quite a conversation piece. Flip it off when all go home.

Your mechanically-minded friends will soon be suggesting ways to make the hand move back and forth or will make other suggestions. I have considered making a longer screw and the clock would be vertical. Then at the top the hand would disengage and drop to the bottom – then start climbing again. The possibilities are endless.

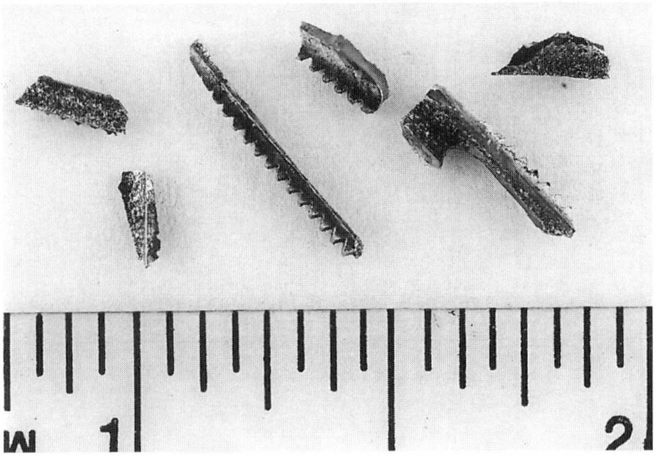

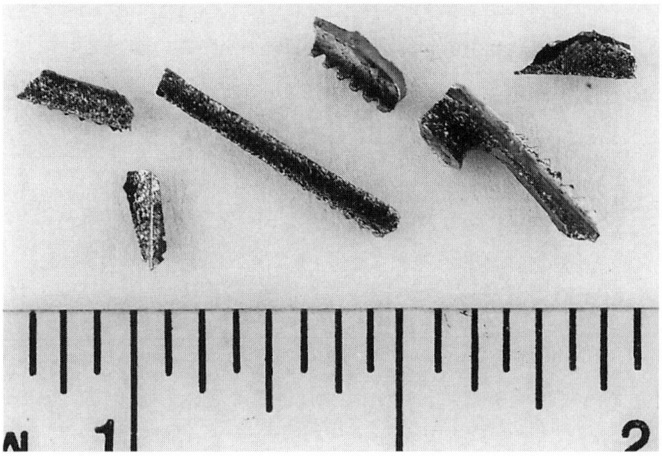

More on EDM

by Jack Lewis
Photos by Author

The photos are of a broken 10-32 tap that was almost completely through a 1/2" plate on a project on which I had spent a lot of time, and it was essential that it be removed. The longest piece is approximately 1/2" long. I bored a hole lengthwise through the tap with my EDM, and picked the pieces out of the hole with an ice pick without damaging the threads. Needless to say, I was overjoyed at getting it out and finding a cheap, easy way of doing so. This is something everyone who does any tapping needs to know about.

I used Hank Meador's basic diagram with the recommended Radio Shack diode, but both light bulbs were 200 watts. Someone had given me a whole box of starting/running capacitors for electric motors, with microfarad ratings ranging from 5 to 35 at 375 volt maximum. As it was a weekend, when most suppliers are closed, I decided to give it a try with what I had on hand. I do not believe the capacitors were electrolytic; I could not find any plus or minus marks to indicate polarity. Not having much confidence that it would work with these capacitors which were so much below the recommended 200 microfarads, I rigged the system up the easiest way possible, using alligator clips on the wires to the capacitors and a piece of dry hardwood 1/2 × 1/2 × 3", one half of the length turned round to go into the drill chuck on my vertical mill. A piece of No. 12 solid copper wire 3" long was clamped vertically to the side of the square portion of the hardwood, and checked with a square on the mill table to be sure it was truly vertical. The top portion of the wire was bent outward at right angles to the wood to accept an alligator clip on the wire marked "tool."

The work with the broken off tap in it was a disk 5-1/2" in diameter and 1/2" thick, and would just fit into the bottom of a cut-off plastic milk jug 2" deep. This was placed into position under the wire electrode, and the mill table adjusted to center the electrode over the tap. Before the final adjustment, the work was attached to the wire marked "work" with a small alligator clip which also came from Radio Shack.

The milk jug was filled with kerosene to make a 1/2" deep cover over the top of the work. The electricity was turned on and the electrode slowly lowered toward the work. Before reaching the work, I noted that bubbles were forming only on the electrode, indicating the polarity to be correct.

When the electrode touched the work, both light bulbs came on full brightness, but dimmed to a low glow when it was backed away, and the crackling of the arc jumping the gap started. I found that by listening to the crackling sounds, which were quite sharp, and by watching the intensity of the glow from lamp "B," I could keep the feed rate at what seemed to be maximum cutting speed. As the worked progressed, a black soot began to form around the cut, which I suppose was composed of metal particles and other byproducts of the process, but which seemed to ionize the arc gap and increase the efficiency of the cut. I never had to retract the electrode and clean out the hole, but was satisfied to just keep going as long as things seemed to be working well.

When the fluid became opaque, I couldn't see anything; I was working by "feel" and by what I could hear and see in the intensity of the glow of the light bulb. I suspect the rate of discharge of the arc was somewhat slower than it might have been using different capacitors, but quite sharp when it did happen, and may have contributed to blowing out the hole and keeping the cut going. My mill has a very slow feed rate which allows quite precise control of the feed and arc length. I wish now that I had noted the original length of the copper wire electrode so I could know how much burn-off it had. Then, in future cuts, I would know when the electrode was completely through the work and could be stopped at the proper point.

I continued the cut until I felt the electrode was completely through the tap before I changed anything or retracted the electrode. The crackling of the arc never stopped, so it must have been going to the sides after the tip cleared the bottom of the hole. On retraction of the electrode, I found the tip to be pointed with noticeable pitting on the sides, even though it seemed to have lost very little diameter.

Upon removing the work from the inky black fluid, I found to my delight the neatest round hole completely through the 1/2" length of the tap and very little larger than the No. 12 wire electrode. An ice pick easily broke out the parts remaining, and are all shown in the photographs. Note on the longest piece the teeth on one side and a portion of the neat round hole on the back.

To those of you who do not possess a mill with a fine feed and wish to try this, it seems possible that a drill press in good shape with a strong return spring and a lightweight extension on the feed handle to increase the sensitivity of feed might work. Needless to say, stay away from the start switch unless you just love excitement!

I would encourage anyone with a broken off tap in a nice piece of work to try this – but be careful! I plan to continue experimenting with this process and will report any improvements which seem worthwhile.

A Simple Timer

by Bill Reichart

Photos by Author

Here is a simple project that all home shop machinists need. I've had this timer for ten or more years, but I never thought of sharing it with other craftsmen. Since building the "Panther Pup" engine, many men have come to my shop and before leaving they all say, "How do you keep track of the hours on such a long running project?"

Actually it's very simple. Just build an hour timer similar to the one I use. This starts with acquiring a 110v hour totalizer. The good news is that these are available from The Surplus Center, P.O. Box 82209, Lincoln, NE 68501-2209. (They advertise in *The Home Shop Machinist*.) They offer an Elapsed Time Meter, their No. 21-1081A, for $6.95. I just ordered one for myself so I can make another timer for my own use. This timer is shown in Photo 1. They also have a toll free number, (800) 228-3407.

1

The timer I built years ago is shown in Photo 2. The hour meter I got back then was rectangular, as you can see in the photo. I made a simple wooden box that is about a 5-1/2" cube. In this box I mounted the timer and a small socket for a 7-watt night light bulb. The bulb pokes through a hole in the front of my box. Next I mounted a push-push switch that controls both the hour meter and the bulb. Attach a cord and plug and you are in business. The bulb serves as a pilot light and a reminder that the meter is running. With this switch, it is so simple to push and the clock starts; push again and the clock stops.

In use, it is simply a matter of recording the numbers on the clock when you start a project. Push on and push off as you progress and weeks, months, or years later you can subtract the starting time from the finish time and you have an accurate record of time on the project.

This gadget is also handy if you do a job for pay and you want to know how much time you had in the total project. Even your wife will borrow it so she can keep a record of the time she has in a quilt or a needlepoint project. And this is why I bought another clock. Now I can time two long running projects at the same time.

The wiring is very simple, but I've included a sketch if it might be required by a non-electrician.

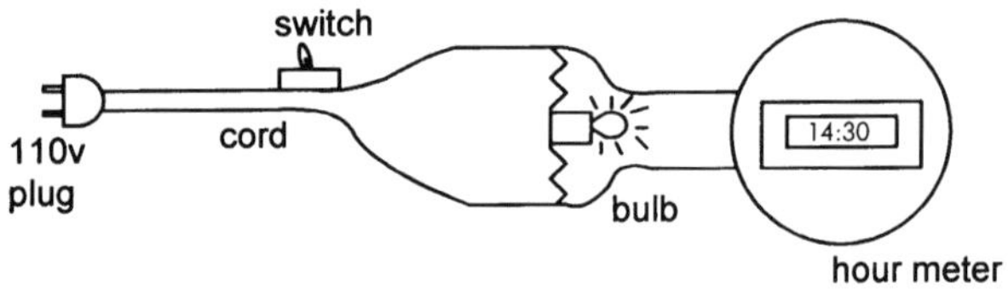

On my unit the front panel is attached with four screws. This makes it easy to open if maintenance should be required. My switch mounts in this front panel.

This hour meter will never replace a stop watch and likewise a stop watch will never replace this hour meter.

2

A Sliding Bevel Gage with Protractor

by Ted Wright

Photos by Author

Whatever you are making from pieces of metal, it will not be long before you will need to measure, mark out or shape parts with an angle on them. A bevel gage and protractor will enable you to carry out these operations, and it is worth remembering that the bevel gage can be set at 90°, thus doubling up as an engineer's square.

Photo 1 shows both a purchased protractor and bevel gage along with my home workshop produced design. The design offered may particularly appeal to those of our hobby with at present limited workshop equipment. Apart from a few simple turned parts, the remaining pieces can be produced with hand tools and are useful exercises in marking out, drilling, filing, producing slots and riveting.

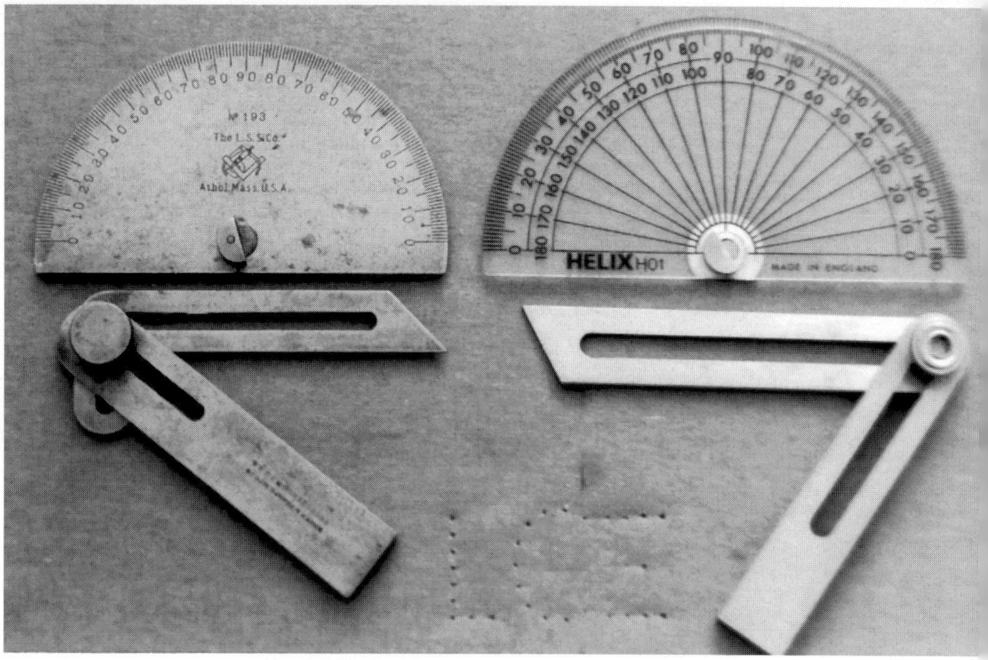

1 ***A commercial bevel gage and protractor together with my home workshop-made unit.***

PART 1

My design is simple in that the protractor is a clear plastic one used by students, draftsmen and designers. It can be purchased from most stationery and office equipment shops. The one I have used has a base length of 4" and all the angular lines and figures are of a fine quality. If treated with respect and stored in a little box, where the base reference cannot be damaged, it should give you years of service.

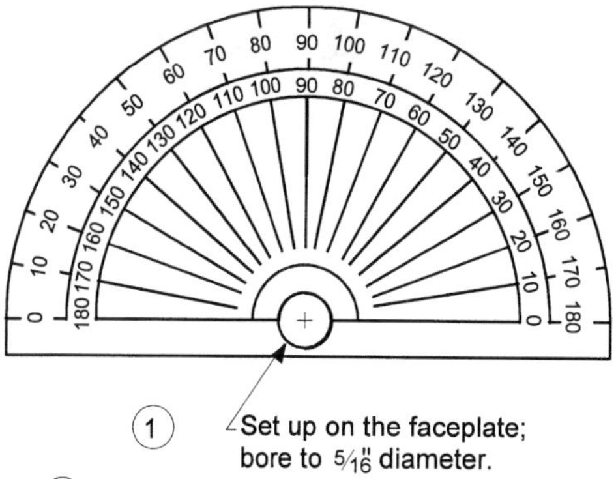

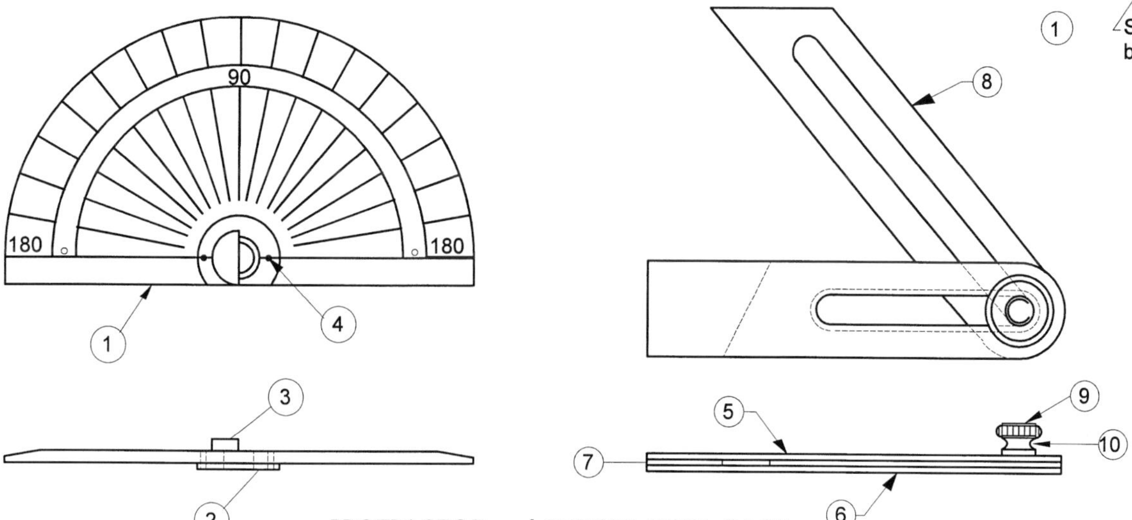

PROTRACTOR and SLIDING BEVEL GAGE

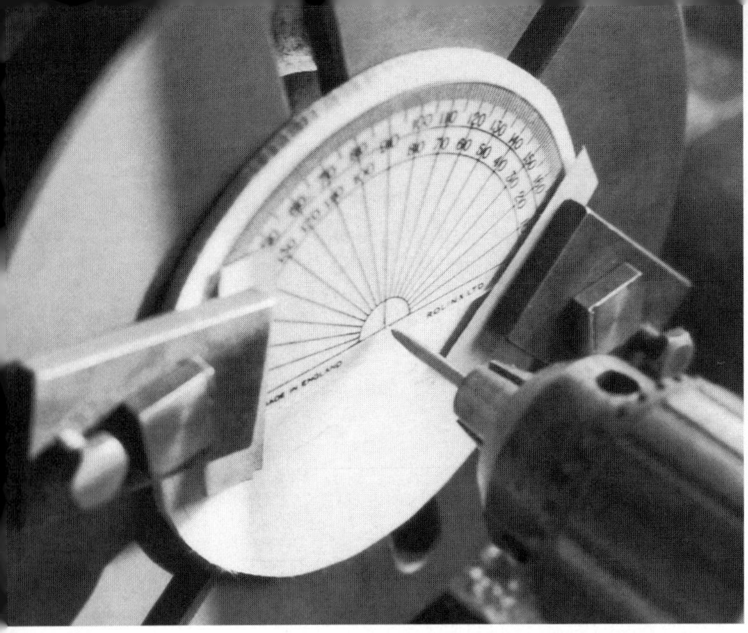

2 *The setup for machining the hole in the plastic protractor.*

3 *A simple filing button.*

A 5/16" diameter reamed or bored hole is required where the 0° – 90° – 180° lines cross on the base line. To my cost I found that the plastic does not take kindly to center punching prior to drilling, so I suggest the following procedure be used. Cut a piece of thin plywood as a backing for the protractor and set up on the lathe faceplate with the crossed reference line somewhere near true. Note the soft card packing also used under the clamps in Photo 2. Set up a sharp center or turned needle point in the tailstock and gently move the crossed lines until true to the point. Take your time and use a magnifying lens. Gently tighten the clamps and, using light pressure, center drill and drill out to letter size N. Finish the hole by either reaming or boring to 5/16".

Before making the turned parts for a protractor and gage, it is worthwhile to turn up a simple double-ended filing button (Photo 3). Chuck a short length of 5/8" diameter CRS, face the end, and turn down to 3/16" diameter for a length just under 1/16". Turn the next 3/8" length down to 9/16" and part off to an overall length of 3/8". Reverse in the chuck and turn down to 1/4" for a length just under 1/16". Remove all sharp edges.

PART 2

Chuck a short length of 5/8" brass with 3/8" protruding. Face the end and center drill, drill right through with a No. 14 drill and finish with a 3/16" reamer or D-bit. Turn down to 5/16", a good fit for the hole in the protractor, and for a length equal to the measured thickness of the protractor. Remove all sharp edges and part off to give a shoulder thickness of 1/32".

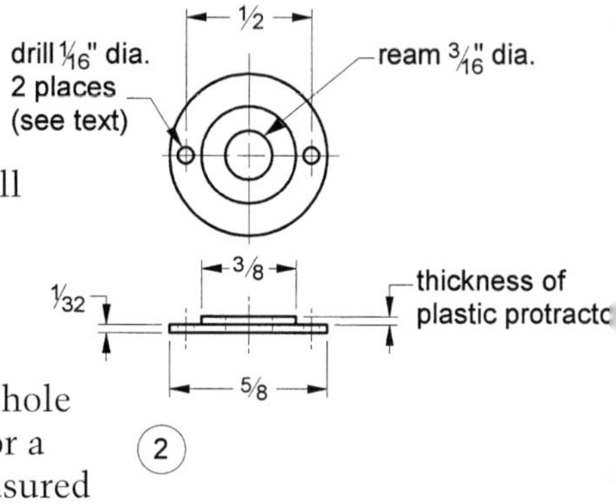

To fit the bush to the protractor, insert it from the back, and where the bush protrudes below the base, scribe a line. Use soft packing in your bench vise when holding this part and file down to just *below* the scribed line. Then drill out two 1/16" holes through both bush and protractor and push in two 1/16" brass pins. Refer to the general assembly drawing. Fix with the super glue if necessary.

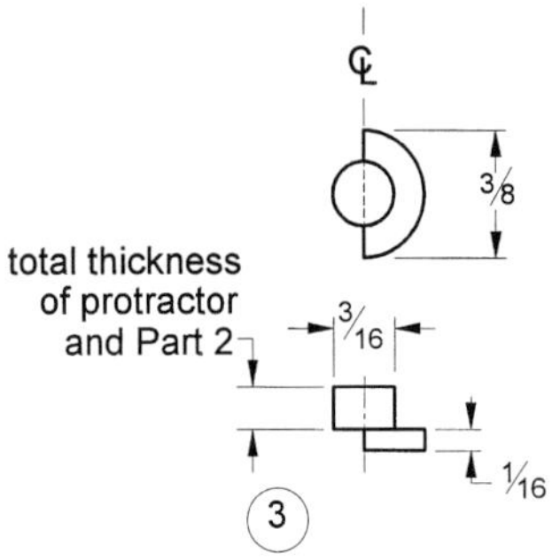

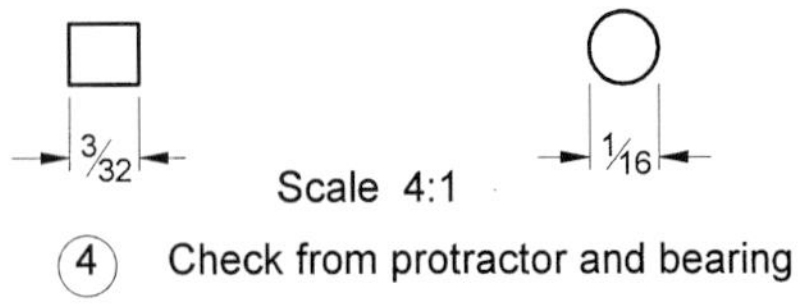

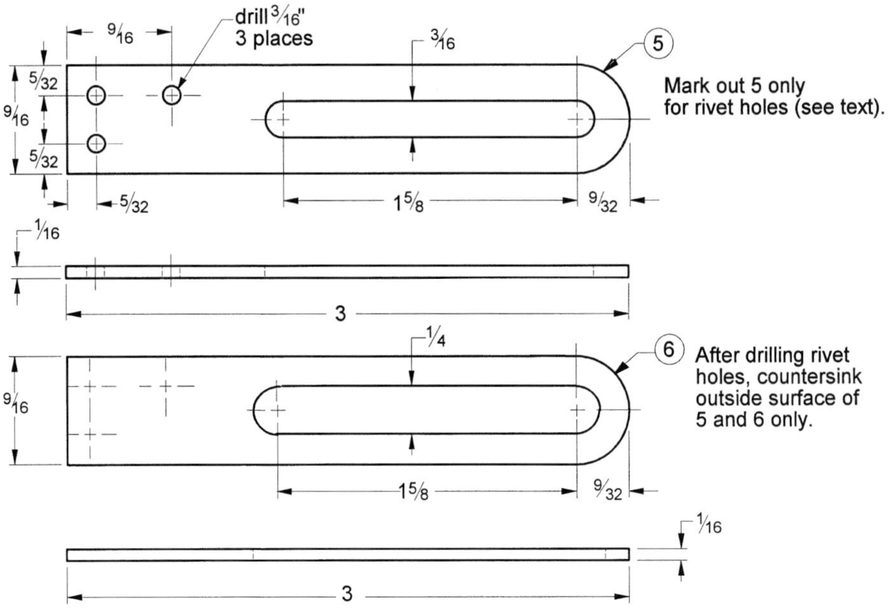

4 ***Roughing out down to a line by filing "vees" first.***

PART 3

Chuck a short length of 3/8" CRS. Face the end and turn down to 3/16" for a length equal to the overall length of Part 2. This should be a good rotational fit in Part 2. Remove all sharp edges, and part off 1/16" from the shoulder.

From the drawing, you will see that part of the 3/8" diameter needs to be cut away and filed down to the center line of the 3/16" diameter. Use a micrometer to check that you remove metal only down to the center line.

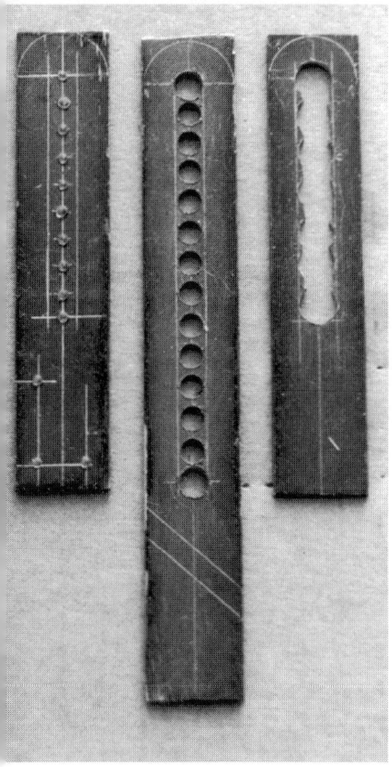

5 ***Marking out, drilling and filing parts of the stock.***

This is only a small part to hold and work on, and will certainly test your filing skills. Luckily, this part can be quickly remade should you by chance remove too much metal when filing. Remove all sharp edges.

PART 9

Face off another short length of 3/8" CRS, and turn down to 3/16" for a length of 7/16". Thread this diameter 2BA or 10-32. Part off 1/16" from the shoulder. This part also requires filing, but I suggest you wait to do this until the stock of the gage is completed.

6 ***Drilling rivet holes in the clamped-up stock.***

PARTS 10 AND 10A

Chuck a short length of 7/16" CRS with 1" extending out of the chuck. Face the end and turn down to 5/16" for 5/32". Knurl, using a straight knurl for the next 3/16". Note that a diamond knurl can be used to produce a plain knurled nut. Refer to the drawing of Parts 10 and 10A.

Center drill and drill No. 23 for 2BA or No. 21 for 10-32 at least 1" deep. Start your taper tap while the part is held in the lathe to ensure a square thread. You can finish tapping later with the part held in the bench vise. If you are making a plain knurled nut, copy Part 10A.

The other version will require a pair of simple curved form tools to complete. Remove sharp edges and part off to length. When complete, I reversed my nut in the chuck and recessed the top surface. I know this will not make it work any better, but it does look like the nuts produced by old-time instrument makers. Hold in the bench vise using soft jaws, and finish tapping the thread.

You will see that for the stock and blade I have specified 9/16 × 1/16" CRS. This is a non-preferred size and you will probably have to machine or file the 9/16" dimension from 3/4" stock. Never mind, it's all good exercise!

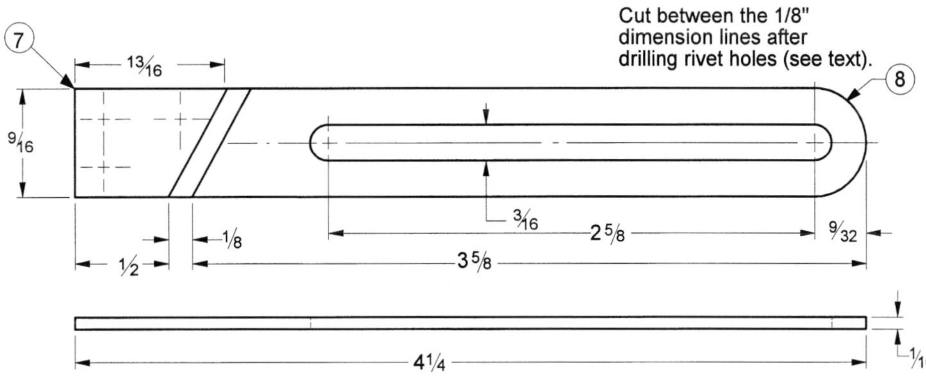

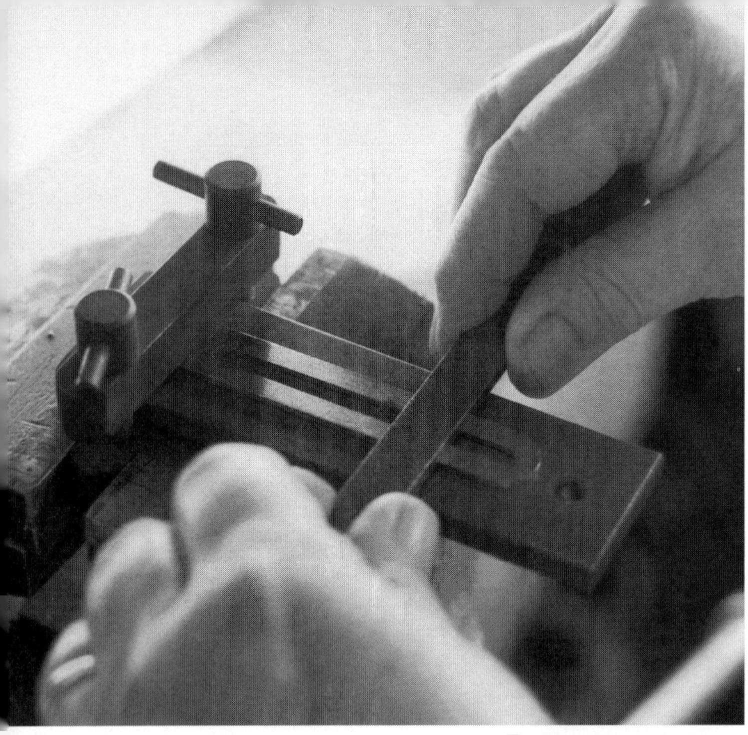

7 *Draw filing the blade.*

8 *If necessary, gently squeeze the stock along its edges to ensure they are parallel.*

Remember, if filing, use the corner of a coarse file to make lots of "vees" down to the line before filing flat (Photo 4).

PARTS 7 AND 8

Cover one side of the blade material with marking out blue (Photo 5) and mark out as the drawing. Note the stock center packing piece at one end with the two 45° lines 1/8" apart. Do not cut off at this stage. Center punch for all the slot holes. Use a 3/16" drill for the two end holes and 5/32" for the rest. Next, file the end of the packings piece square, and use your filing button to round off the end of the blade (Photo 6).

PARTS 5 AND 6

Blue and mark out two lengths of 9/16 × 1/16" CRS for the stock. File one end square on each part. Center punch all the slot hole positions and also the three rivet hole positions on one part only. Drill out the slot holes. Note that one will finish 1/4" wide, and the other 3/16". Finish off the rounded ends using the filing button as a guide. File out the two slots to your marked lines. I used a small square and flat file to produce the slots. Check that either a 1/4" diameter or 3/16" diameter drill will slide from one end to the other. Remove all sharp edges.

On a flat surface, clamp the blade (with the packing piece) and the two parts of the stock together using a toolmaker's clamp. Check that the square end of each part is level, and that the rivet hole center punch marks are showing on the top surface. Drill the rivet holes 3/32" (Photo 6).

Now take the blade and carefully saw down between the two lines that are 1/8" apart. Again, remove any sharp edges and also finish all the flat surfaces by draw filing with a dead smooth file (Photo 7).

Before riveting, cut out a piece of thick, good quality paper, the same size as the packing piece, and insert this also between the outside parts of the stock. You can now rivet the three parts of the stock together using 3/32 × 1/4" soft iron countersunk head rivets.

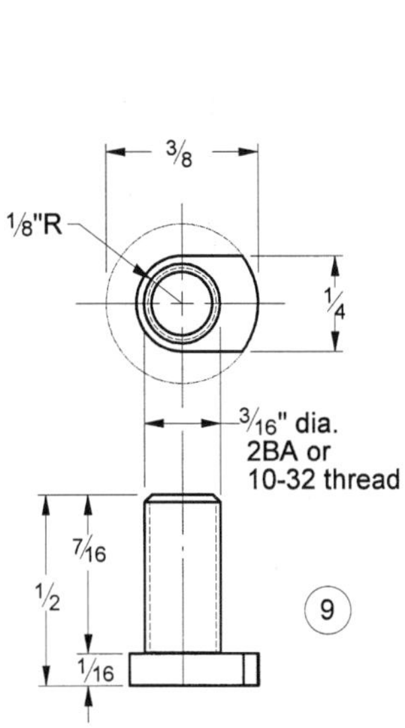

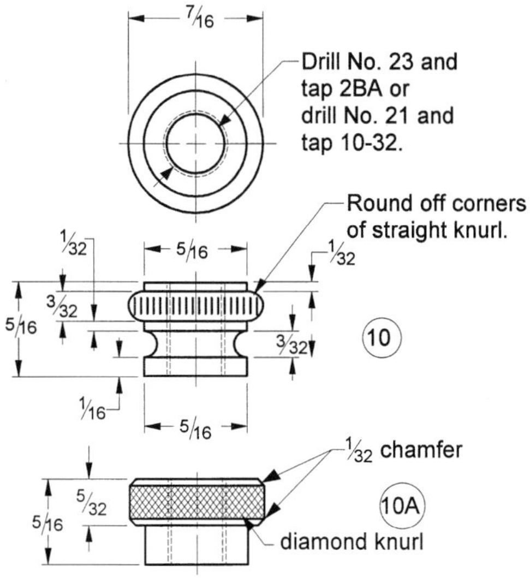

9 ***Hold the file down flat when filing the rivet heads.***

Do not hammer too hard when riveting; the rivets are very small and do not require much force to flatten them into the countersinks. If you find that this operation has moved the two outside pieces of the stock slightly out of parallel, very gently squeeze them back in the bench vise while holding them at an angle (Photo 8).

Be careful when you file away the rivet heads; they should not show when finished. To start, hold the file as shown in Photo 9, and keep the file down flat in the work. Finally, drawfile all the outside surfaces of the stock and again just remove any raised sharp edges.

Finish off Part 9 now. As you file away the head, check it in the stock until it will slide right to the end of the 3/16" wide slot.

Assemble the stock blade, lock screw and knurled nut. Check that the blade is free to move to any position along both slots. Ease any tight spots, if necessary.

Finally, make up a simple storage box to protect your well made bevel gage and protractor (Photo 10).

BILL OF MATERIALS

PART NO.	DESCRIPTION	SIZE	NO OFF	MATERIAL
1	Plastic protractor	4" base	1	
2	Bearing	5/8" dia.	1	brass
3	Locating spindle	3/8" dia.	1	CRS
4	Locking pins	1/16" dia.	2	brass
5	Stock side plate	9/16 × 1/16"	1	CRS
6	Stock side plate	9/16 × 1/16"	1	CRS
7	Stock packing piece	9/16 × 1/16"	1	CRS
8	Blade	9/16 × 1/16"	1	CRS
9	Locking screw	3/8" dia.	1	CRS
10	Knurled locknut	7/16" dia.	1	CRS

10 ***A simple storage unit to protect the bevel gage and protractor.***

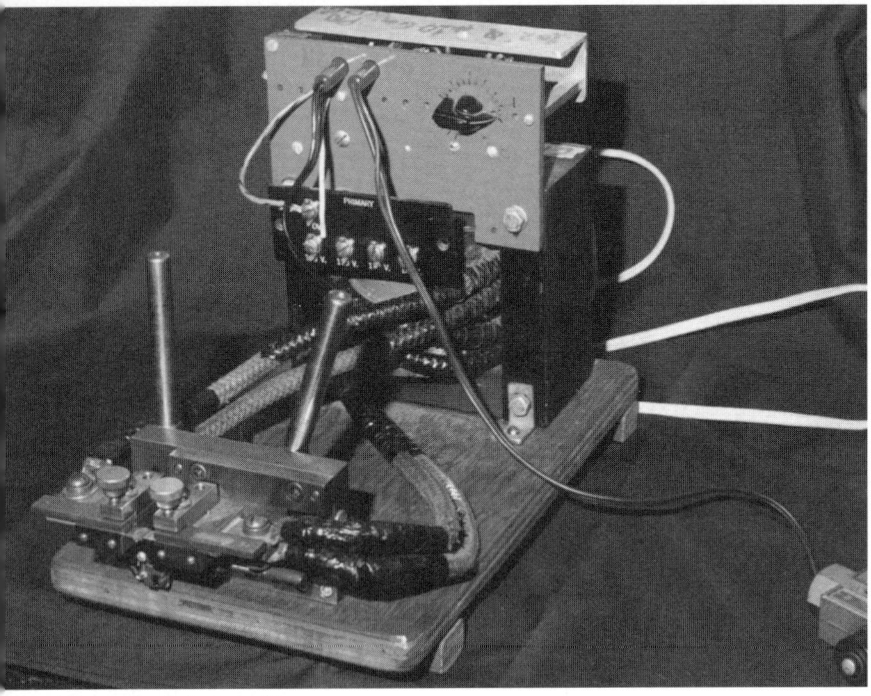

The author's butt welder.

A Small Shop Butt Welder

by Wolfgang F. Habicher

Photos by Author

I built my little shop butt welder some years ago. The original purpose for this welder was to butt weld stainless steel tubes to the hot cap of an experimental hot air engine. This it did very well. With a little modification, the welder subsequently proved suitable for welding of band saw blades, extensions to drills up to 3/8" diameter, and clevis ends to rods for various locomotive valve gear.

I've tested sample welds for strength by bending, and found the joints to be excellent. However, the welding of saw blades was tricky, and I found it necessary to make weld data sheets listing the weld parameters used for a particular blade roll, to enable me to recreate the successful setups.

The construction of the welder was quite straightforward: A 1 kilowatt transformer with its 115 VAC primary coil wound with No. 10 wire was modified by stripping off the secondary winding and replacing it with two turns of heavy welding cable with its rubber insulation stripped off. The core was large enough to accept two cables in parallel.

The heat control is a simple triac phase controller built to plans from a General Electric SCR Manual. It works well and has never given any trouble after I figured out the appropriate rating. Until that time, I burned out three triacs! They were a little expensive at that time.

Here is what went on: The inrush current at maximum setting turned out to be 40 amps at 115 volts! No wonder the 15 and 25 amp triacs I had used failed. I bought the heaviest triac available at that time, rated for 40 amps at 400 volts. This device has withstood the test of workshop use and time. After switch-on the current rapidly drops to about 30 amps, which a 20 amp circuit breaker can easily supply for the short period necessary to weld.

For best results, a direct line to the service panel wired with No. 12 – or better yet, No. 10 – cable would be the way to go. In the absence of this, a kitchen appliance outlet or the washing machine outlet would suffice. This should be fused with a 20 amp slow blow fuse or circuit breaker. (I've used both methods in different residences).

The heavy secondary winding supplies 2.75 volts AC at up to 1250 amps. With extended use the secondary cables and clamping jaws get rather hot. This tool is not suited for continuous use, but with, say, 15 seconds on and 2 minutes off, it will live quite happily without injuring itself or the neighborhood.

The photographs show the general arrangement and clamp assembly clearly. On the transformer is mounted the heat controller (knob on the right) and the plugs for

The clamp arrangement.

The clamp arrangement showing insulation of the left-hand side clamp.

the limit switch (this switches off the triac control voltage after the upset stroke) and the "weld start" button (heavy limit switch on the right). The aluminum angle toward the back is the heat sink for the insulated stud-mounted triac. The two rods sticking up from the clamp assembly are handles to squeeze the jaws together for upsetting the joint; this works well with heavier workpieces. For welding saw blades, an accurately controlled force is necessary and I use a spring-type fish scale for this purpose.

The transformer size is 5-1/4" high, 6-1/2" wide, and stacked 2-7/8" thick. The winding core size is 2" wide, 3-5/16" high, and 2-7/8" thick. The clamping jaws are brass with copper lugs for the cables. The right side jaw is on a small precision slide I salvaged from some redundant gages over 25 years ago! Note that the left side jaw is insulated from the backplate.

The clamping arrangement is a little crude, but it is versatile and different setups are quickly arranged. However, in the near future (after I've completed Robert Langlois's EDM machine), I shall fit an adjustable force spring arrangement to control the upsetting force; for small workpieces, this is necessary for consistent results.

This little device could also serve as a spot welder. The stainless steel ash pan for my 3/4" scale Canadian Pacific Hudson locomotive was spot-welded together with this machine. I've also welded connections onto heating elements where these had broken off. I'm sure other uses will be found in time. The welder was not difficult to construct and makes tasks possible that would be difficult, if not impossible, to do by any other means. It has become pretty well indispensable in my workshop.

A Homemade Die Filer

by Melvin L. Kalb
Photos by Author

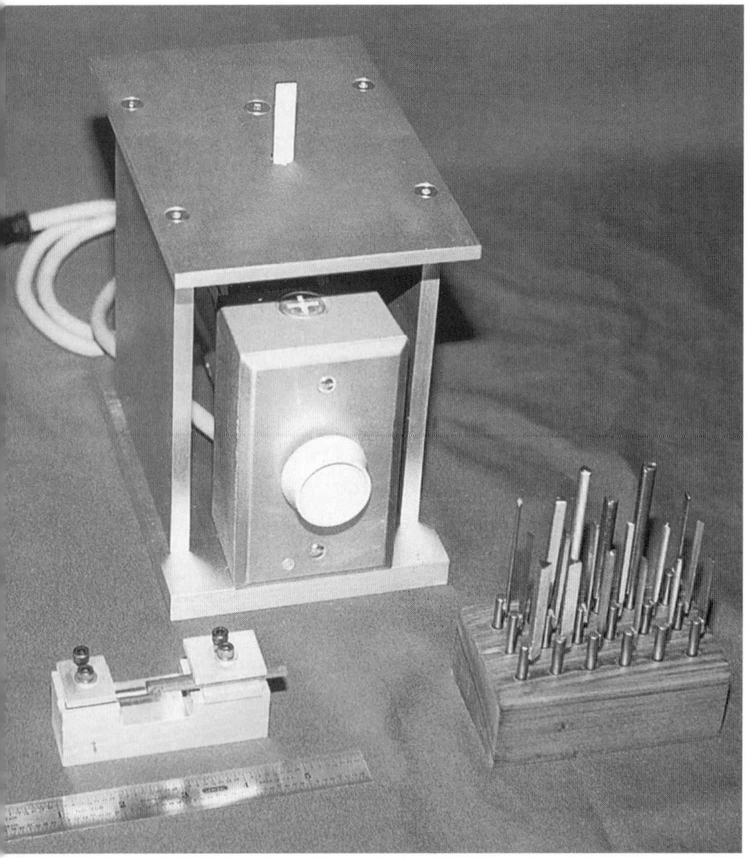

1

Some months ago I started a project that required a die filer, and because I didn't have one I decided to see what I could make. I reviewed a number of designs I had, but none seemed to click. I let the project slide.

Then I "found" a hand-held sabre saw I had stored on a back shelf. I thought it might have the right stuff. I made no changes to the saw except to remove the sole plate, which I replaced with one of aluminum. The saw was mounted upside down on this plate and became the top working surface (Photos 1, 2, and 3). Steel would be better for this piece because it wears better, and one of these days I will change it. For the frame, I used aluminum from my scrap box, which I bolted together with socket head bolts. The hole near the top of the side plate gives access to a setscrew, which secures the files to the sawshaft.

Commercial die files are hard to find, expensive, and not quite what I needed. Since I would be doing only light work I thought I would try to make my own. I had several sets of needle files and other small files in the shop, but they cut with a "push" and die files cut on the "pull." Then I thought, why not turn the files upside down and soft solder them onto drill rod?

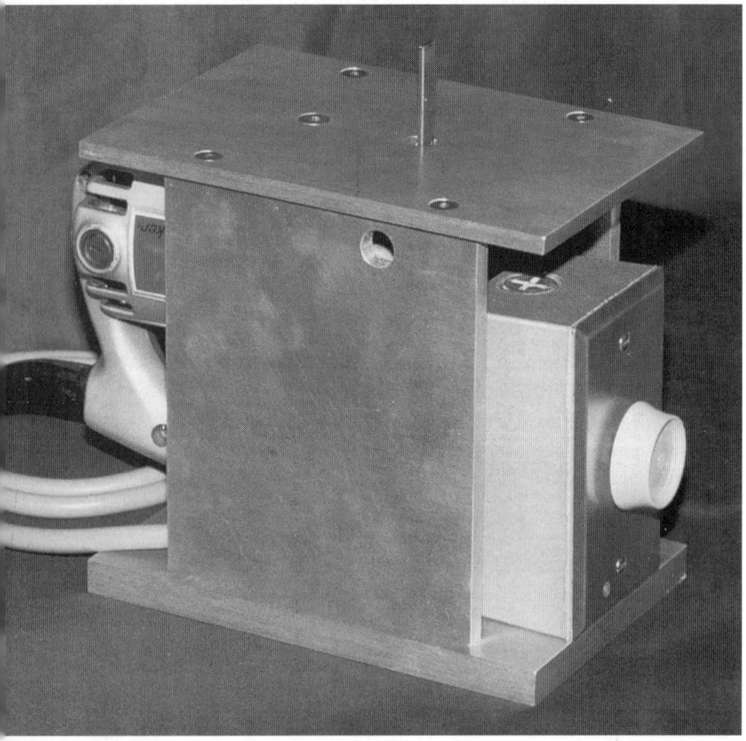

2

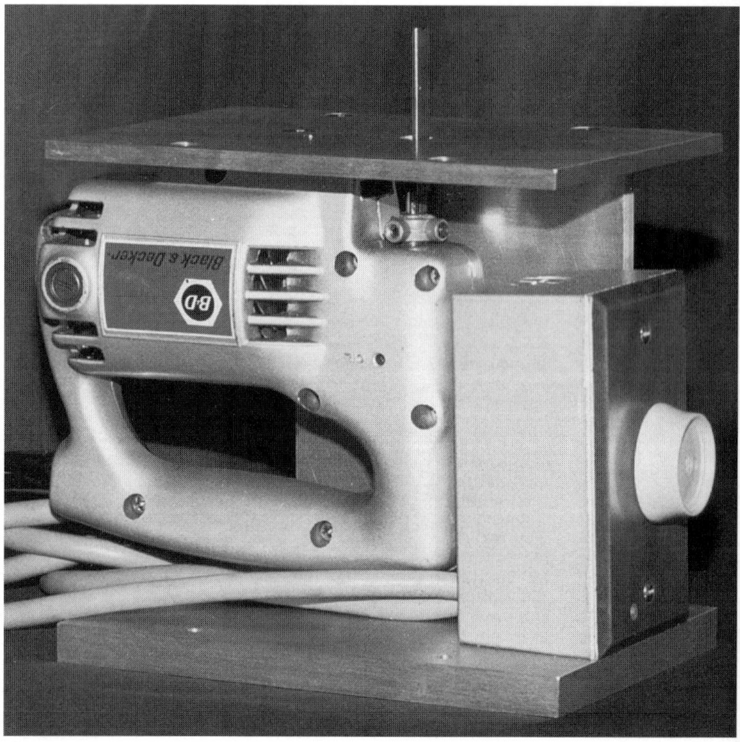

3

Silver solder would be better, but heat destroys the hardness of files. Soft solder is not considered to be a good structural material but in thin sections can be fairly strong. I thought I would give it a try.

My first attempt was a 1/4" wide flat needle file that I trimmed to 2-1/2". I milled a flat on a piece of 3/16" drill rod (the shaft on my saw is drilled .190") to fit the length of the file. I used lead free, silver bearing solder (.98 Sn/.02 Ag) which, when used with an acid flux, adheres readily to steel and has a bit better strength than lead/tin solder. I made a fixture to hold and align the rod and file while soldering. I was pleased with the results (Photo 4).

I continued to try ideas for mounting the files and realized that because the parts I would be filing would be less than 1/2" thick, I didn't need long files. The saw has a stroke of 3/4", so a file length of about 1-1/2" seemed right. With that in mind, I tried soldering only the bottom 1/4" of the file to the drill rod. This worked surprisingly well, especially for square, round, and triangular shapes which are thicker. Long, thin files still need to be supported their full length. My collection of files is shown in Photo 5.

When I first turned on the filer, the speed was much too fast. In order to slow it down, I put a solid state, incandescent lamp dimmer in series with the motor. The saw uses a brush-type motor so the speed is almost infinitely variable. Under load the filer slows down a bit, but for my purpose has worked fine. A receptacle box and cover plate enclose the dimmer.

I didn't think I would have that much use for a filer but now use it frequently. For the few hours it took me to put together, I have been well rewarded.

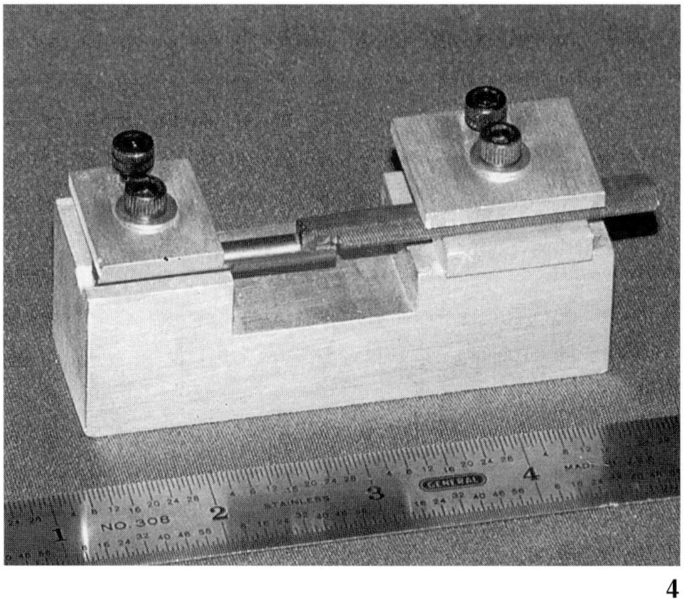

4

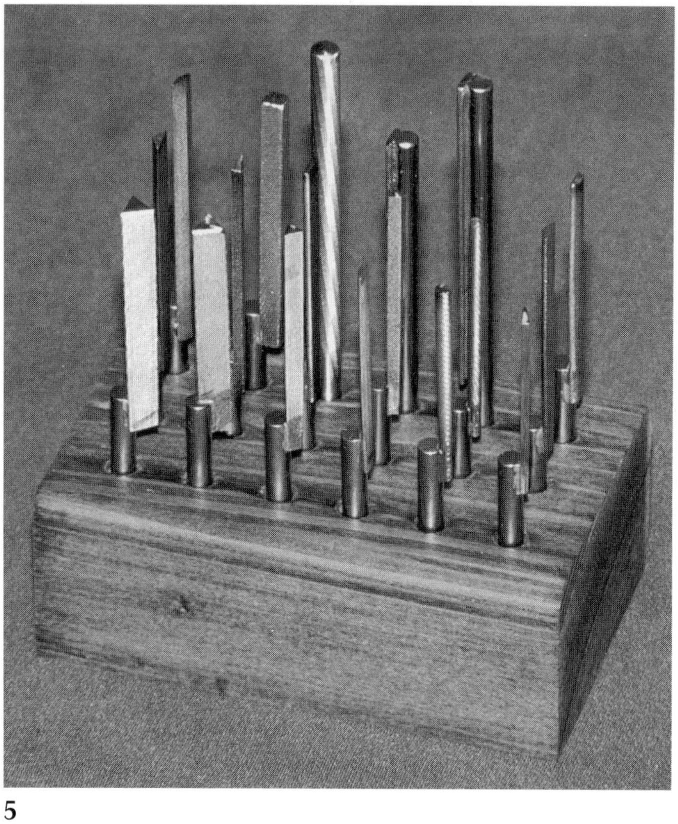

5

An Indicator Stand

by Earl L. Bower

Drawing by Author

If, like me, you have trouble reading dial calipers or depth mikes, you will like this easy project. It should take only one Saturday to complete and it will make it easier to read those depths and steps. All you need is a piece of flat 0-1 tool steel, a dial indicator, and several pieces of round 0-1 tool steel.

To start the project, measure the indicator stem and get the diameter and length. This is very important. You want a snug fit on the diameter and the height of the base must allow you to zero the indicator. Once you have these dimensions, you can start the base.

Start the base by laying out the profile. If you have a milling machine or a milling attachment for your lathe, you can machine the contour easily. Otherwise, you will have to drill and saw to the hole on both sides and then file the contour. The radius on the ends could be ground on a bench grinder.

The next step is the most important. You want the hole to be at exactly 90° to the bottom of the base. If your lathe has a small enough bore to allow the 3" dimension to span, then use the faceplate and two clamps to center the base. Once you have the base centered, drill and ream the bore for a snug fit on the indicator stem. Also machine the height of the base to allow approximately 1/32" clearance between the indicator dial face and the top of the base. After machining, lay out two holes 90° apart on the base top, as shown, for 6-32 setscrews. This will allow you to lock the indicator to the base, and should not permit your setting to change.

While you are making chips, this would be a good time to machine the three extensions you will need. These extensions will allow you to measure in 1" increments to a total depth of 4" deep. This should be more than enough depth for the hobbyist.

Start by turning the end to .112 × 3/16" long and then using a threading die for 4-48 thread the 3/16" length. Using a file, make an undercut at the shoulder to just below the thread depth. This must be done or the extension will not screw in completely and will give you an improper reading.

Once you have the thread cut, reverse the piece in the chuck and thread the opposite end the same way. Now take the piece out of the chuck and cut one end 1" +.015" and cut the other end 2" +.015". Return the piece to the chuck and thread one end as above. Then cut this piece to 3" +.015". Now using a collet, if available, cut each respective piece to exactly 1", 2", and 3".

To use the base, screw the tip of your choice in the indicator, and insert the indicator into the base. Then set the base on a flat and parallel surface. Adjust the indicator up or down to zero it, and snug up the setscrew. Do not overtighten the setscrews. This will only cause the indicator shaft to drag and hang up. You need to tighten them only enough to hold the indicator in place.

Now the indicator is ready to use. Just set the base on the surface you want to measure from and read the dial.

Depending on your type of indicator, you may have to re-mark the dial to read direct. This is easily done by using a typing eraser and either a fine-tip marker or rub-on transfer numbers.

To measure to the bottom of a hole deeper than 1", use the extensions between the indicator shaft and the tip you use. If you do not want to make the extensions, several tool companies sell them. You will have to use one, two, or three 1" extensions for depths. They do not supply the extensions in any sizes except 1" long.

The tips can be purchased, as well. If you change tips after the initial setup, you will have to re-zero the indicator.

Have fun making this base. I know you will find pleasure in using this tool in the shop. You may also want to make a nice wooden case for keeping it in when not in use.

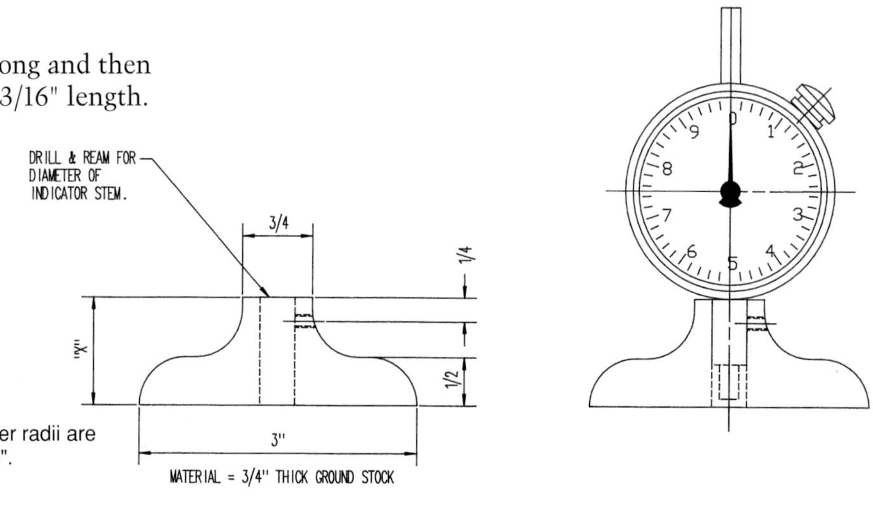

All corner radii are to be ½".

More on Three-phase Converters

by Robert J. Matthys

Steve Acker's article on three-phase converters ("A Simple Phase Converter," *HSM,* January/February 1991) contained much useful information, but there is more to the three-phase converter story. First, you do not have to use a separate three-phase idler motor to run a three-phase motor from a single-phase power line. You can use the three-phase motor on your mill or lathe (or whatever) as the idler motor. This reduces the converter cost considerably, but at the same time restricts the converter to driving just one motor load. If you have two or more three-phase motors to drive (two in my case, a mill and a lathe), then a separate idler motor for the three-phase converter makes sense.

Second, adding two run capacitors to the converter will increase its horsepower, and help balance and equalize the voltage and phasing of the individual phases of the three-phase converter. The converter will work without the run capacitors, but two of the individual phase voltages will be down 15% to 20% without them, and the power capability of the converter is reduced 25%. The run capacitors are separate from the start capacitor, and have about one-seventh the capacitance of the start capacitor. The run capacitors are permanently connected across the two idler motor phases that are not excited by the incoming single-phase power line.

CAPACITORS

The size of the start capacitor varies with the motor size in horsepower. The start capacitor should be about 70 microfarad/HP when using the oil-filled paper-type capacitor. It apparently takes more capacitance with the electrolytic type, as Acker found the start capacitor should be 100 microfarad/HP when using the electrolytic type of start capacitor. In practice, the start capacitor capacitance is not critical. If the start capacitance is half of the above values, the motor just takes longer to come up to speed. If twice the above values, the motor comes up to speed very quickly but with a heavier starting current that may dim your workshop lights for an instant.

What does the start capacitor really do? Well, one-phase and three-phase electric motors will both run on single-phase power, but they won't start on single-phase. They both need at least two-phase excitation to start rotation. A capacitor will shift the phase, and when tied to a second motor winding will provide the two-phase excitation needed for starting the motor. After the motor gets up to speed, the start capacitor can be removed from the circuit, and the one-phase or three-phase motor will continue to run on single-phase power.

The size of the run capacitors also varies with the motor size in horsepower. The two run capacitors should each be about 12 microfarad/HP. The capacitance values may be different for the two phases. The capacitance values are not critical and are determined by trial, adding or subtracting capacitance until the voltage across the two motor phases is equal to the line voltage across the third motor phase.

During the startup of a 220-volt converter, the line-to-line voltage on two of the three phases of the idler motor will rise to 330-370v AC. This means that both the start and run capacitors must be rated at 330-370v AC (or greater) for a 220v converter.

The oil-filled paper-type makes the best start capacitor. It has an essentially unlimited life, and comes in an oval-shaped metal can. However, it takes a bank of three to nine of them (depending on their capacitance) tied in parallel to get enough capacitance for motor starting purposes. This is due to the limited capacitance available in a single capacitor. Used oil-filled paper capacitors are available at electronic surplus stores for $2 to $10 each, depending on size.

Because the start capacitor is used only during the motor startup, the cheaper (and lower quality) non-polarized aluminum electrolytic type can also be used as a start capacitor.

Unfortunately, electrolytic capacitors have a limited life, because their internal chemicals evaporate over time. They must be periodically replaced (every four to ten years or so). They also have high internal resistance losses, which cause high internal heating in AC voltage applications, and which pretty much limits their AC use to intermittent duty things such as motor starting.

The run capacitors MUST be the oil-filled paper type, as the electrolytic type will not stand up to continuous AC duty.

Do not buy old or surplus electrolytic starting capacitors, as there may be little capacitance left in them due to chemical evaporation over time (old oil-filled paper capacitors are okay). Electrolytic capacitors can be identified by their round, black plastic cases. New electrolytic capacitors cost \$10 to \$15 at motor supply stores.

RELAYS

It takes two relays to operate the converter. One relay connects the start capacitor to the idler motor during startup, and is operated by pushing the momentary "START" button. The second relay applies the incoming 220v single-phase power to the idler motor. This relay has an extra lock-up contact to keep the relay closed (and keep the 220v coming to the motor) after the momentary START button is released.

Relays are rated in terms of the maximum contact current and motor power they can safely handle. The 25 amp (1-3 hp) and 30 amp (1-4 hp) sizes are the ones to use. Unused and slightly used relays are available at \$5 to \$10 each from electronic surplus stores and used motor outlets.

MOTORS

If you use your three-phase load motor as the idler motor, then its characteristics are already fixed. If a separate idler motor is used, it should be somewhat bigger than the load motor, as the idler motor does not operate efficiently and it must also carry the electrical losses in the load motor, which are not included in its horsepower rating. Steve Acker uses a two-times bigger ratio for idler motor horsepower to load motor horsepower, which is fine, but a one-and-a-half times ratio will also work. As usual, nothing too critical.

A new 1-3 hp motor will cost \$100 to \$300. Three-phase motors are the simplest and most reliable of any motor type, and they last a very long time. Consequently, there are lots of used three-phase motors available at \$25 to \$50. If the motor you buy has been used a lot, its shaft bearings may have gotten noisy, and you may want to replace them to reduce the noise level if it's to be located inside your shop.

CONVERTER CIRCUIT

Refer to the schematic of a typical converter circuit, showing the two relays, the idler motor, and the start and run capacitors. The idler motor shown can be either a separate three-phase motor or the actual three-phase motor you want to drive. The schematic also shows a momentary "STOP" switch, that opens the lock-up contact on the second relay and makes the relay drop out, disconnecting the 220v single-phase power into the idler motor. The relay coil connections are for 220v AC coils. 110v AC coils can be used instead by reconnecting the relay coils across the 110v line.

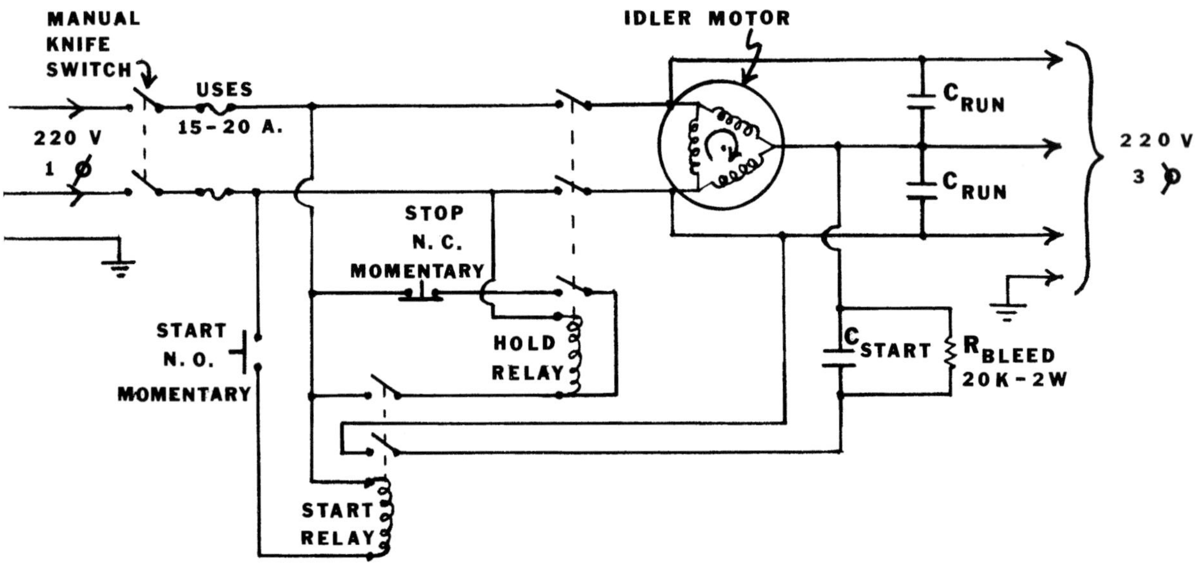

The momentary START and STOP switches are commonly sold as a pair in a small metal box ($5 used, $15 new), so they can be put in a convenient location for the operator. They usually have one normally open contact on the START button, and one normally closed contact on the STOP button, just as is shown in the schematic. Other contact arrangements can be had, but are not as widely available.

SAFETY

After starting the idler motor, the start capacitor retains an electric charge of up to 500v DC. If the start capacitor is the electrolytic type, the charge will bleed down to zero volts in a short while through internal leakage. If the start capacitor is the oil-filled paper type, the 500v DC charge will take a very long time to bleed off because the internal leakage of the capacitor is extremely low. So, for safety reasons, an external bleed resistor is placed across the oil-filled paper-type start capacitor. A 20,000-to-100,000 ohm two-watt resistor will do the job nicely. The run capacitors do not need bleed resistors as they are shunted by the idler motor windings which will bleed the run capacitors down to zero volts.

Also for safety reasons, all of the electrical components should be enclosed in a *grounded* metal box. Ground the motor frames, too. Use heavy connecting wires (No. 12), and fuse both sides of the 220v line. Safety first. Remember that 220v is more than enough to kill you if you make a serious mistake.

ELECTRICAL SCHEMATIC OF ONE-PHASE TO THREE-PHASE CONVERTER

R_{BLEED} is Not Needed if C_{START} is electrolytic type,

C_{START}= 70-100 Mfd/HP, C_{RUN}= 12 Mfd/HP

HOBBY PROJECTS

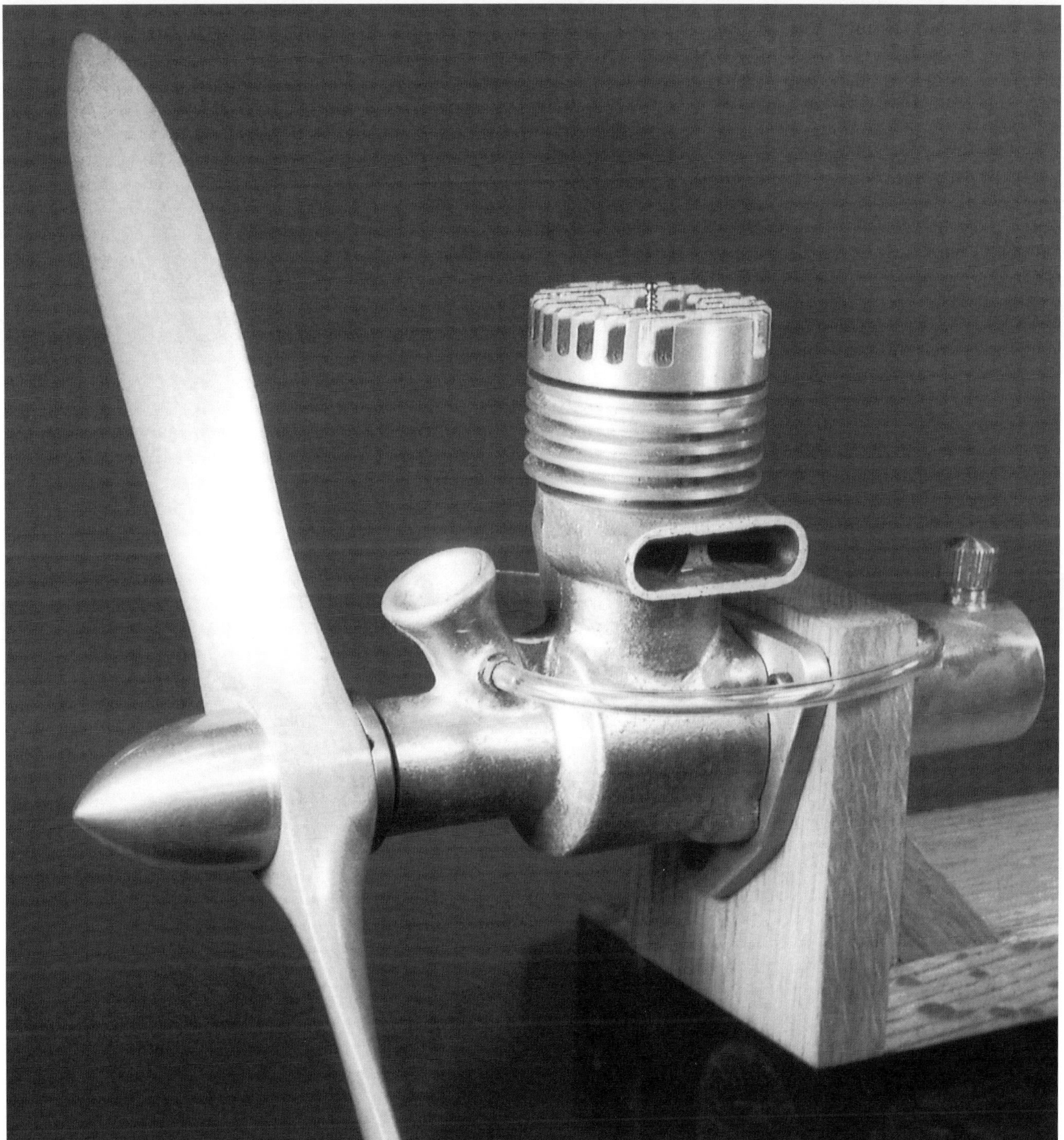

The Titan .60 engine is shown. Note the ample exhaust port and cooling fins.

Building the Titan .60

by George Genevro

Photos by Author

One of the more challenging experiences for a craftsman is machining, fitting, assembling, and running a model engine. What had been an inert conglomeration of bar stock, castings, and fasteners has now been transformed into an active, operating mechanism. Most important, after the engine is completed, is the realization that you are the one who performed this complex and interesting series of tasks.

Building any model engine, regardless of whether it is a simple single-cylinder, two-stroke cycle type or a complex multi-cylinder project, requires determination, patience, a desire to expand one's skills and learn new ones, and the ability to attend to the details that insure success. Taken singly, the machining and fitting operations are not difficult. If each part is well made and the relationships between the operating elements of the engine are understood, the engine will work well and be a source of pride and deep satisfaction.

Typically, the successful completion of a builder's first engine is a major stepping stone to more complex projects and the development of a greater range of machining skills. Of probably equal importance is the expansion on one's technical skills and knowledge in such areas as basic metallurgy, heat treatment, fits and tolerances, applied mathematics and other related topics. One also learns that parts that are "good enough" or "nearly okay" will never be really satisfactory. The self discipline learned in remaking parts that are not really right is valuable and is the hallmark of the good craftsman.

None of this can happen, however, until the first machining setup is made, the first chips are produced and the satisfaction of making a good – really good – part is experienced. Now is the time! Sharpen the tool bits and drills, lay out the measuring tools and let's go to work.

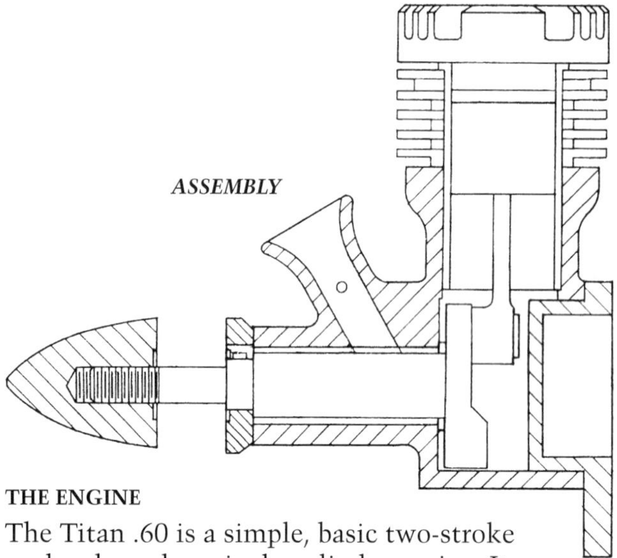

ASSEMBLY

THE ENGINE

The Titan .60 is a simple, basic two-stroke cycle, glow plug, single-cylinder engine. It was designed to be built using the machine tools usually found in home workshops, or in industrial arts or vocational education shops in schools. The number of machining operations has been held to a minimum commensurate with the production of a reliable, durable

The right side view of the completed engine showing the flared intake and needle valve.

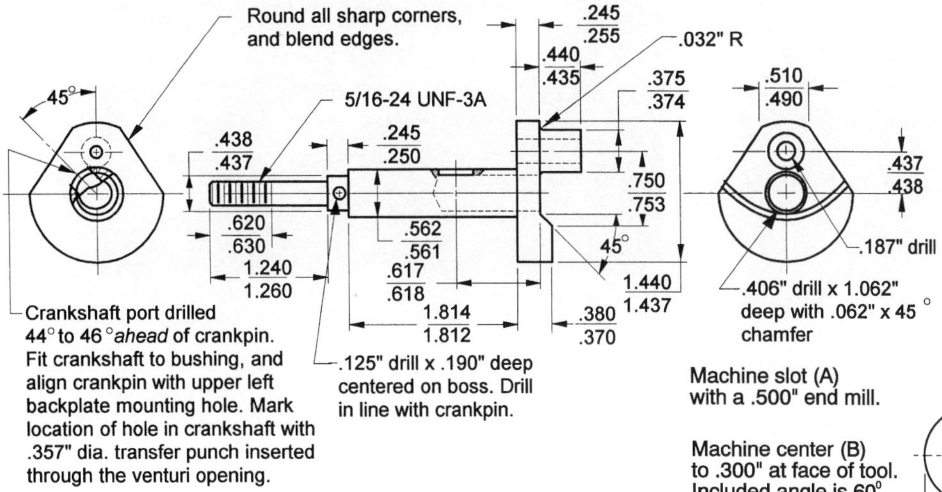

(1) CRANKSHAFT
AISI - C - 1018
Alternate material:
low alloy, low carbon steels

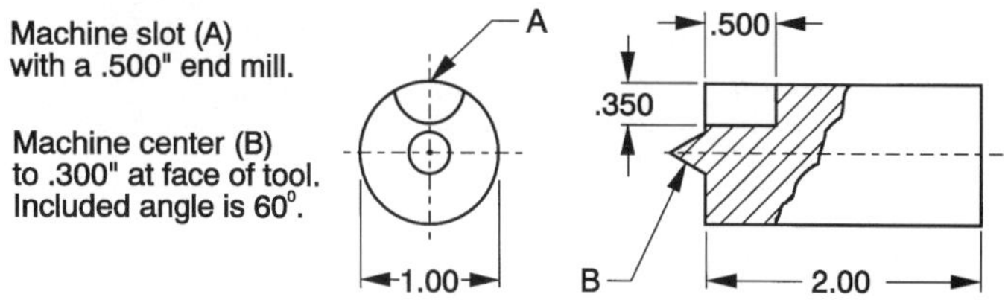

TOOL A - DRIVE CENTER
low carbon steel

engine. In the design process, attention was paid to the geometric relationship of the various parts to minimize the possibility of errors resulting from complex setups. Another major consideration was the choice of metals used and the metallurgical relationship between the major working parts.

While the number of special fixtures and jigs recommended may seem to overshadow the construction on the engine itself, in the long run time will be saved, fewer errors will be made, and the tools will be available to fellow engine builders. Organization of a joint effort involving two or more builders is an excellent way of reducing the time each person spends building tooling.

CRANKSHAFT (Part 1)

The crankshaft is machined from a steel billet and incorporates the rotary valve. If the builder has heat-treating and grinding facilities available, a medium carbon steel such as AISI-C-1050 may be used instead of the AISI-C-1018 specified. Other alloy steels such as AISI-E-4130 may be used if desired.

PROCEDURE

1) Face the ends of the billet to length and center drill each end. On the crankpin end, deepen the .125" diameter center drill hole to a depth of .750". This will be the reference point required for drilling the center hole needed when the main bearing diameter and threads are finished.

2) Accurately mark the center of the crankpin location using a vertical mill or other available means and center drill it.

3) Offset the billet in a four-jaw chuck, align the crankpin center hole with a wiggler, and machine the crankpin and counterweight area. Drill the crankpin and polish it.

4) Chuck the billet in an accurate three-jaw chuck or in a four-jaw chuck, align the previously deepened .125" hole, and center drill it.

5) Mount the crankpin end of the billet in a three- or four-jaw chuck and machine the main bearing and prop shaft areas about .100" oversize.

6) Machine the special drive center noted as Tool A. After chucking it in the three-jaw, take a light clean-up cut on the 60° point to assure accuracy.

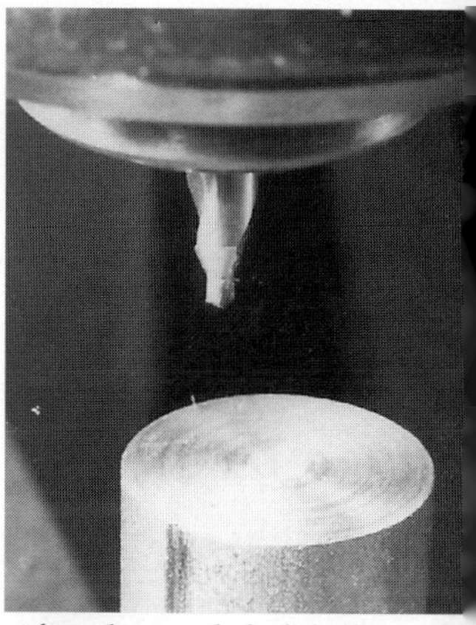

After the crankshaft billet has been machined to length and each end is center drilled, the crankpin location is established and center drilled.

The crankshaft billet is offset in the four-jaw chuck, and the center of the crankpin is located using a dial indicator and a tool with one end radiused about .060" and the other end center drilled.

7) Set up the crankshaft between centers and place soft aluminum tubing over the crankpin to protect it as it is driven by Tool A. Machine the main bearing, prop washer shoulder, and cut the 5/16-24 UNF 2A thread for the spinner. Polish the main bearing journal and the thrust face.

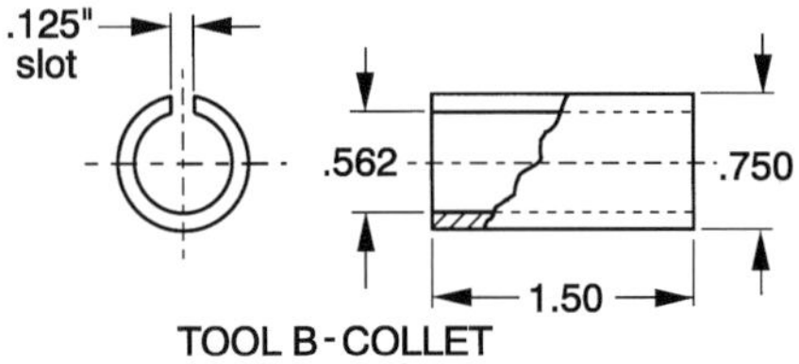

TOOL B-COLLET
aluminum or brass

8) Mount the crankshaft in a three-jaw chuck, protecting the journal with Tool B or by some other means. Drill and ream the gas passage and chamber or radius the outlet.

The finished crankpin is polished and drilled, leaving a small fillet at the juncture of the crankpin and the crank cheek.

9) Drill the .125" diameter hole for the prop washer drive pin.

10) Machine the cutaways for balancing either by gripping the crank cheek in the milling vise, or by using Tool B to protect the main bearing journal while the crankshaft is held in the vise.

Except for the rotary valve port in the main bearing journal, the crankshaft is now complete. The rotary valve port will be marked on assembly and

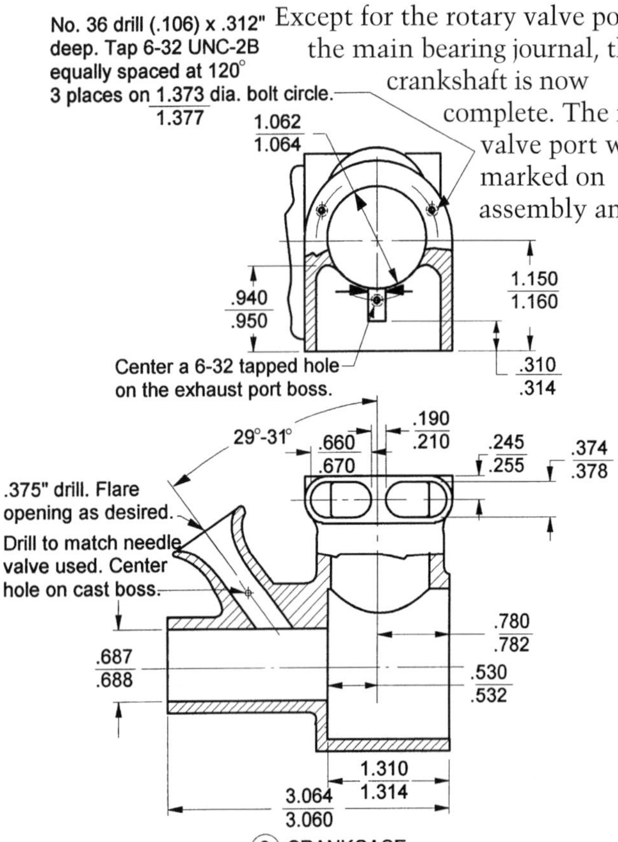

(2) CRANKCASE
aluminum alloy casting

The main bearing area is rough-turned while the end of the crankshaft billet is held in a three-jaw chuck.

drilled at that time. Deburr all sharp edges with a fine file and abrasive cloth. Protect the crankshaft from corrosion.

CRANKCASE (Part 2)

The crankcase is cast of No. 713 aluminum. This is an aluminum-zinc alloy sometimes

The balancing cutaways on the crank cheek are machined on the milling machine and all sharp edges are rounded.

Using Tool B to protect the finished main bearing journal, hold the crankshaft in a three-jaw chuck to drill and ream the gas passage.

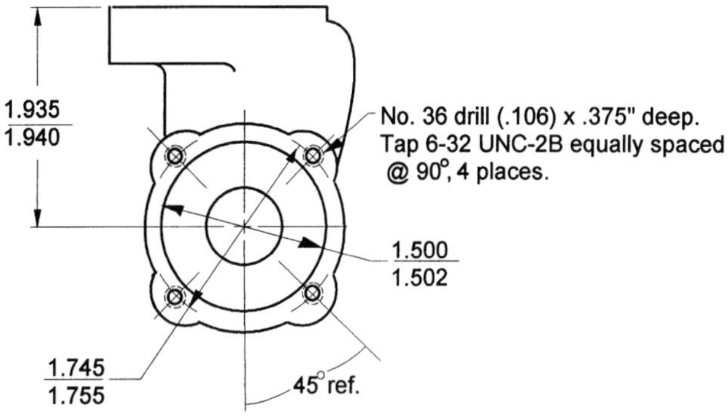

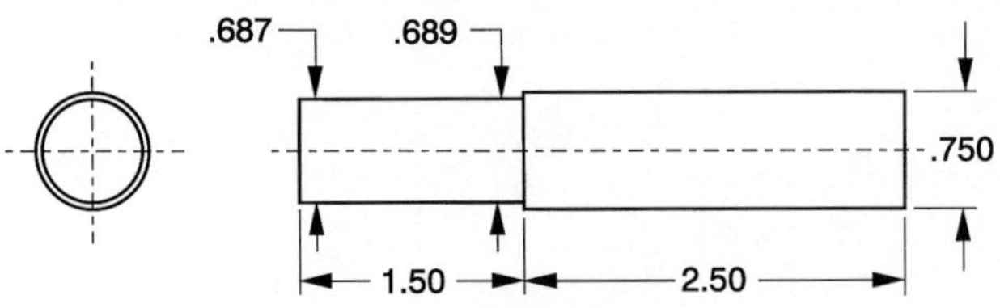

TOOL C-STUB ARBOR
low carbon steel

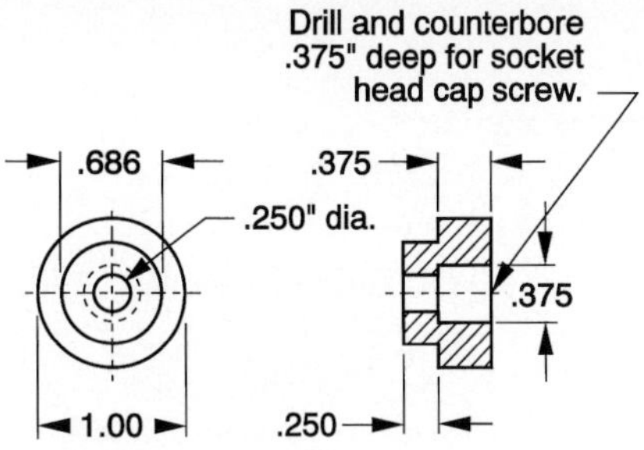

TOOL D2-CRANKCASE HOLDING TOOL
aluminum

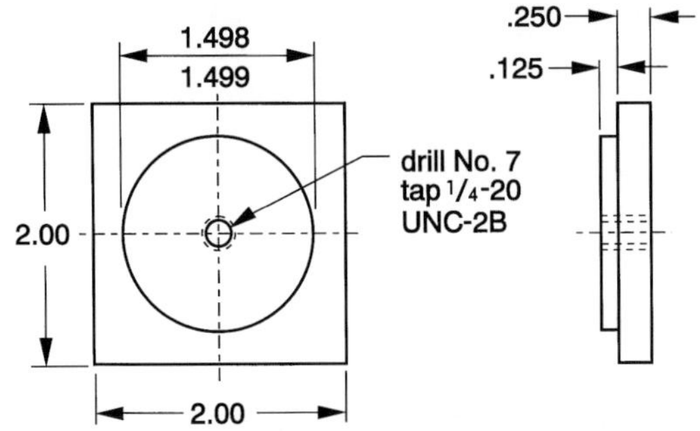

TOOL D1-CRANKCASE HOLDING TOOL
aluminum

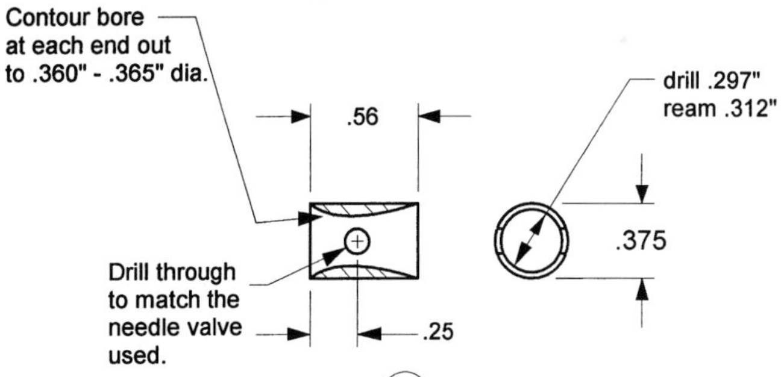

(2-1) VENTURI
yellow brass

known as *Tenzalloy.* It is naturally aged for maximum strength and machines cleanly and easily. For best results, use a cutting fluid specially compounded for aluminum, particularly for drilling, reaming, and tapping operations.

PROCEDURE

1) File away any burrs on the casting, paying particular attention to the parting line area opposite the intake venturi.

2) Chuck the casting by the round main bearing housing in a three-jaw chuck and align the rear of the crankcase as evenly as possible.

3) Face the back of the case just enough to clean it up and recheck the alignment.

4) Bore the case to within .100" of the finished diameter and face the inside and back to within .050" of the finished size.

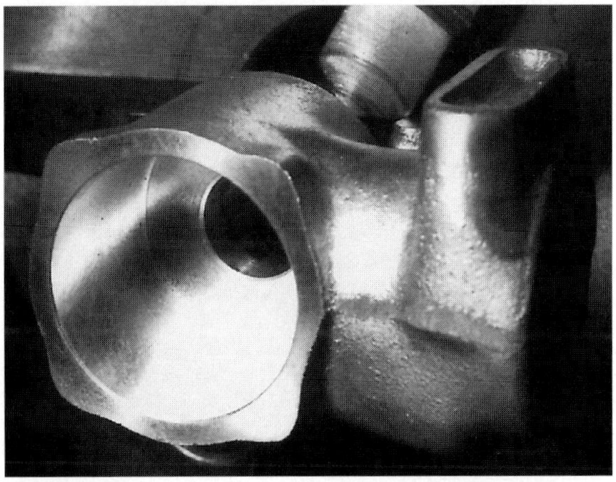

The crankcase casting is deburred and held by the straight main bearing portion. All boring, facing, and reaming operations are done at this time to insure concentricity.

5) Using the largest boring bar that will fit in the cored hole for the main bearing housing, bore it to within .015" of the finished size. At this point, the hole should run true, and therefore, be concentric with the crankcase bore.

6) If the proper reamer is available, ream the main bearing housing at slow speed using cutting fluid. Carefully oilstone the entering edges of the reamer to assure a smooth hole.

7) Bore the crankcase to size, do the final internal and external facing operations, and chamfer the edges of the holes.

NOTE – The essential crankcase and main bearing housing lathe operations are now done. If a powered milling attachment is available, the top of the casting may be faced and bored before it is removed from the chuck.

8) Machine a stub arbor (Tool C) to hold the casting while the front end is faced to length.

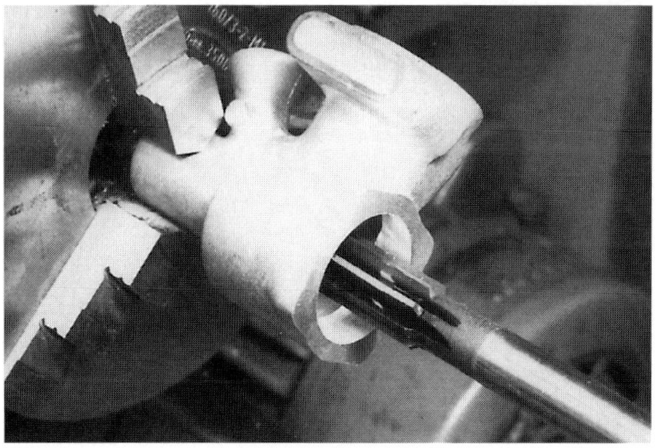

The crankcase main bearing housing is reamed to size to accept the pressed-in main bearing sleeve.

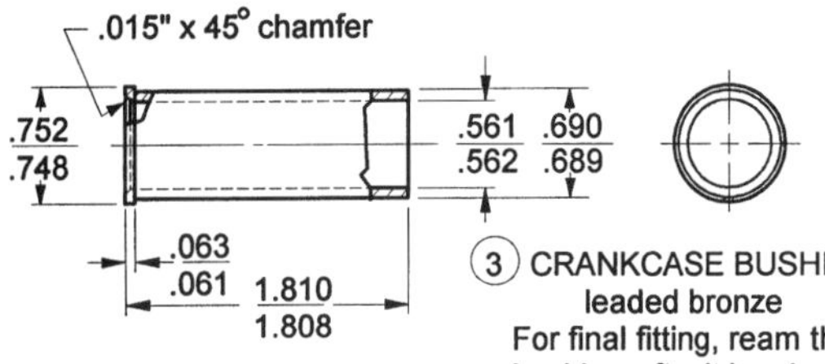

(3) CRANKCASE BUSHING
leaded bronze
For final fitting, ream the bushing *after* it has been installed in the crankcase casting.

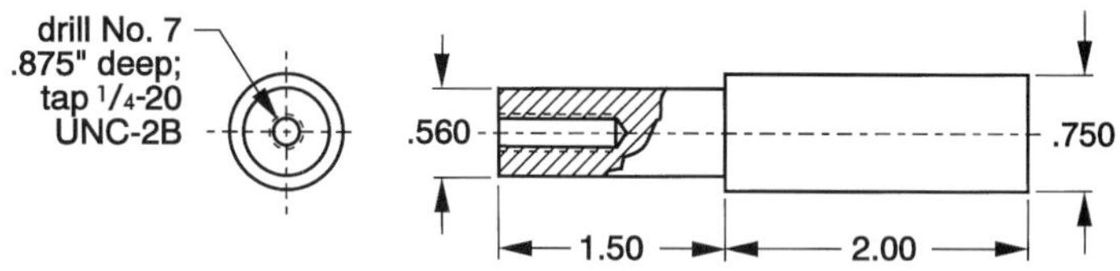

TOOL E - BUSHING ARBOR AND INSTALLATION TOOL
low carbon steel

9) Make Tools D-1 and D-2 as shown in the Special Tools Addendum. These are used to protect and help align the partly machined crankcase casting in the four-jaw chuck, if a lathe is used to face and bore the upper part of the crankcase. The tools are installed on the crankcase with a socket head cap screw of the proper length.

NOTE – If a vertical milling machine is used to face and bore the upper part of the crankcase, it will not be necessary to make Tools D-1 and D-2. Regardless of which method is used, make sure the bore is at 90° to the crankshaft center line. Any irregularity at this point will result in undue wear and friction.

10) Face and bore the upper part of the case and chamfer the edge of the hole.

11) Machine the exhaust port opening using a ball end mill or a regular two-lipped end mill with the corners of the cutting edges well rounded.

The major machining operations on the crankcase are now complete. Secondary operations such as installing the main bearing bushing, drilling, reaming, and flaring the venturi hole, drilling the intake for the needle valve, and drilling and tapping various small

The crankcase is held on Tool C while the front of the main bearing housing is faced to length.

The top of the crankcase is machined to the proper height before the boring operation is performed.

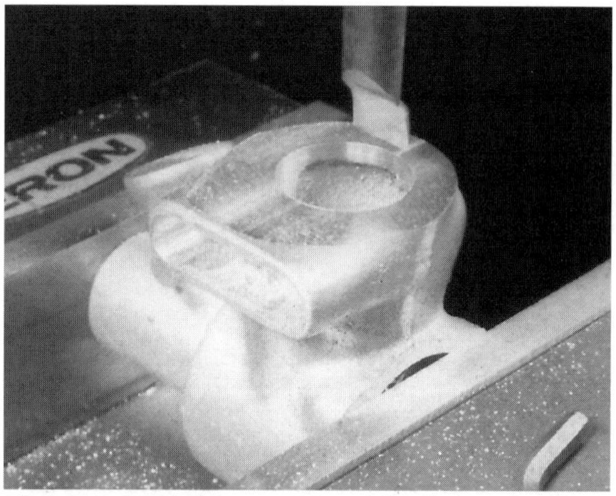

The recess for the lower portion of the sleeve is bored using an offset boring head.

The exhaust port in the crankcase may be machined with a ball end mill or with a regular two-lipped end mill with generously rounded cutting edges.

holes will be covered later in the Miscellaneous section. Machining of the venturi insert (Part 2-1) is also covered in the Miscellaneous section.

MAIN BEARING BUSHING (Part 3)

The bushing is machined from a durable copper, tin, and zinc alloy containing about 2% lead. When properly finished, it is very compatible with the polished steel journal of the crankshaft. Standard bronze bushing stock may also be used.

PROCEDURE

1) Machine the inside of the bushing first. It should be faced, center drilled, and drilled in increments to within about .060" of the finished size. Then it should be bored to straighten the hole.

2) If the proper size reamer is available, the last .010" can be removed by reaming. Lightly oilstone the blades of the reamer to make sure there are no burrs that could cause scratches in the bore.

3) Face to length and chamfer each end of the bore.

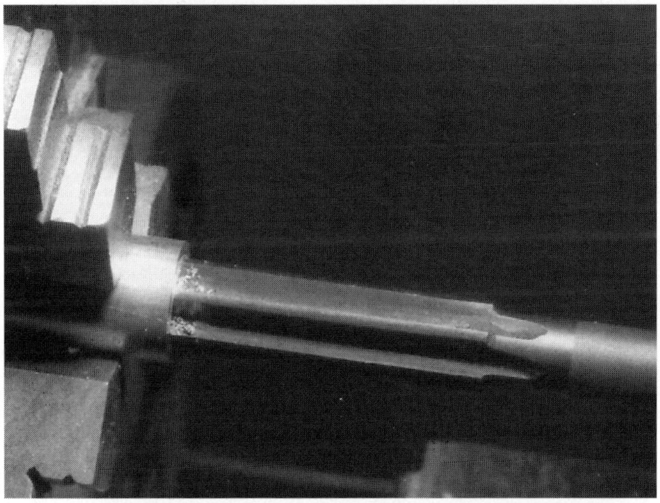

The main bearing sleeve is drilled and reamed to size using a reamer with carefully oil-stoned cutting edges to produce the best possible finish.

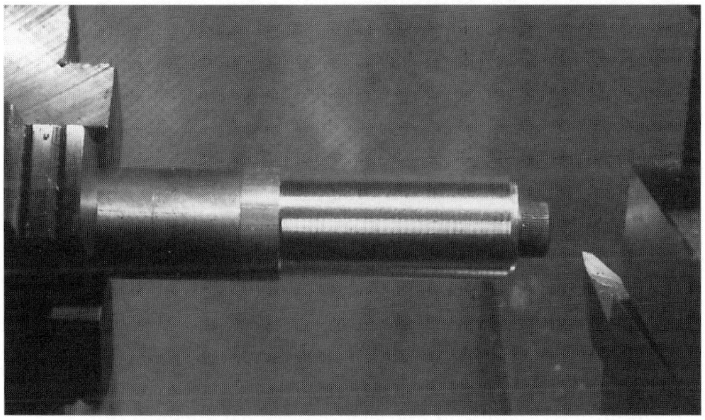

The outside of the main bearing sleeve is machined while being held on Tool E.

4) Tool E, which will be used to hold the bushing while the outside diameter is machined and to install it in the crankcase, should be made at this time.

5) To machine the outside diameter, use a mandrel or Tool E to assure that the outside diameter and the bore of the bushing are concentric. After the outside diameter has been machined to the correct size, taper the forward end (opposite the flange) slightly with a file to ease installation.

The bushing is now complete except for the operations that must be done after it is installed.

BACKPLATE (Part 4)

The backplate is cast of the same alloy used in the crankcase and head. It closes the back of the crankcase, and also serves as a rugged radial mount for the engine.

All of the facing and turning operations on the front face of the backplate are done at one setting while the casting is held in the three-jaw chuck.

All of the facing and boring operations on the back face of the backplate are done at one setting, using a thin strip of soft aluminum or copper between the front boss of the backplate and the chuck jaws. The outer bolt circle may be scribed at this time, if desired.

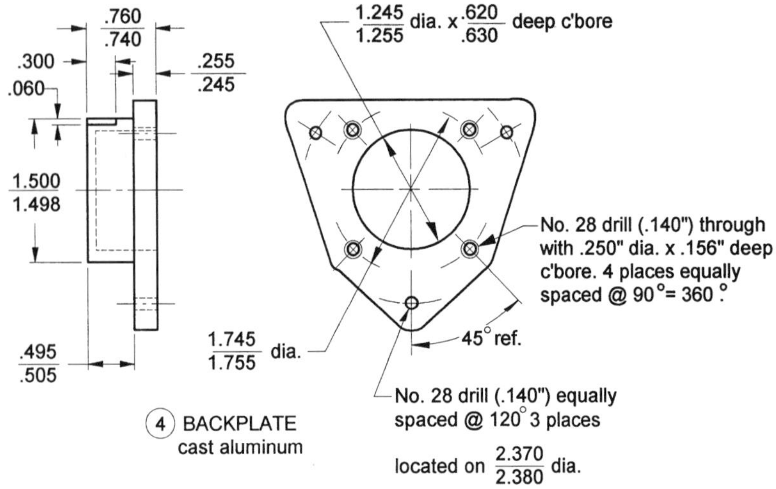

PROCEDURE

1) Deburr the casting and chuck it by the triangular portion with the boss exposed. Allow about .060" of the material to extend from the chuck jaws, and align the casting in the chuck so the face of the casting runs as true as possible.

2) Turn the boss to about .060" over its finished diameter.

3) Face both surfaces, initially removing a minimum of material from the face adjacent to the chuck jaws. Machine the face of the boss to its proper length.

4) Turn the boss to the specified diameter.

NOTE – If you choose to use an O-ring rather than a gasket to seal the crankcase/backplate joint, secure a 1.500" diameter O-ring no larger than .060" in cross-section. Then machine a groove about .065" wide and .050" deep at the rear of the boss on the backplate, and deburr the edge of the groove. You will find this makes a neat, effective seal.

5) Remove the backplate from the chuck, and place the machined boss in the chuck, using aluminum sheet or shim stock to prevent marring the surface. The rear portion may be faced and bored to size at this time.

6) The location of the three radial mount holes may be established now.

7) Machine the flat at the top of the backplate boss to clear the piston at bottom dead center.

The perimeter of the triangular part may be finished by filing, belt sanding, disk sanding, or any combination of these methods. With the exception of drilling operations that require a drill jig we will make later, the backplate is now complete.

Castings for the Titan .60 may be obtained by writing to George Genevro, 2050 Christina Street, N.W., Salem, OR 97304.

A small flat area is machined on the top of the back plate boss for piston skirt clearance at bottom dead center.

All facing and turning operations on the bottom face of the cylinder head are done while the casting is held in a three-jaw chuck.

CYLINDER HEAD (Part 5)

The cylinder head is cast of the same zinc-aluminum alloy used in the other castings. It should be machined to the dimensions shown, and may be altered later to change the compression ratio, if desired.

PROCEDURE

1) Deburr the casting, chuck it by the larger diameter and align it with enough material protruding from the chuck jaws to allow the necessary facing operations.

2) Machine the boss that extends into the cylinder to the specified diameter and depth.

3) Center drill the boss, drill through with a No. 3, and tap with a 1/4-32 UNS-2B tap using tapping fluid.

4) Machine the combustion chamber cavity, and deburr the area where it intersects the glow plug threads.

5) Using Tool F to protect the surface of the .937" diameter boss, chuck the head casting, turn the outside diameter, face the top of the head, and chamfer or radius the outer top edge of the head as desired.

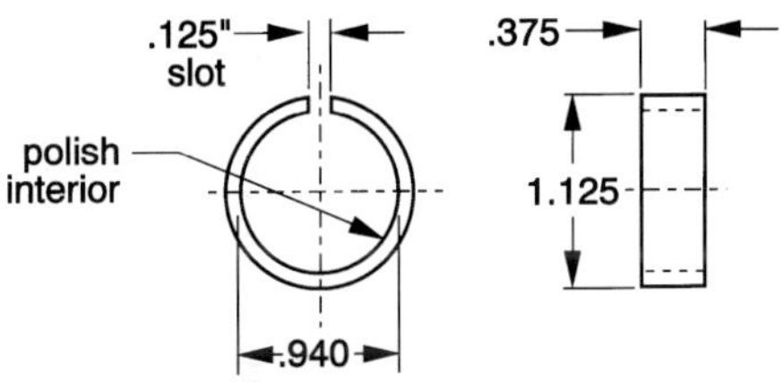

TOOL F - HEAD PROTECTOR
low carbon steel, brass or aluminum

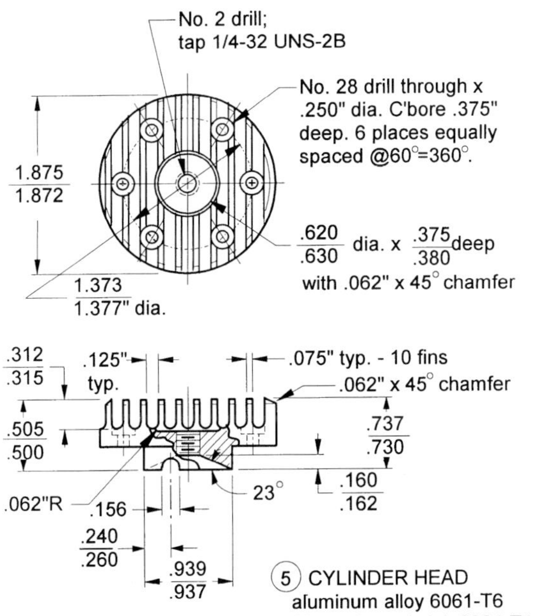

(5) CYLINDER HEAD
aluminum alloy 6061-T6
Alternate material: 2024-T4

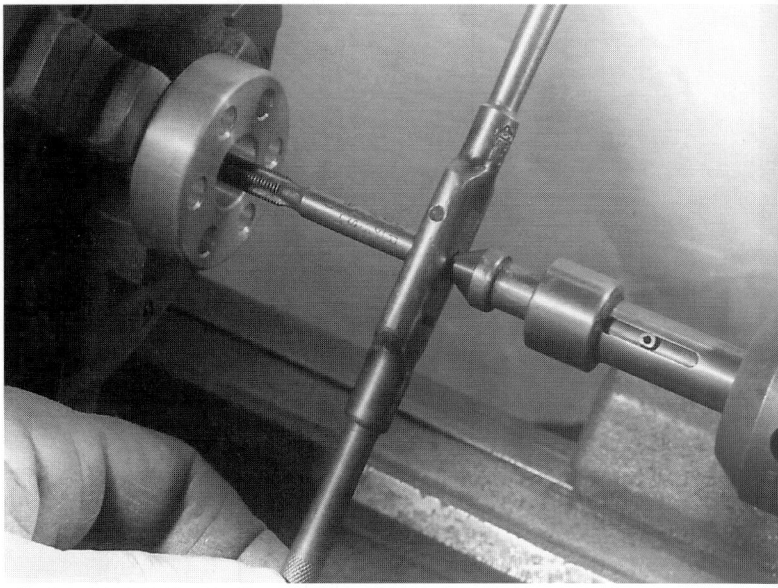

The head is held by the lower boss using Tool F when tapping the glow plug hole. Care must be taken in aligning the tap with the hole.

6) Counterbore the glow plug recess to the specified depth, and chamfer or radius its intersection with the top of the head as desired. Be sure the counterbored area where the glow plug gasket will seat is flat and free of burrs.

7) Using Tool Q, mount the head in the milling machine with the .937" diameter boss uppermost. Using a .125" ball end mill, machine the clearance groove for the piston baffle.

The remaining operations on the cylinder head, such as drilling and counterboring the screw holes and machining the fins, will be covered in the Miscellaneous section.

The recess in the lower boss of the head that accommodates the piston baffle is cut with a .125" ball end mill. Tool Q is used to hold the head in the milling vise.

The fin blank is bored for a shrink fit on the larger outside diameter of the cylinder sleeve.

The cylinder fins are machined as the fin blank is held on Tool H.

5) To facilitate the fin spacing operation, set the compound rest so its travel is parallel to the ways. With the carriage locked, the compound rest feed dial can be used to establish the position of the tool bit for cutting each fin.

6) For best results, grind a roughing tool and a finishing tool for cutting the fins. The roughing tool should be about .090" wide at the nose and should be tapered about .005" per side, making the end opposite the cutting edge about .080" wide. The corners of the nose of the tool bit should be slightly rounded.

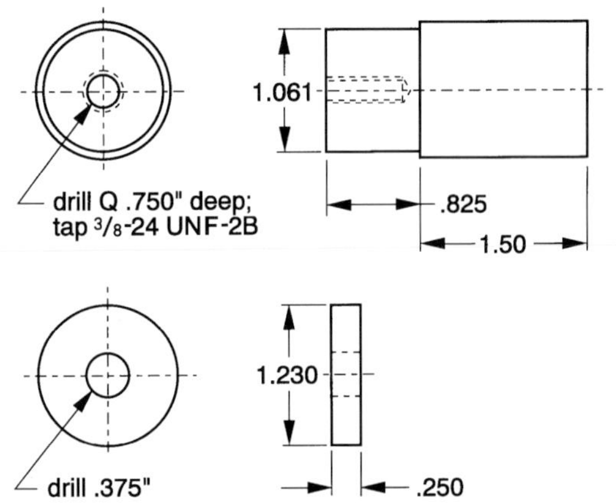

TOOL H -ARBOR AND WASHER
low carbon steel

7) With the part securely mounted on Tool H, rough out the fins to within about .010" of the finished depth, using an appropriate cutting oil.

8) Grind and oilstone the finishing tool to the specified width and finish the fins to their full depth. The outer edges of the fins can now be radiused or chamfered as desired.

The remaining operations on the cylinder fins will be done at assembly and are described in the Miscellaneous section.

The depth of the bores in the bottom of the piston must be held to close tolerances. Note the shape of the boring bar used to cut the recess between the underside of the head of the piston and the top of the piston pin bosses.

The partly machined piston is ready to be cut off by sawing or with a parting tool.

PISTON (Part 8)

The piston is made of high-strength, fine-grained cast iron. This dense yet slightly porous material, which incorporates a minute amount of free carbon in flake form, is ideally suited for a lapped piston running in a honed steel sleeve.

PROCEDURE

1) Chuck the piston blank securely so about 1" of material extends beyond the chuck jaws. Face the end and center drill it. This faced surface is the reference point for the depth of the cavities that will be machined in the piston.

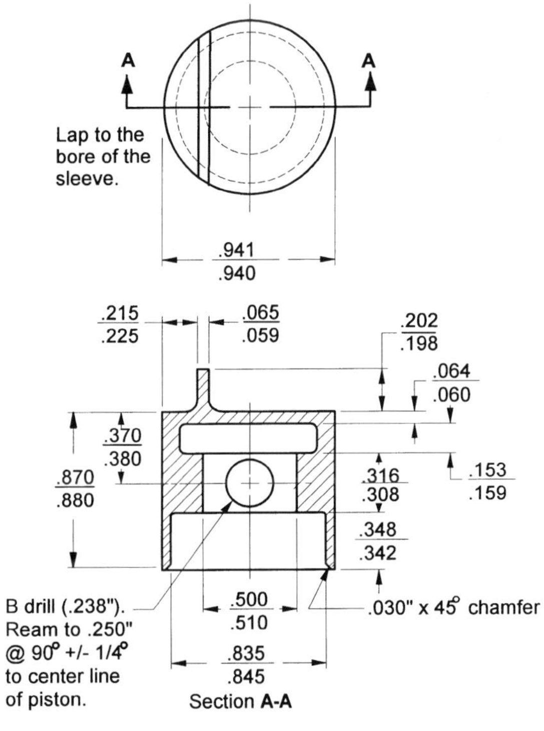

(8) PISTON
gray cast iron

2) Drill a .500" hole to the specified depth and square off the bottom using a small boring bar or a .500" end mill held in the tailstock chuck. A .500" drill ground square on the end or a .500" counterbore with the pilot removed may be used.

3) Machine the larger recess to the specified diameter and depth and chamfer the lower edge of the bore .030" wide at 45°.

4) To machine the recess above the piston pin bosses, it is probably more expedient to make a special boring bar from a .375" square high speed steel tool bit. It is easily mounted, strong, and less cumbersome to use in tight quarters than a small boring bar with a clamped-in tool bit.

5) Bore the recess above the piston pin bosses, taking care to measure all depths accurately.

6) Machine the outside diameter of the piston to .950", and turn the area immediately above the body of the piston to .920" diameter for a distance of .250".

7) Remove the piston from the chuck and cut off the excess material, making sure to leave enough of the .920" diameter area for the baffle.

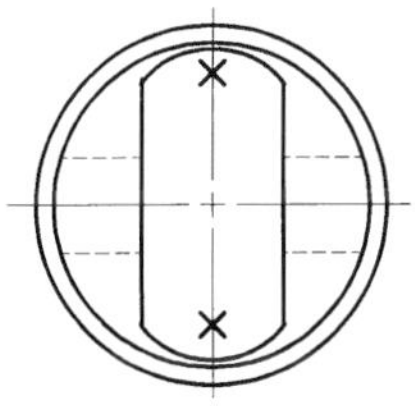

A view of the underside of the piston showing areas (X) machined away to lighten the piston and reduce engine vibration.

8) To hold the piston in the milling vise while machining the baffle, Tool I should be made. This tool is also used to hold the piston while excess material is machined from the area around the piston pin bosses to lighten the piston.

The piston baffle is machined using Tool I to hold the piston in the milling vise. An end mill with the corners radiused about .015-.020" should be used.

The drilling and reaming operations should be done at one setting. Lightly oilstone the cutting edges of the reamer to insure an accurate and smooth hole.

With the piston held inverted in Tool I, machine away excess material adjacent to the piston pin bosses.

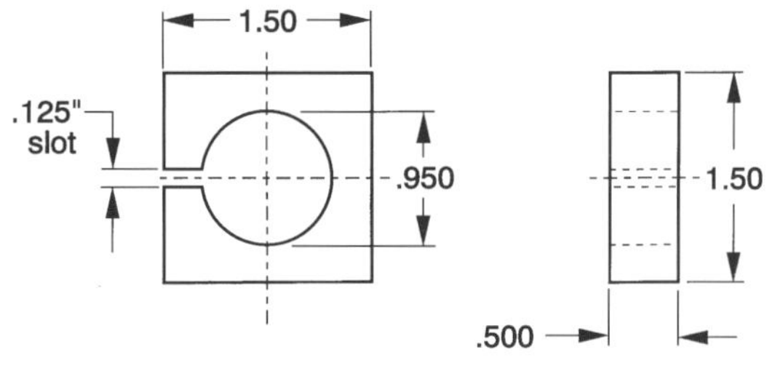

TOOL I - PISTON FIXTURE
aluminum

9) Machine the baffle, taking care to maintain the specified piston head thickness. Remove all sharp edges from the baffle.

10) Locate the piston pin hole, center drill it, and drill with a letter C drill (.242").

11) Place the piston in Tool I in the milling machine vise with the open end up and mill the area adjacent to the piston pin bosses as desired. The piston should be made as light as possible without sacrificing strength to reduce engine vibration.

12) To relieve any stresses that might have been inherent in the cast iron, or possibly introduced by the machining operations done so far, the piston should now be heated to a dull red heat with a propane torch, held at this temperature for at least several minutes, and allowed to cool slowly with a small can or other container over it.

13) To properly hold the piston while the outside diameter is being machined, it is necessary to make yet another simple tool. Tool J-1 is a piece of steel bar stock that has been machined so the inside diameter of the lower part of the piston is a snug slip fit on it. Its companion piece, Tool J-2, is used to allow the tailstock center to support the piston by applying light pressure on the head. The outside diameter of the piston should now be turned to .940".

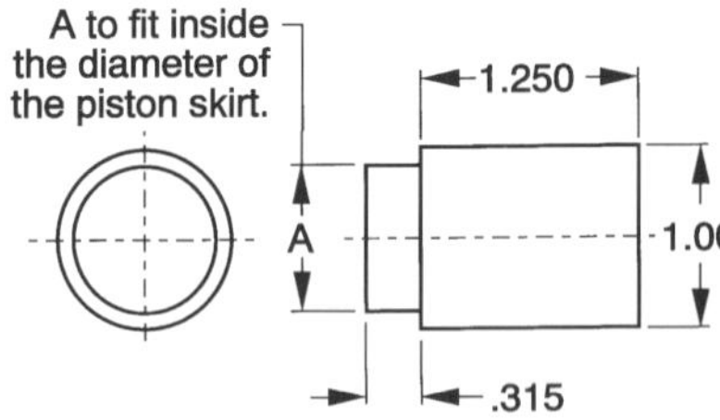

TOOL J1 - PISTON ARBOR
low carbon steel

After the piston has been stress-relieved, the outside diameter is machined using Tools J-1 and J-2.

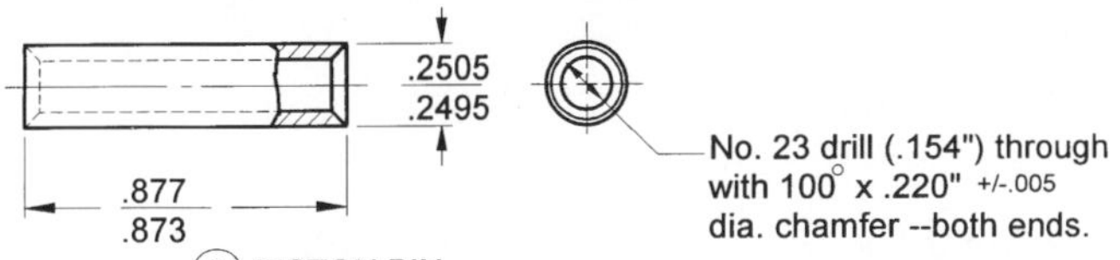

(9) PISTON PIN
drill rod
1 required

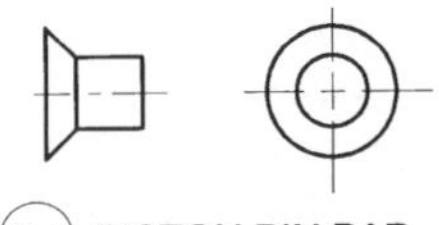

(9-1) PISTON PIN PAD
aluminum rivet (AN 426AD5-4)
1. 2 required
2. Pads to be pressed into each end of piston pin. Size the pin and pad assembly to the overall length of .930" / .925".
3. Scale: 2/1

14) Ream the piston pin hole with a hand reamer to .250", and lightly deburr the edges of the holes.

The piston is now complete.

PISTON PIN (Part 9)

The piston pin is made of .250" diameter drill rod and may be used without heat-treatment.

PROCEDURE

1) Chuck the drill rod, face it, center drill it, and drill through with a .125" drill, followed with a .154" drill. Chamfer the hole at a 100° included angle to accommodate the rivet which is used to form the piston pin pad.

2) Face the piston pin to length and chamfer the other end.

3) Cut the rivets to length and press into place. The exposed rivet head should be contoured, and the assembled piston pin should be machined to an overall length of .925" to .932".

The piston pin is now complete.

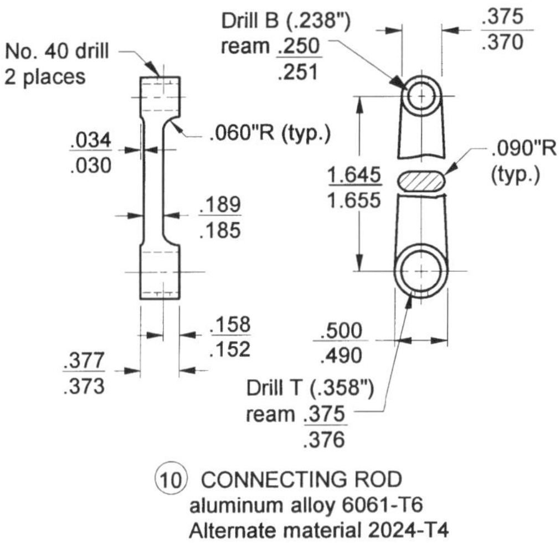

(10) CONNECTING ROD
aluminum alloy 6061-T6
Alternate material 2024-T4

CONNECTING ROD (Part 10)

The connecting rod is machined from aluminum bar stock, and while 6061-T6 is the recommended material, other strong aluminum alloys may be used. The major factors to consider in making this part are the alignment and center-to-center distance of the bores, and the internal finish.

The first operation in making the connecting rod is to drill and ream the bores. The table feed screw is used to accurately locate the bore centers.

Excess material is milled from the sides of the connecting rod with an end mill.

PROCEDURE

1) After the layout is complete, place the rod billet in the milling vise. Drill and ream both bores using cutting fluid.Chamfer the edges of both holes about .030" at 45°.

2) Mill away the excess material between the bores on both sides of the rod.

3) Using .125" dowel pins or similarly sized rod, place the rod blank in the milling vise using the dowel pins to locate the bores of the connecting rod relative to the top of the

The connecting rod is held in the milling vise and located with pins while the edges are machined.

vise. The excess material on both sides of the rod can now be machined off.

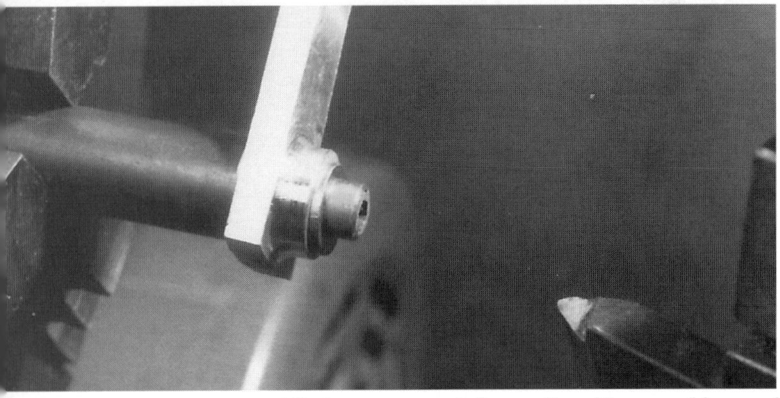

The bosses around the small and large end bores of the connecting rod are machined while the rod is held on Tool K.

4) A simple double-ended stub arbor (Tool K) is placed in the three-jaw chuck or a collet, and an 8-32 socket head cap screw and washer are used to mount the rod on the tool. The excess material can now be turned off the bosses at either end of the rod. This will leave only a small amount of material on the ends of the rod to be removed with a file or disk sander.

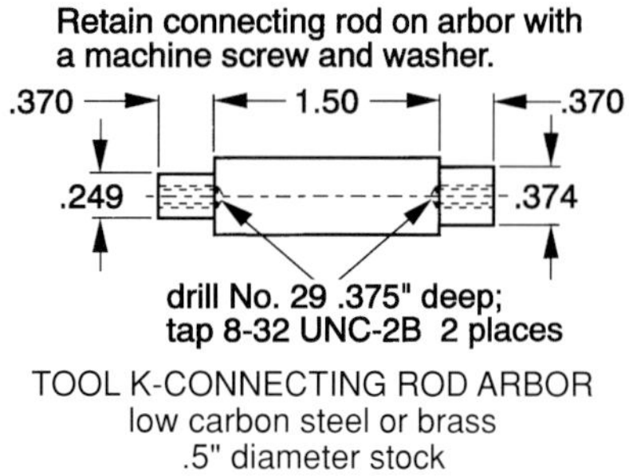

5) Oil holes should be drilled at the top and bottom ends of the rod using a Number 40 drill. Chamfer the holes with a small center drill or countersink.

6) All sharp edges should be rounded.

That completes the connecting rod.

The crankshaft, connecting rod, wrist pin, piston, and main bearing sleeve are ready for final finishing operations and assembly.

PROPELLER DRIVE WASHER (Part 11)

The propeller drive washer is made of low carbon steel bar such as AISI-C1018 or AISI-12L14.

PROCEDURE

1) Chuck the bar stock and machine the propeller drive face of the washer. Drill and ream the .312" hole and break all sharp edges.

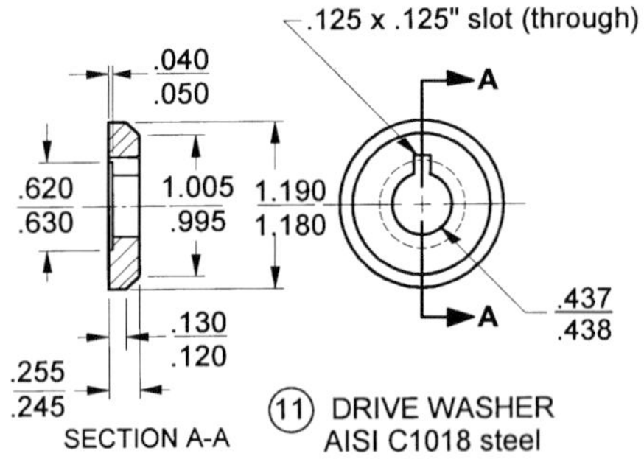

2) Knurl the face of the washer with a single straight knurl, if one is available. The drive face of the washer may also be serrated by other means if desired.

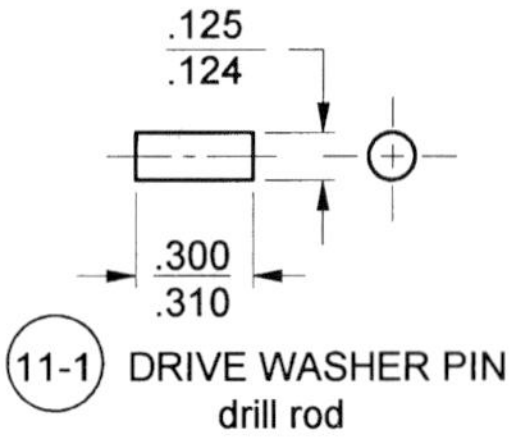

3) Machine the taper on the rear face of the washer and cut it away from the bar with a parting tool.

4) Using a .125" end mill, machine the recess in the bore of the washer for the drive key. If an end mill is not available, the recess may be cut with a small file.

The propeller drive washer is knurled by usinq a sinqle straight knurl. It may also be serrated by other means, if desired.

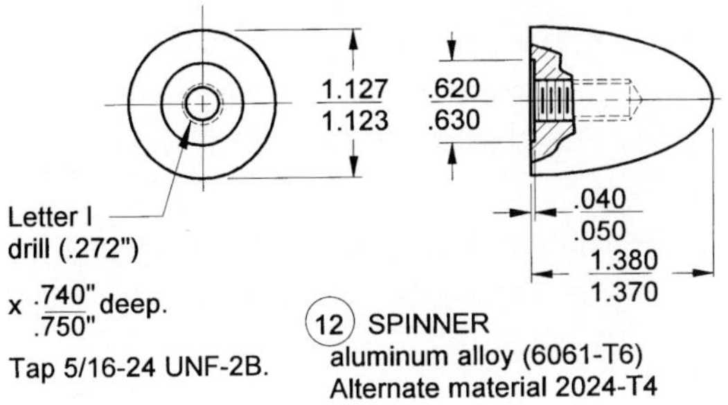

(12) SPINNER
aluminum alloy (6061-T6)
Alternate material 2024-T4

SPINNER (Part 12)

The spinner is machined from 6061-T6 aluminum bar stock and adds a nice finishing touch to your engine. Concentricity between the internal thread and the outer surface is an important consideration in making the spinner.

PROCEDURE

1) Chuck the bar stock. Face one end and drill with an I drill (.272") for the 5/16-24 UNF-2B thread.

2) Tap the hole, starting with a plug or starting tap and finishing with a bottoming tap. Support the shank end of the tap with a spring center or with the tailstock center.

3) Locate and drill the .187" cross hole used to install and remove the spinner.

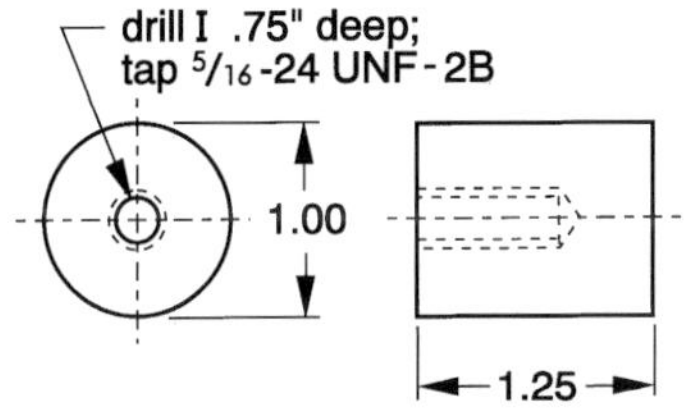

TOOL L-SPINNER ARBOR
low carbon steel
OPTION: use the blank end of TOOL G.

4) To hold the spinner securely while it is turned to the desired external shape, yet another simple tool should be made. To make Tool L, chuck a piece of 1" round steel bar stock, face it, drill it about .750" deep with an I (.272") drill, and tap it with a 5/16-24 UNF tap. (If desired, the blank end of Tool G may be drilled and tapped and used in place of Tool L).

5) Using a 1.25" long piece of 5/16-24 UNF threaded stock, mount the spinner blank securely on the tool in the chuck and turn it to the desired shape. Finishing operations on the spinner may be done while it is still mounted on the holding fixture.

The spinner blank is tapped in the lathe. The axis of the tapped hole must be perpendicular to the face.

The spinner is contoured while it is held on Tool L.

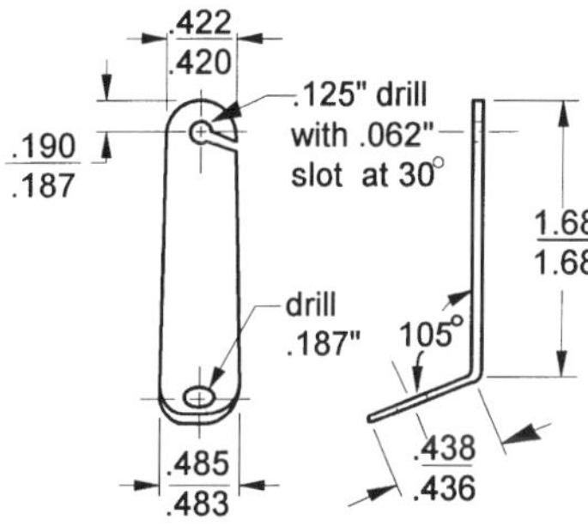

(13) NEEDLE VALVE CLIP
brass or stainless steel
Dimensions may change, depending on the type needle valve used.

The holes in the drill jig for the cylinder, head, and upper face of the crankcase may be located and drilled using a rotary table. Hole spacing and concentricity with the bore are of the utmost importance.

PREPARATIONS FOR FINAL MACHINING AND ASSEMBLY OPERATIONS

At this point, the major machining operations are complete and you have a collection of parts, most of which will require additional secondary operations before they can become parts of the operational engine. Before any of these tasks are performed, you will need to make additional tooling to position the necessary holes accurately in mating parts. While it may seem expedient to simply lay out and drill the holes in the various parts, making drill jigs saves time in the long run, and greatly reduces the chance for errors, and may provide new machining experiences. Also, if you decide to make additional parts, such as cylinder heads for experimental and development purposes, interchangeability is assured.

DRILL JIGS

Only two drill jigs are required. One is for the four holes at the rear of the crankcase and for the corresponding holes in the backplate. The other one will be used for the six holes in the cylinder head, the corresponding holes in the top fin of the cylinder/sleeve assembly, the three holes that go through all the fins, and the three tapped holes in the top surface of the crankcase. To use the two basic drill jigs for all of the operations listed above, several accessories such as Tools M-2, N-2, and N-3 must be made. These are detailed in the drawings for Tools M and N.

The crankcase/backplate drill jig (Tool M) is made from .375" thick mild steel plate. The holes may be located by use of a rotary table, if one is available, or by using the jaws of a four-jaw chuck as reference points. The drill jig may be left soft or carburized, or hardened drill bushings may be made and installed.

PROCEDURE (Four-jaw chuck layout method)

1) Locate the center of a 2" square piece of .375" thick steel plate and center drill it in the drill press.

2) Set up the stock in a four-jaw chuck and align it so the face of the stock and the center drilled hole run true.

3) Bore it to 1.501" and break all sharp edges. Chamfer one edge of the bore .060" at 45°.

4) Using the dial on the cross-slide or other means, locate the 1.750" bolt circle and scribe it with a sharp tool.

5) Bring one chuck jaw to a position 45° from the vertical axis and cut a piece of scrap stock to an exact fit between the jaw at the lower left (facing the chuck) and the ways or carriage.

6) Turn the sharp tool sideways in the toolholder, scribe a line across the bolt circle, and repeat the process for the other three holes. Be sure the scribing tool is on center.

7) Remove the part from the lathe, carefully prick punch the intersection marking each hole location, and then center drill with the smallest available center drill.

8) At this point, decide whether you will use the drill jig in the soft condition, carburize it, or make drill bushings to be pressed in. If you decide to either carburize it or use it in the soft condition, drill the holes with a No. 36 drill which is the tap drill size for a 6-32 screw. If you choose to use drill bushings, drill with a No. 26 drill, ream the holes .156" in diameter, and chamfer the edges. Machine the drill bushings from .187" drill rod, making them .375" in overall length. Leave a head about .060" long, machine the outside diameter to .157", and drill with a No. 36 drill. They can now be hardened and installed.

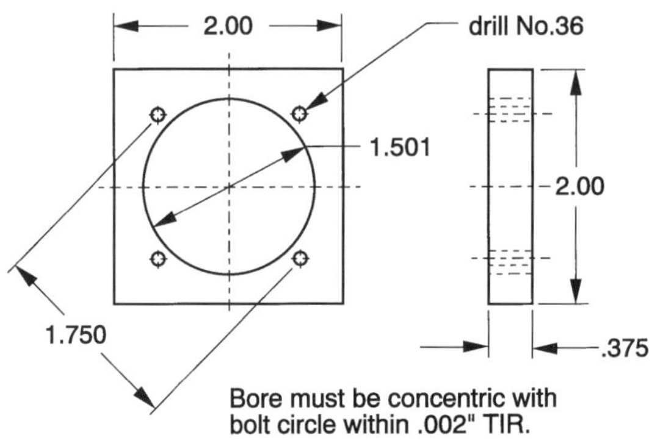

TOOL M1 - DRILL JIG
low carbon steel

9) Make the adapter (Tool M-2) that will align the drill jig with the crankcase bore.

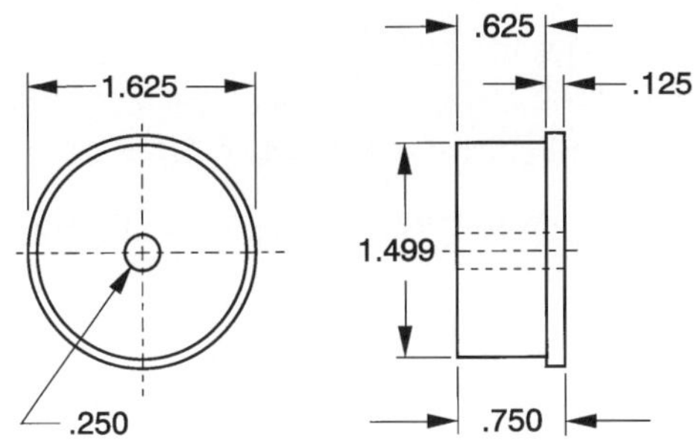

TOOL M2 - DRILL JIG ADAPTER
aluminum
Use to align TOOL M1 to crankcase bore.

PROCEDURE (Rotary table layout method)

1) After completing steps 1 through 4 above, remove the part from the four-jaw chuck and mount it on a 1.500" centering stub on the rotary table.

2) Prick punch the intersection of one of the diagonal lines scribed for the initial layout and the bolt circle, and use a wiggler to bring that point directly under the center line of the milling machine spindle.

3) Center drill, drill as desired, move the rotary table 90° and repeat the process until all four holes are drilled (and reamed, if necessary).

4) Chamfer all holes and break all sharp edges.

CYLINDER HEAD/FIN/CRANKCASE DRILL JIG

This drill jig (Tool N-1) is made from .375" mild steel plate or bar stock 2.00" in diameter. As in the case of the crankcase/backplate drill jig, the hole layout can be done on the lathe using a three-jaw chuck or by means of a rotary table. The jig may also be left soft, carburized or fitted with drill bushings, as desired.

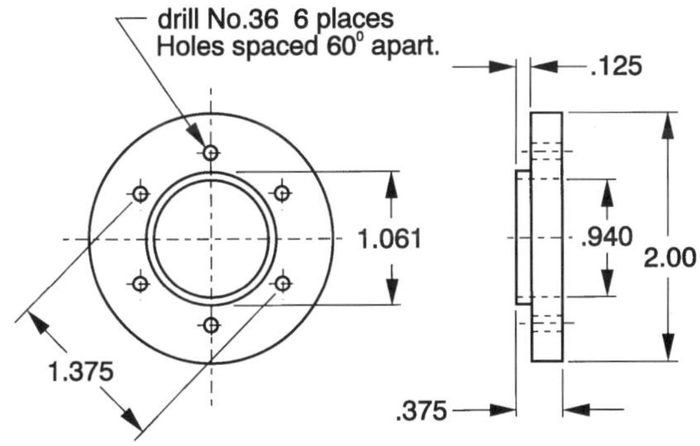

TOOL N1 - DRILL JIG
low carbon steel

PROCEDURE (Three-jaw chuck layout method)

1) Chuck the material in a three-jaw chuck, and face both sides so the blank has parallel faces and is .375" thick.

2) Bore through to .940" and chamfer the edge of the bore .040" about 45°.

3) Face as indicated, leaving a 1.061" boss.

4) Using the cross slide dial, locate the 1.375" diameter bolt circle and scribe it with a pointed tool.

5) Bring one chuck jaw to the level position at the front of the machine, and cut a piece of scrap to an exact fit between the chuck jaw and the ways or the carriage.

6) Using a sharp tool mounted sideways in the toolholder and positioned exactly on center, scribe all the way across the previously marked bolt circle. You should now have two hole locations marked 180° apart.

7) Repeat the process with the other two jaws of the chuck. You should now have six marks on the bolt circle spaced 60° apart.

8) Prick punch each of the hole locations, center drill it with the smallest available center drill, and drill with a No. 36 drill. If you have chosen to use drill bushings, drill with a No. 26 drill and ream to .156".

9) Make the adapter (Tool N-2) that will align the drill jig with the cylinder bore. The 1.061" diameter boss on the drill jig will align it with the bore at the top of the crankcase.

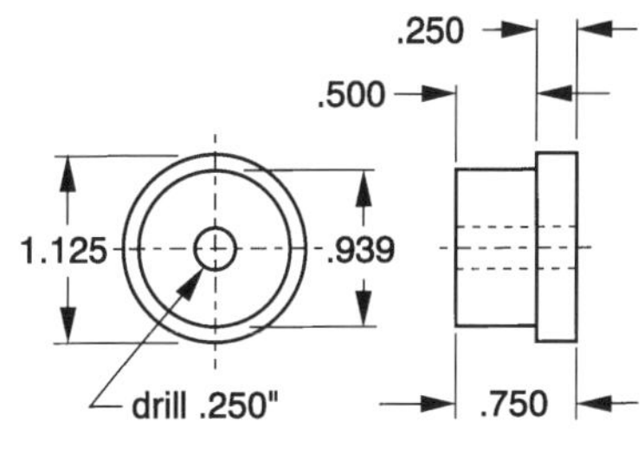

TOOL N2
low carbon steel

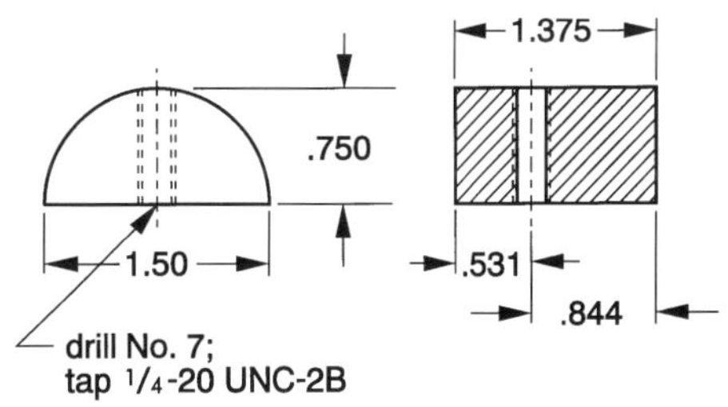

TOOL N3
aluminum

10) Make Tool N-3 to allow the drill jig to be securely clamped to the top of the crankcase when drilling the three holes at that position.

PROCEDURE (Rotary table layout method)

1) Machine the part as indicated in steps 1 through 5 above, including marking one hole location on the bolt circle and prick punching it.

2) Set up the part on the rotary table, using a centering stub .939" in diameter.

3) Use a wiggler to bring the hole location under the center line of the spindle. Center drill and drill the hole as desired. Advance the rotary table 60° and repeat the process until all six holes are drilled (and reamed, if necessary).

HOLDING FIXTURE

Unless an angle vise is available, the process of drilling, reaming, and flaring the intake will be made much easier by making Tool P. With this tool, you will have the option of using a milling machine or drill press to drill the hole, and the crankcase can be held securely and safely without danger of distorting it.

1) Scribe diagonal lines to locate the center of the 1.75 × 2.00" face of the aluminum billet and prick punch it.

2) Chuck the billet in the four-jaw chuck with the 1.75 × 2.00" that was marked exposed, and center it with a wiggler.

3) Face it, leaving a boss 1.499" in diameter and .190" long.

4) Remove it from the chuck, locate the hole that will be tapped, and drill with a No. 7 drill. Tap the hole 1/4-20 UNC-2B.

5) Scribe the 30° reference line on the side of the billet, as shown.

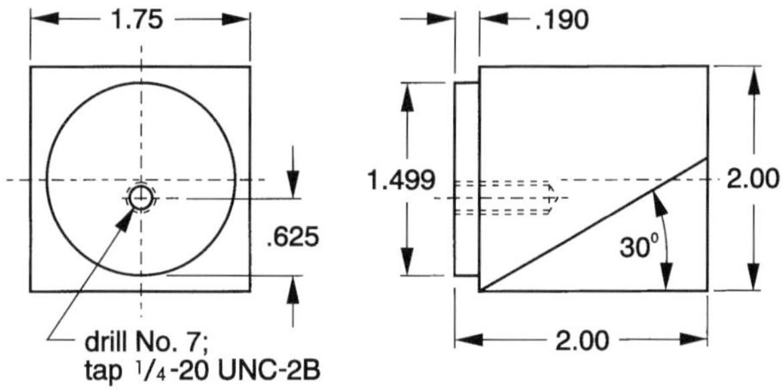

TOOL P - CRANKCASE FIXTURE
aluminum

The basic drill jig (Tool N-1) and the necessary accessories are shown.

The three radial mount holes are spaced 120° apart on the backplate casting.

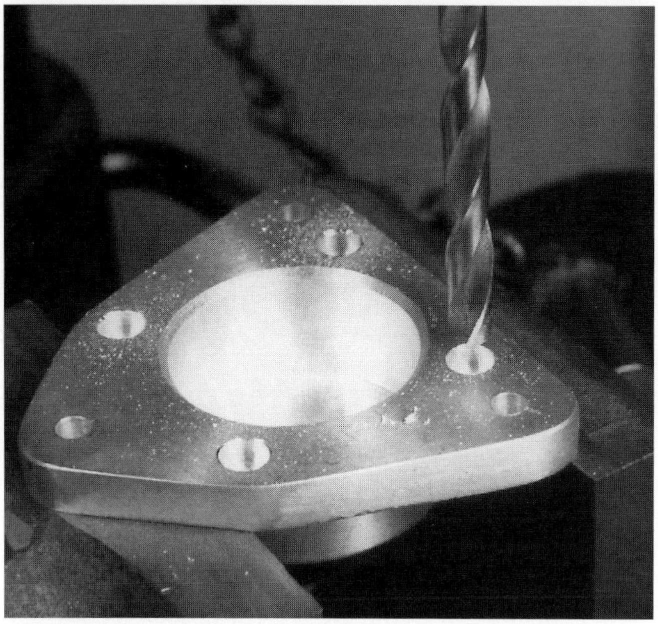

The holes for mounting the backplate to the crankcase are counterbored to accept the heads of the socket head cap screws.

FINAL MACHINING OPERATIONS AND SUB-COMPONENT ASSEMBLY

You are now ready to do those critical operations that will convert the individual parts you have machined into an operating engine. Care and patience at this stage will ensure satisfying and rewarding results for your machining venture.

DRILLING AND TAPPING OPERATIONS

Since all of the socket head cap screws used to assemble the engine are 6-32 in size, all of the holes in the drill jigs are the tap drill size for this thread. Holes in the parts that will be attached to the tapped parts will be drilled out to the body drill size after they have been drilled to the tap drill size while attached to the drill jig.

PROCEDURE (Backplate)

1) Align the drill jig (Tool M-1) with the backplate, clamp it, and drill the four holes.

2) Enlarge the holes to body drill size with a No. 28 drill.

3) Using a .250" counterbore, recess the holes at the back of the backplate for the heads of the socket head cap screws. (If a closer fitting counterbore is desired on both the backplate and the cylinder head, a letter A or a 15/64" drill may be used. It should be followed with the same size drill ground with a square end to properly finish the bottom of the hole).

4) Drill the three mounting holes at the perimeter of the backplate, using either a No. 28 drill for 6-32 screws or a No. 19 drill for 8-32 screws.

PROCEDURE (Crankcase)

1) Using the drill jig and the aligning tool (Tools M-1 and M-2) held in place by a through bolt and washers, drill the four holes at the rear of the crankcase that will be tapped for the backplate retaining screws.

2) Tap the holes with a 6-32 tap, using a tap alignment tool which is simply a piece of .500" square steel bar with a No. 28 hole drilled through it perpendicular to one surface. (Use of a tool of this type for any tapping operation is highly recommended as a means of insuring straighter tapped holes and greatly reducing tap breakage.)

3) Locate the drill jig (Tool N-l) on the top of the crankcase using Tools N-2 and N-3, and drill three No. 36 holes 120° apart. Be sure one hole is centered over the web in the exhaust port.

4) Tap the holes with a 6-32 tap using the tap alignment tool, and deburr.

PROCEDURE (Cylinder head)

1) Clamp the drill jig (Tool N-1) to the cylinder head using washers and a 10-32 bolt and nut. Drill the six holes and remove the drill jig.

2) Enlarge the holes to body drill size for 6-32 screws with a No. 28 drill and counterbore the holes for the heads of the socket head cap screws. Be sure the counterbore is deeper than the fin depth.

3) Carefully deburr all holes and chamfer the upper edge of the counterbored hole .030" with a 60° or 82° countersink.

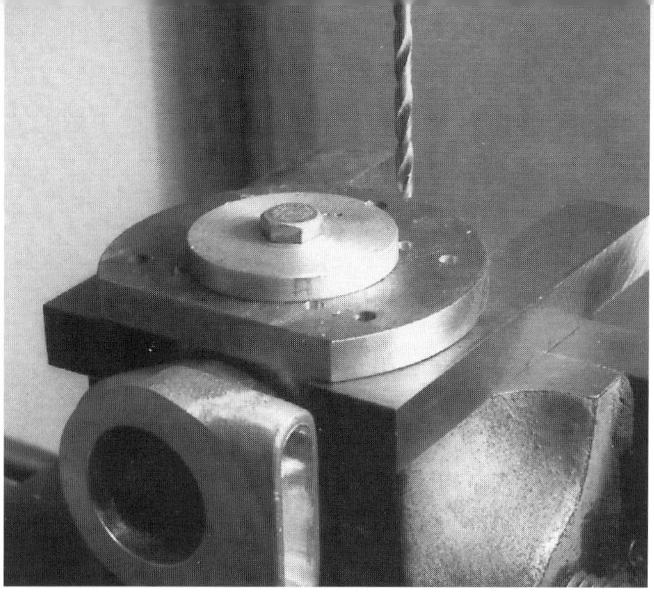

The four holes for mounting the backplate to the crankcase are drilled using a variation of Tool M-1 and Tool M-2 held together with a bolt and washer.

The six holes in the cylinder head are drilled using Tool N-1. Note the retaining bar used to hold the drill jig firmly in place.

4) Using Tool O, mount the head in the milling machine vise. Regardless of whether you are using a horizontal milling machine and a .125" thick milling saw or a vertical milling machine and a .125" end mill to cut the grooves between the fins, you will be using the cross-slide to position the head for proper fin placement and spacing. Therefore, place the head in Tool O so a line running through two diametrically opposed mounting holes is at a 90° angle to the face of the column of the machine.

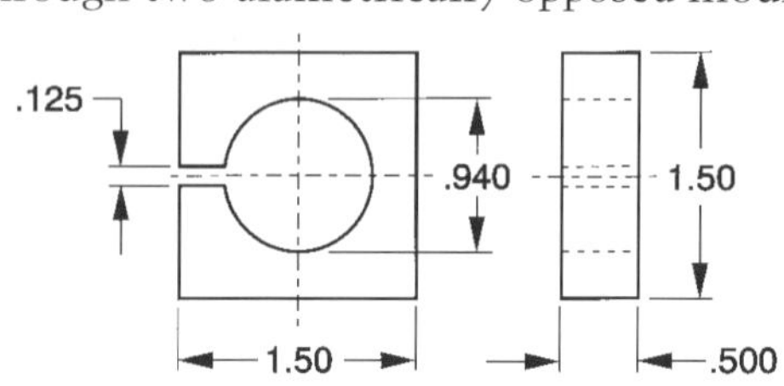

TOOL O - CYLINDER HEAD HOLDER
aluminum

5) Since the outside diameter of the head is 1.875", each of the ten fins will be .075" thick. Since each space is .125" wide, the total ten fins (10 × .075" = .750") and nine spaces (9 × .125" = 1.125") will equal 1.875". Spacing must be done with care to have the first and last fin be equal in thickness.

6) To start the fin cutting operation, move the end mill or milling saw so it just touches the side of the head, and zero the cross-slide dial. With the cutter set to the proper depth, move the head to one side and advance the cross-slide screw .200" (.125" for the space and .075" for the fin). Cut the fins, moving the table .200" each time for spacing.

7) Deburr the fins with abrasive cloth and a small hand scraper.

The fins are cut using a .125" milling saw. The head is held in the vise with Tool O and a spacer of equal width is used at the other end of the vise jaws.

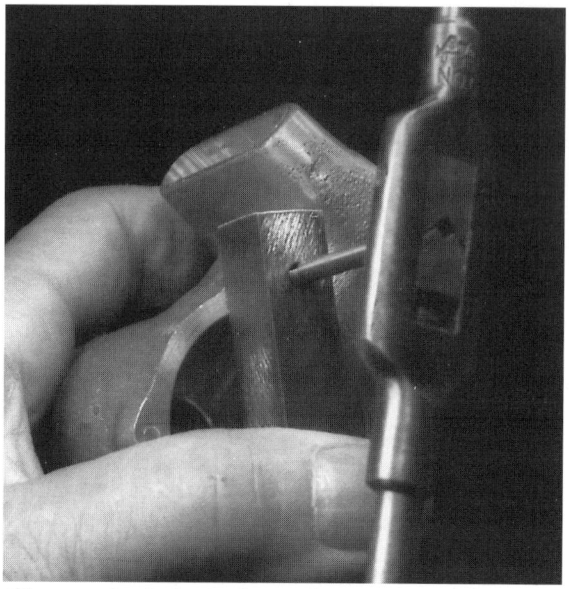

When tapping holes in the crankcase, use of a simple tapping guide made of .500" square cold rolled steel will help prevent tap breakage and misaligned threads.

After a needle valve has been selected. The pads at the side of the intake passage should be milled to size and drilled.

The three holes at the top of the crankcase are drilled using Tools N-1, N-2, and N-3.

PROCEDURE (Cylinder/fin assembly)

1) Attach the drill jig (Tool N-1) and the aligning tool that locates it relative to the bore (Tool N-2) with a through bolt and a nut and washers. Be *sure* one hole in the drill jig is aligned with the center web of the exhaust port in the cylinder sleeve.

2) Drill three holes 120° apart completely through the fins. One of these holes must line up with the center web of the exhaust port.

3) Drill the remaining three holes only through the thick top fin.

4) Remove the drill jig and enlarge the three long holes to body drill size (No. 28 drill) for 6-32 screws.

5) Tap the three holes in the top fin using the tap aligning tool. The cylinder/fin assembly is now complete except for the honing operation to fit the piston.

The heated aluminum fins have just been assembled onto the chilled cylinder sleeve.

FINAL CRANKSHAFT/ CRANKCASE OPERATIONS

This is the first of the important processes in which parts are mated to each other in their final working relationship. Exercising extra care at this point can be a major factor in producing a more attractive and satisfying engine.

PROCEDURE

1) Make sure the crankcase and the bushing are free of burrs, and press in the bushing using Tool E. If desired, the crankcase may be heated to about 300° F, to ease installation.

2) Ream the bushing to fit the crankshaft.

3) Attach the crankcase to Tool P and machine the intake passage as desired. The hole may be drilled and reamed to .312" diameter for initial running and enlarged to .375" later, if desired. An alternate method is to drill and ream the intake passage .375", and to make a venturi insert (Part 2-1), as shown on the plans. This is to be a slip fit in the intake passage and is held in place by the needle valve spray bar.

4) Machine and drill the pads on the side of the venturi to accept the needle valve that has been selected.

5) Carefully deburr the crankcase bushing where the venturi hole intercepted it, and recheck the fit of the crankshaft in the bushing.

6) Install the crankshaft, position the crankpin adjacent to the upper left backplate mounting hole (looking forward), and mark the intake passage hole location on the crankshaft journal with a scriber or a transfer punch.

The drill jig (Tool N-1) has to be carefully located on the cylinder sleeve/fin assembly to assure that the center web in the exhaust port will line up with the rib in the center of the exhaust opening in the crankcase casting.

All of the holes in the cylinder fins are drilled using Tools N-1 and N-2. Three of the holes will be drilled only through the thick fin at the top and tapped while the other three holes are drilled through all the fins and are enlarged to clearance size.

A brass lap may be used for the final fitting of the piston and cylinder sleeve. Patience is important at this point.

7) Remove the crankshaft from the crankcase, drill the intake port hole and carefully deburr it. This establishes the basic intake port timing, and its duration can be extended by simply increasing the diameter of the hole or altered and extended by machining away material from the leading or trailing edge of the hole.

8) Install the propeller drive washer and check for end play. If the end play is less than .005" or more than .015", adjust it by machining either the end of the crankshaft journal or the end of the crankcase. The crankcase and crankshaft are now complete and ready to accept other components when finished.

FINAL CYLINDER/PISTON OPERATIONS

Fitting the piston to the cylinder sleeve is one of the most critical operations in building an engine. Since no piston rings are used, the piston must be fitted closely to prevent combustion pressure leakage, and yet must move up and down in the cylinder with a minimum of drag. It is a painstaking job but one that must be done well.

A trial assembly of the engine, less cylinder head, indicates that all parts fit and that the crankshaft, rod, and piston move freely.

PROCEDURE

1) Since the piston has been normalized and turned round, no more material should be removed from it. It must be free of burrs, and the edges of the piston pin holes should be chamfered with a hand scraper.

2) If possible, contact a local auto engine rebuilder, and have the sleeve honed with an automotive piston pin hone until the piston will go partway into the lower end of the bore. To get a free running fit that will hold compression when the engine is hot, the lower end of the bore should be about .0001" looser than the upper end. This is referred to as "choke" in a cylinder bore, and is common practice in air-cooled engines.

3) If a fine hone is available, use it for the final fitting. An alternate method is to make a split brass or cast iron lap, and do the final fitting using a mixture of rouge and oil, washing the piston and sleeve frequently.

4) If desired, you may do the final fitting by mounting the piston on the connecting rod, and placing a wooden dowel in the lower end of the rod. Using the finest jeweler's rouge available and light oil, move the piston in the sleeve with a combined linear and rotary motion until the desired fit is achieved. Patience at this point will reap great rewards.

FINAL ASSEMBLY

Now that all secondary operations have been completed, you are ready to assemble the engine. Lubricate all parts with light oil during assembly.

PROCEDURE

1) Insert the crankshaft into the crankcase and place the connecting rod onto the crankpin.

2) With the crankpin at its uppermost position, install the piston and piston pin,

The engine mount has been made and coated with polyurethane varnish. The fuel tank is mounted by means of a stud that goes through the mount and a nut.

The engine is bolted to the mount using 6-32 socket head cap screws and aircraft-type self-locking nuts.

noting the position of the baffle on the piston relative to the bypass.

3) Slip the cylinder sleeve/fin assembly over the piston and into the crankcase. With the cylinder held in place, the crankshaft should rotate freely as the piston moves up and down. If the side of the connecting rod touches the intersection of the lower and upper bores of the crankcase, disassemble the engine, remove material as necessary, clean, and reassemble.

4) Install the cylinder head, being sure to tighten the three long cap screws evenly. Rotate the crankshaft to check for interference between the piston baffle and the head.

5) Install the backplate, needle valve, prop drive washer, a propeller with a diameter of 12" or 13" and a pitch of 6", and the spinner.

MOUNTING AND RUNNING THE ENGINE

The engine mount shown on the plans and in the photos proved to be ideal for running the engine. It is strong and convenient to use, especially if the tank shown is made and installed. Take care to bolt the engine securely to the mount, using aircraft-type elastic self-locking nuts and washers. The wooden mount should be securely held in a vise or clamped to a bench.

Regardless of the type of tank you use, be sure the top of the tank is slightly below the level of the needle valve. Install a long reach glow plug and use a 1.5 volt dry cell or a nickel-cadmium battery pack, as recommended by your local hobby shop. Use commercial glow fuel with about a 10% nitromethane content. If degummed castor oil is available, about 5% by volume may be added for the break-in runs.

Initial needle valve settings will have to be determined by experimentation, depending on the type of needle valve used. The engine should be primed by squirting a small amount of glow fuel into the cylinder through the exhaust port with the piston at bottom dead center. A few drops squirted into the intake passage will often make starting easier.

SAFETY NOTE

It is strongly recommended that a leather glove be worn when starting the engine. Paint

Note the relationship between the position of the tank and the needle valve in the venturi. The only things needed now to run the engine are fuel and batteries for the glow plug.

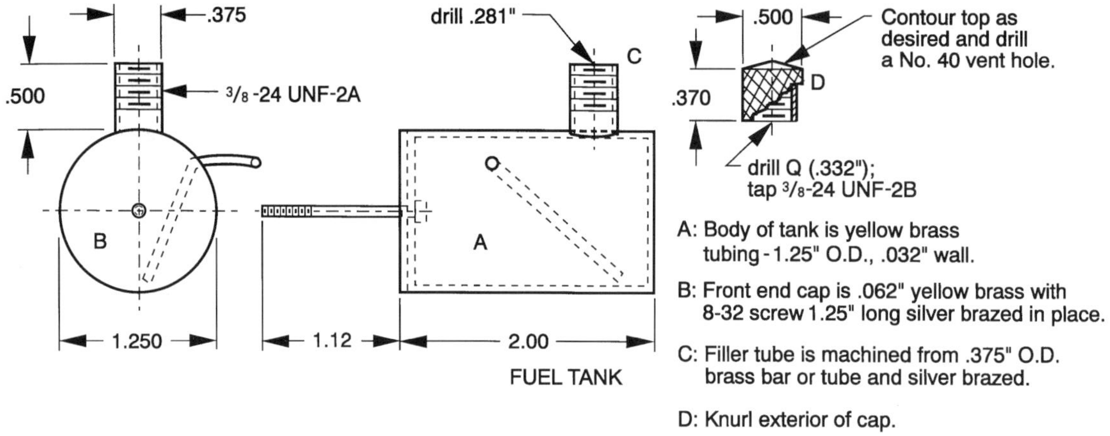

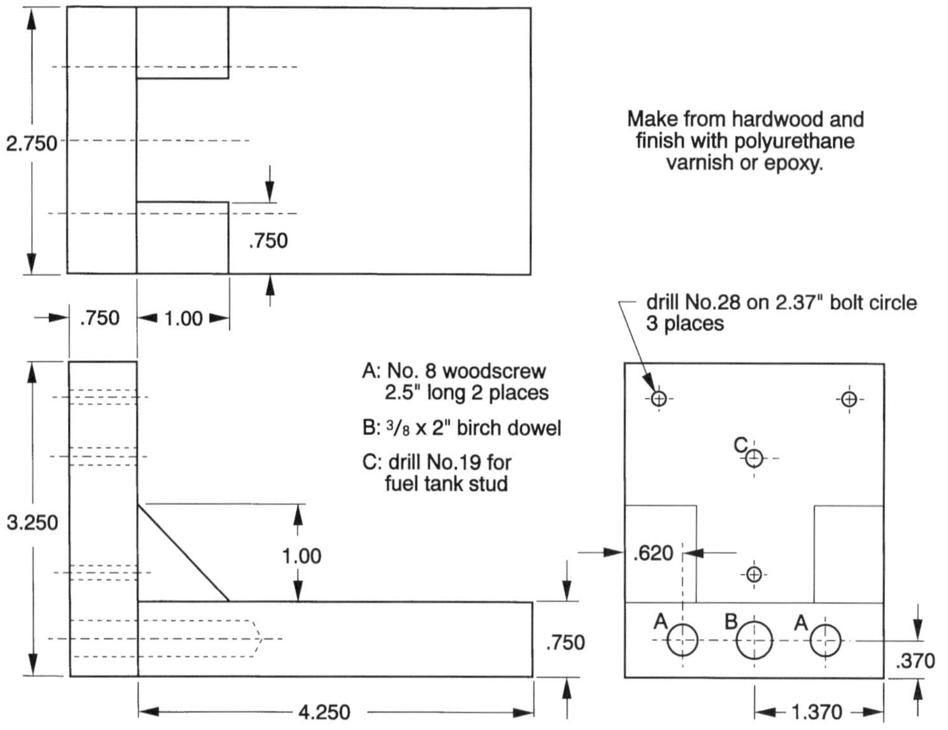

the tips of the propeller white for a distance of about 1". This will make the propeller disk much more visible when the engine is running. Run the engine in a well ventilated space to avoid an accumulation of exhaust fumes.

CONGRATULATIONS!

You have completed a complex project – a series of demanding tasks that had to be done well if the engine was to be successful. You are now ready to break in and enjoy your engine. Use it safely, keeping in mind that a propeller rotating at high speed demands your undivided attention and respect.

If you are inclined to experiment, there are many changes you can make to your engine. Some of these involve reworking existing parts, but others require making new and slightly altered components. In the search for improved performance, your imagination and capacity for research are the major factors.

IT RUNS! Note the secure mounting in the vise.

FOR THE BEGINNER

The Walking Staff

by Don Titus
Photos by Author

The Walking Staff is a pretty straightforward project any hiker or photographer will find useful in the field. The only tools you'll need are a hacksaw, file and a lathe; since the parts that need to be turned are small, even a small Unimat lathe will work. Most of the materials for the project can be purchased from a local hardware store and photo retailer, with the remainder coming from your scrap box.

This is a great project for some of the folks who have written in asking how they should get their feet wet. Even if you make a mistake and need to remake a part, you'll be out only a couple of bucks. This project is geared toward the novice, but with a little creativity, a seasoned machinist can add some custom touches that only experience and craftsmanship allow. The walking staff makes a great gift for the up-and-coming photographer or outdoor enthusiast.

The only item that might be a little difficult to round up is the clamp; all the other parts came out of the scrap box, my junk drawer, or the local hardware store. If you're really creative, the clamp can be fabricated from aluminum or brass, then polished. I decided to use one made by Bogen for use on their tripods; it's a nice clamp, comes painted black, is reasonably priced (about $4.99) and is available at most camera suppliers that handle the Bogen tripods.

PARTS 1 AND 2

When I started this project, the first decision I had to make was whether I wanted the staff to collapse in two or three sections. For strength and convenience, I decided to use the two-section design; this allows the staff to collapse to about 3 feet. If collapsed length is a concern, then you might want to consider making the shaft in three parts, though this will reduce the strength slightly, require an additional clamp, and necessitate purchasing an additional piece of tubing the proper size. Weight is always a concern when you are out hiking or picture-taking for the day, so I decided to use aluminum for the shaft and steel for the end fittings. If you want to fancy things up a bit, you could use brass for the end fixtures or go to aluminum to further reduce weight.

After a little checking, I found the local hardware store carries a limited stock of steel and aluminum in a variety of shapes. They had aluminum tubing in three sizes, and the 3/4" OD tubing was a prefect slide fit inside the 7/8" OD tubing. Had I decided to make a three-section staff, the 5/8" OD fit nicely into the 3/4" OD tubing. Make sure the tubing slides together smoothly before buying it; sometimes the tubing in the racks gets bent or dinged up in shipping, or the wall diameter is such that one piece will not fit inside the next larger size. The tubing at the hardware store was available in precut lengths of either 6 or 10 feet, so I bought a couple of 6-footers.

Before you start cutting, you must decide how long the staff needs to be when fully extended; this will depend primarily on the height of the user and personal preference. I'm 5'9" tall and wanted to keep the length manageable for packing purposes, so I settled on an extended maximum length of 6 feet. If you are making your staff in two parts, take the length you decided upon and add 12". Then divide this figure in half. In my case, the total length would be 84" (72" plus 12") divided by 2, so each part was 42" long. This allows for a 6" overlap where the parts slide together, and provides for a good stable joint with little flex from side to side. I suggest you make the shaft parts a little longer than you think you will need; they can always be trimmed at a later date.

The staff is designed in such a manner that it can be shortened at any time in the future if you find it is too long, thereby custom-fitting it to the user. Cut two pieces of tubing this length, one from each piece of tubing you are using; these will make up the upper and lower shaft pieces of the staff. If you are making your staff in three sections, take the length you decided on and add 18", then divide this figure by 3. You would then cut three pieces of tubing this length, one from each size you are using. Make sure your cuts are straight; the handle bushing and pointed tip butt up against these surfaces. If they are a little crooked, you can touch them up with a file. Deburr the inside and outside ends of all the pieces of tubing, check the slide fit, then set them aside.

1 ***All the parts and material, except the tubing, for building the walking staff.***

2 ***Turning Part 3 on my 9" South Bend lathe.***

While you're at the hardware store, pick up a package of the rubber feet like the one shown in Photo 1. They come in a variety of sizes, and you'll need one to fit the size tubing that will end up as the lower leg of the staff.

PART 3

The scrap box produced a piece of steel round stock about 1.00" in diameter and 6" long. Make sure the piece you choose is a little longer than you need to allow enough room for the lathe chuck to hold the piece securely. The round stock was chucked up in the lathe three-jaw chuck, faced, and turned to be a snug fit in the 7/8" tubing for a length of about 1.00". Be sure to keep the step that forms at 90° to the body of the part so you will get a nice mate with the end of the tubing. When this is done, you can remove the excess material with a hacksaw, or part it off in the lathe so it has a total length of 1.50". Either way, you will need to rechuck the part on the area you just turned, and turn to a diameter of 0.875" to match the outside of the larger piece of tubing. Do not remove the part from the lathe; you will now need to face the end, center drill the part, and drill and tap the threads for the 1/4-20 threaded rod. After tapping this hole, I used a file to clean up the sides of the bushing that would be exposed. If you want to get fancy, this area could be rounded or coved. Let your imagination run wild. If your idea doesn't work, you are only out your time. For the time being this part is finished, so set it aside.

PART 4

You can make the tips to fit your needs. I decided to make a pointed tip that would grab in hard-packed snow or ice, for those times when I foolishly wander out in Minnesota's Winter Wonderland. There is nothing more exhilarating on a crisp, sunny, winter morning than starting down a hill only to discover the warm sun has

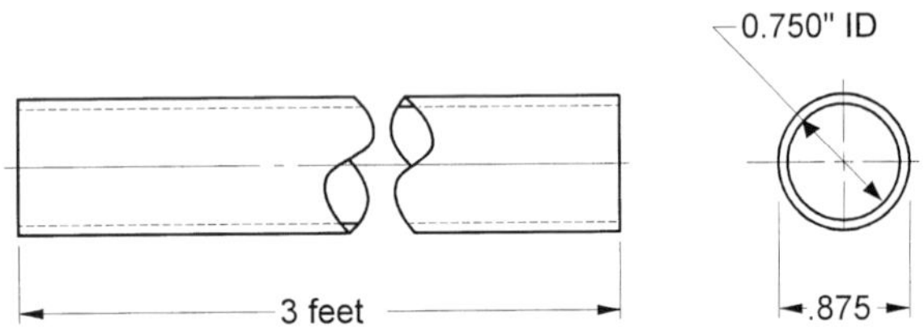

PART 1
aluminum tubing

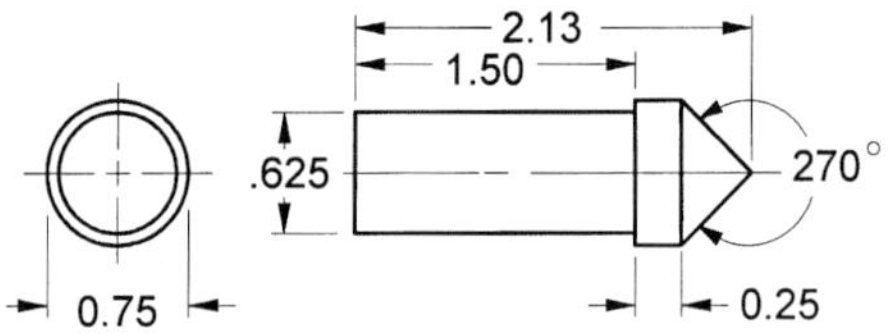

PART 4
steel

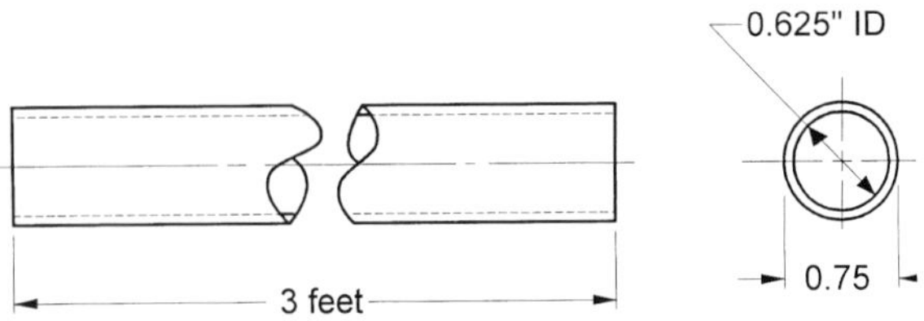

PART 2
aluminum tubing

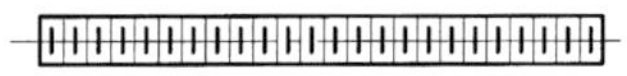

1/4-20 threaded rod

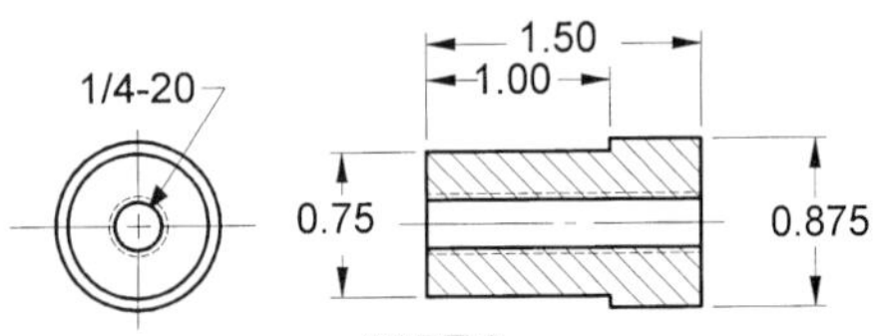

PART 3
steel

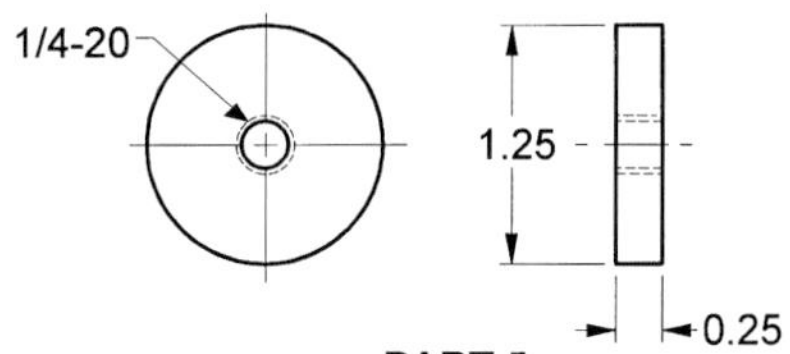

PART 5
steel

3 ***My 10-year-old son has shown an interest in metalworking and talked me into helping him build a walking staff for himself. This is a great project for the novice.***

turned the hard-packed snow to glare ice and there is nothing in sight to hold onto. When this occurs, the trip down the hill is usually much quicker than you would like, especially while trying to juggle a couple of cameras, lenses and a tripod. Don't get me wrong – it would take a lot to get me to move out of Minnesota with its abundance of wildlife and change of seasons. I love it here!

You'll also find the pointed tip helpful when hiking in an area where the soil has a lot of clay, and it has rained recently. After a moderate rain, the clay quickly fills the cleats in boots, making the navigation of trails with moderate grades quite exciting. For this part, I decided to use a piece of hot rolled round stock about 3" long and 0.875" in diameter that I found in the scrap box. It was chucked up in the lathe, the end was faced, and it was turned to a snug fit in the piece of tubing that forms the lower leg for a length of about 1.50".

As above, make sure the shoulder is square with the portion just turned to give a good mating surface when the parts are assembled. When this is done, the part is turned around in the chuck, the end faced, and the OD turned down to match the OD of the tubing. Set your compound slide at 45°, and cut the angle forming the point. Use a file to clean up the sides that are exposed and the point. There is no need to clean up the shaft as the rough surface left by turning will give epoxy a better surface to grip to. I had given some thought to hardening the tip and then decided this would be an unnecessary step. If you prefer a hardened tip, consider using drill rod or tool steel when making this part; it will be easier to harden and provide better durability.

PART 5

I planned to use the walking staff as a monopod for my camera when I didn't feel like lugging along a tripod. This created the need for a lock ring in the design. The lock ring keeps the knob and/or a camera from unscrewing from the threaded portion of the staff. If you are making strictly a walking staff and never plan to use the threaded portion to hold a camera, the lock ring is not necessary. The lock ring was made from a piece of 1.25" hot rolled round stock I had leftover from a previous project. You can use aluminum or brass, if you prefer. Size isn't really important as long as the lock ring provides a good gripping surface. The dimensions given in the drawing are just for reference.

Mount the material you have chosen for the lock ring in the lathe chuck, face the end, and take a few shallow cuts to true it up. Center drill the end, then use a live or dead center in the tailstock to support the outer end of the piece. Besides providing stability, this support also helps to keep the part from walking out of the chuck while knurling.

Replace the cutting tool with a knurling tool, and knurl an area about 0.25" or slightly wider. For those who have never tried knurling, try it; it's not a difficult procedure but it does take a bit of patience. I would like to point out that I am not an expert when it comes to knurling, although I do have a few suggestions that might be useful for the beginner.

If this is your first attempt at knurling, you might want to start with either aluminum or brass for the lock ring; these materials are easier to work with and will provide more than adequate strength. The speed at which you set your lathe when knurling depends on the material you are using, the size of your equipment, and your setup. When knurling steel, I prefer to use the back gears on my South Bend lathe to set the speed at the second slowest setting. When knurling brass or aluminum, I get better results by bringing the speed up a few steps. Once

4 ***Another up-and-coming hobby machinist at work!***

you move the knurling tool into the part and it starts to cut, try not to back it off until you're finished knurling. A knurling tool usually has two small unprotected wheels spinning at a relatively high rate of speed. Give them the respect they deserve. Due to their design, those little wheels will grab anything close to them; they especially like brushes, shirt sleeves, hair and fingers, and they are not partial!

For the novice, don't be afraid of the knurling tool – just be aware of what you are doing and take the necessary safety precautions. Over the years, I have found a squirt bottle seems to work well for applying cutting fluid when knurling, and you'll want to use plenty of cutting fluid to wash out the chips that accumulate. If you decide to use a brush to apply cutting fluid, DON'T get it anywhere near the knurling tool. This advice comes from learning my lesson the hard way.

When knurling steel, I've used standard cutting oil with good results; aluminum or brass seem to do better with a lighter weight fluid like *WD-40.* When you finish knurling, drill a hole approximately 0.50" deep using a No. 7 drill bit, then thread it 1/4-20, as shown in the drawing of Part 5. At this point, take time to bevel the edges of the area just knurled to give it a finished look, and remove any sharp burrs raised by knurling. I have a lathe bit that has the tip rounded. By slowly moving this into the knurled edge, it forms a nice coved effect or, if you prefer, you can use a file and just roll the edge. Now use a parting tool or hacksaw to remove excess stock and bring the part to its proper length.

For those of you who do not have a knurling tool, here is another option. Cut a narrow strip of paper that will go around the outer edge (circumference) of the finished lock ring. Lay the strip out flat on a table and mark an equal number of graduations on the strip. If your lock ring had a diameter of 1.25", you would end up with a strip of paper about 4.25" long. Lay out marks at .25" intervals on the paper strip using a pencil. Use a black magic marker or toolmaker's die on the outer edge of the lock ring. Tape the strip around the lock ring, and use a scribe to transfer the marks from the paper to the ring, then remove the paper strip. Using pads in your vise, clamp the ring with the marked edge slightly above the vise jaws.

Using a round file (I like chain saw sharpening files as they cut fast), start at one of the marks and cut a round-bottomed slot about .125" deep across from side to side. Then rotate the ring in the vise to bring up the next mark and file a slot. Keep this up until you have gone all the way around the lock ring. Remove the ring from the vise, and see if all the cuts appear to be the same depth. If not, touch up those that need it, keeping in mind this isn't a precision process by any means. This will provide a gripping surface that will work almost as well as knurling! If no round file is to be found, use a triangle file and cut slots; even a hacksaw will work in a pinch. You're just trying to improve the gripping surface.

PART 6

My junk drawer produced a nice black, hard-plastic shift knob I decided to use. These are available from your local auto parts store for just a few bucks, or you can make whatever suits your fancy! This particular knob was threaded 3/8-18, so I made a bushing with these threads, and drilled and tapped it 1/4-20 so it would be the proper size. I used *Loctite* to hold the bushing in the knob so it wouldn't come loose.

ASSEMBLY

With all the parts finished, all that is left is to assemble them. I decided to use epoxy to hold the bushing in the top of the staff and the pointed tip in the other end. I prefer to use the slow setting epoxy rather than the five-minute type. It seems to provide a better bond and appears more durable.

Use a piece of 120-grit wet or dry sandpaper to scuff the inside of the tubing being used for the upper portion of the staff. Clean the bushing (Part 3) to remove all traces of oil. A little lacquer thinner, *Gumout* or alcohol works great for this. Mix up your epoxy according to the instructions. Spread some on the bushing and then around the inside of the tubing. Push the bushing into the tubing, wipe off any excess, then set this assembly aside to dry as per the instructions on the epoxy usually 12 to 24 hours. Make sure to stand the assembly on the bushing end so any excess epoxy doesn't run down the inside of the tubing. This would create a real problem when the other piece is slid up inside.

When this is dry, cut a piece of 1/4-20 threaded rod approximately 2.25" long. Put *Loctite* on a 1.0" section of the threads and screw it into the bushing. Allow this to dry the recommended period of time. If *Loctite* isn't available, you could use epoxy for this, also. Use some sandpaper, and scuff inside one end of the tubing being used for the lower leg of the staff. Degrease the portion of the point (Part 4) that goes into the tube, and mix up some more epoxy resin. Apply it to the part and inside the shaft, slide the parts together, wipe off the excess, and stand on the point until dry.

The rubber foot is used when the point isn't needed, and protects the point from damage when using the walking staff on hard surfaces. I decided to cut a small metal disk out of brass (about 0.125" thick) to put in the bottom of the rubber foot to keep the point from pushing through the rubber. This is optional. After everything is dry, usually overnight, all that's left is to install your clamp or clamps, slide the tubes together, slip the rubber foot over the pointed end, and screw the lock ring and knob to the top, and you're finished. If you don't need the knob to be removable, you won't need the lock ring, and can use *Loctite* or epoxy to keep the knob from unscrewing. I decided to wrap the upper portion of the handle to provide both a better gripping surface and some insulation when the mercury drops below zero.

5 ***My son with the walking staff he built with very little help.***

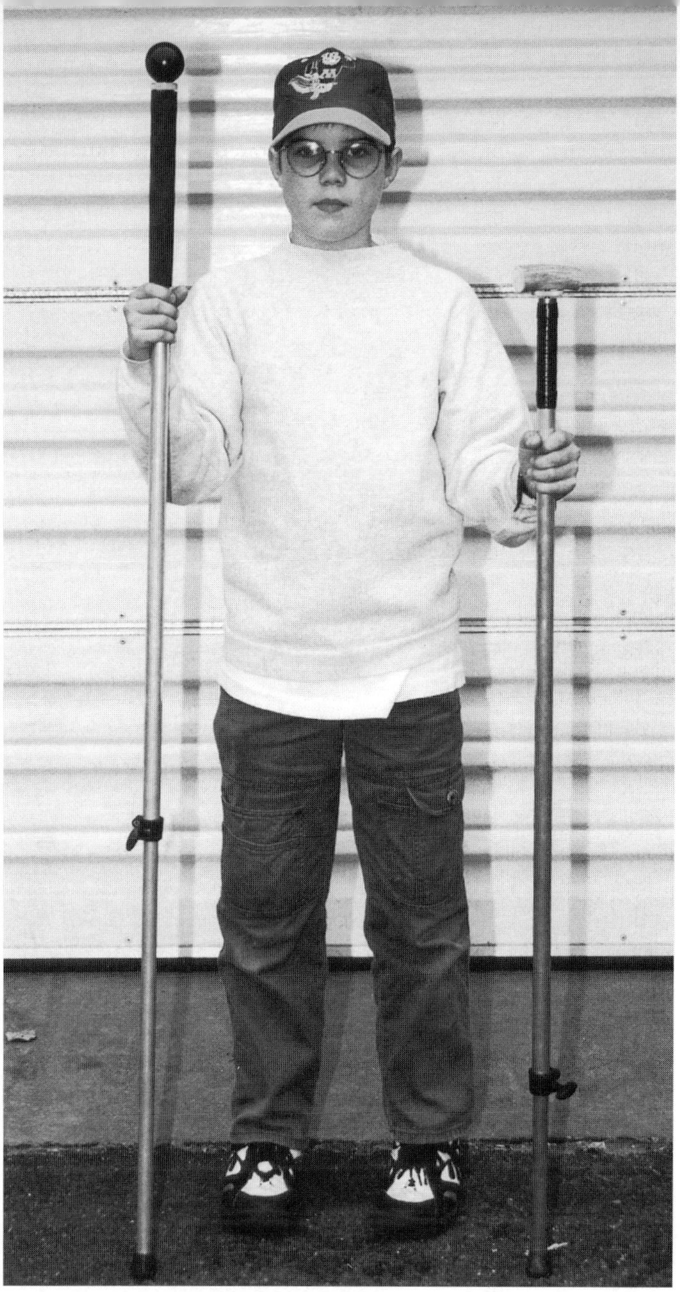

6 ***Our pair of finished walking staffs.***

POSSIBLE OPTIONS AND MODIFICATIONS

Some modifications you might want to consider when building this project would be to use a durable plastic like *Delrin* and make a plastic tip for the other end of the lower leg. By making it the same size as the OD of the tubing, you would be able to turn the lower leg around when you want to go from a non-marring end to the pointed end. The lower leg has ample room to store waterproof matches, a fishhook and line, or other items that might come in handy, if you are forced to spend an unscheduled evening in the woods.

Consider interchangeable tips – using a threaded bushing instead of the pointed tip. Then make your tips so they screw into this bushing. Then your tip options are unlimited. Instead of a black plastic knob, you could make a hollow knob with a screw-on top where additional items could be stored. The possibilities are limited only by your imagination!

The walking staff is very handy for probing areas in which you really don't want to stick your hand, and for retrieving items just out of reach, such as apples in trees or items in the water. I made a short leather strap that works as a sling for carrying the staff across my back when I don't need it; or, if you prefer, make a wrist strap that fits under the lock ring. I think anyone who spends time outdoors will appreciate the versatility of the walking staff.

BILL OF MATERIALS

PART NO.	QTY.	MATERIAL	SIZE
1	1	aluminum tubing	3' × 0.875" OD (0.75" ID)
2	1	aluminum tubing	3' × 0.75" OD (0.625" ID)
3	1	steel	1.50 × 0.875" round stock
4	1	steel	2.13 × 0.75" round stock
5	1	steel/aluminum/brass	1.25 × 0.25" round stock
	1	threaded rod	¼-20 × 2.25"
	1	rubber foot for chair leg	
	1	clamp	

A Steam/Air Engine for Fun

by George Kerekgyarto

Photos and drawings by Author

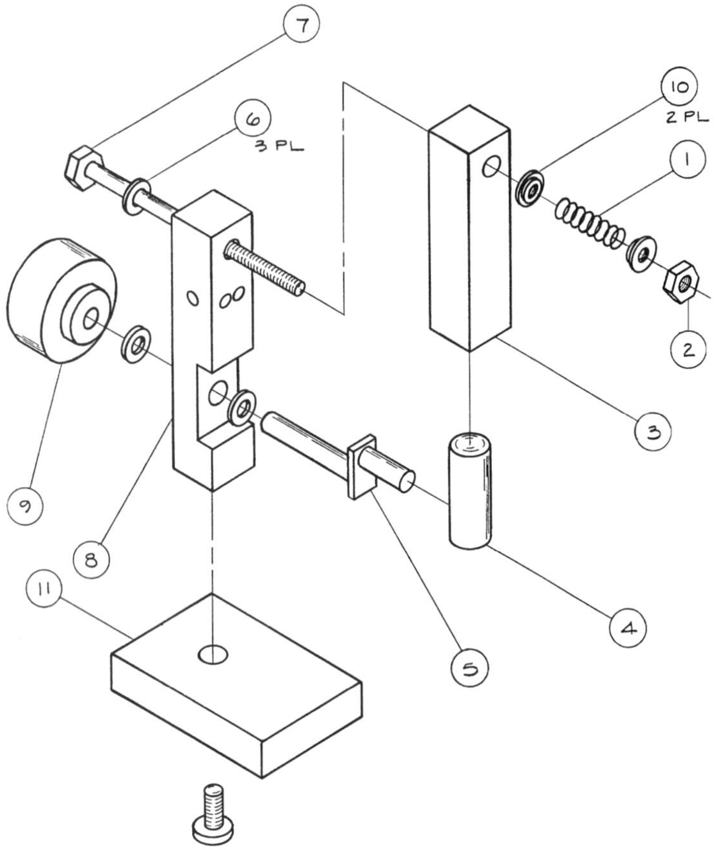

Figure 1

Most home shop machinists do jobs that are unique to the equipment they have available. The oscillating steam engine shown in the photos is simple, yet challenging to machine using basic machinist tools. The Bill of Materials has been developed to take advantage of standard manufactured stock. However, modifications can easily be made depending on the availability of materials. If suggested material sizes are not available, more machining will be required, which may be the case for most home shop machinists. The following instructions will provide the necessary direction, yet allow for your own creative methods to surface in machining the steam engine to the required tolerances.

The steam engine can be machined by the most inexperienced machinist, and provide a challenge to the most experienced one. Boiler designs can be adapted to this plan quite easily and provide challenging fabrication techniques. The engine design permits steam or compressed air to be used as a fuel source. This particular model steam engine has run for up to three hours using *Sterno* as a heat source to create the steam in the boiler. All parts presented in these plans are machined except the bolt, spring and nut. These may be bought at the local hardware store or can be machined for fun, except for the spring.

The following procedures will aid the home machinist in completing this engine:

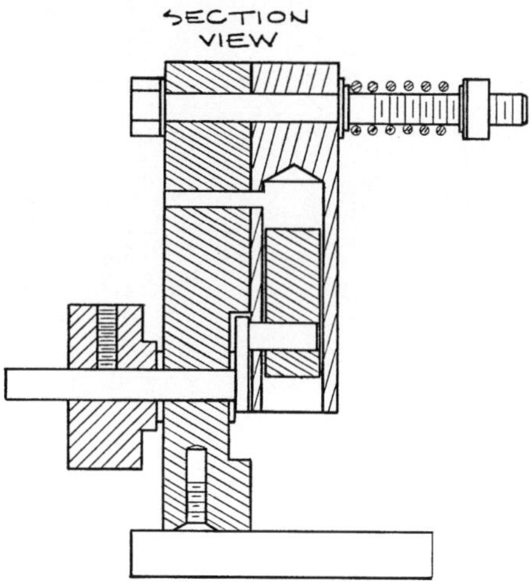

Figure 2

MACHINING THE CYLINDER AND THE STANDARD

Machine the cylinder and the standard to square within .005" on a mill, and to the exact length using a lathe (Figures 3 and 4). If standard stock is used, little machining will be required at this point. On a lathe, drill the hole for the piston (Figure 3). Machine the steps in the standard and the cylinder on a vertical or a horizontal mill (Figure 4).

Stack the cylinder and the standard together in a proper match (one on top of the other) and using a drill press, drill through both pieces, insuring alignment for the bolt. Now drill the intake and exhaust ports (Figure 4, Section A-A). In a mill, machine the cylinder and the standard to the exact required size.

Clean off burrs with a file, wrap parts and set them aside.

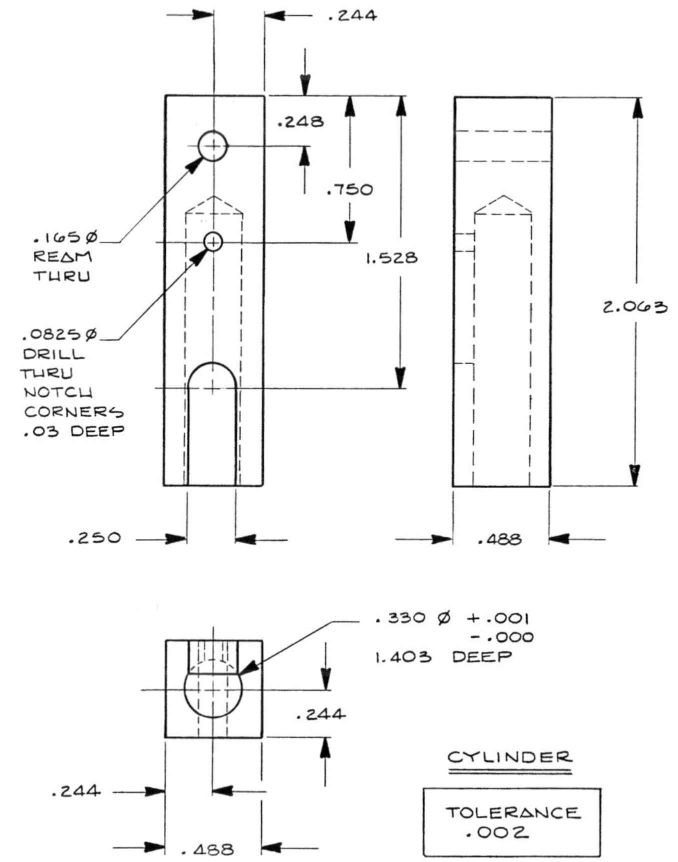

Figure 3

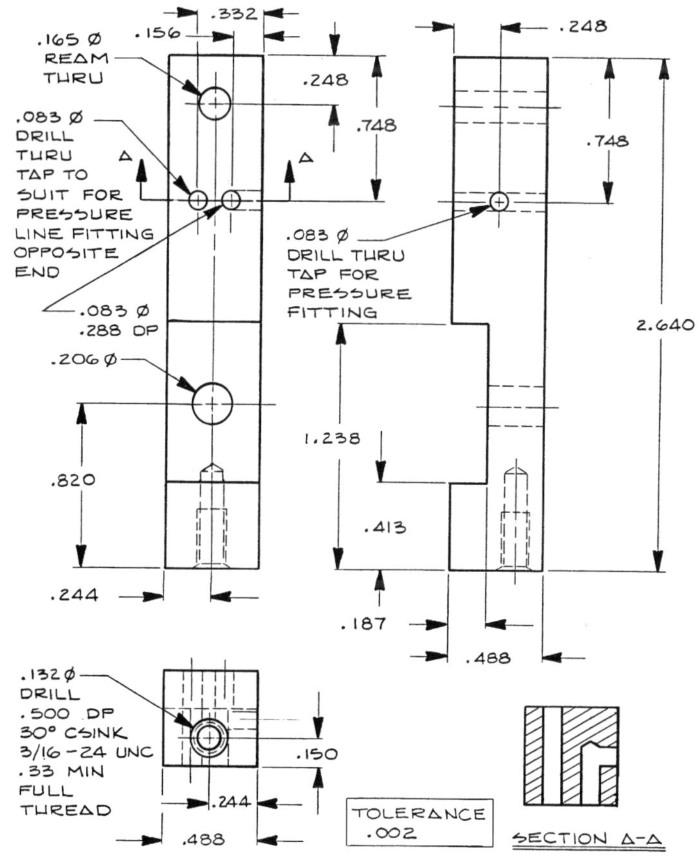

Figure 4

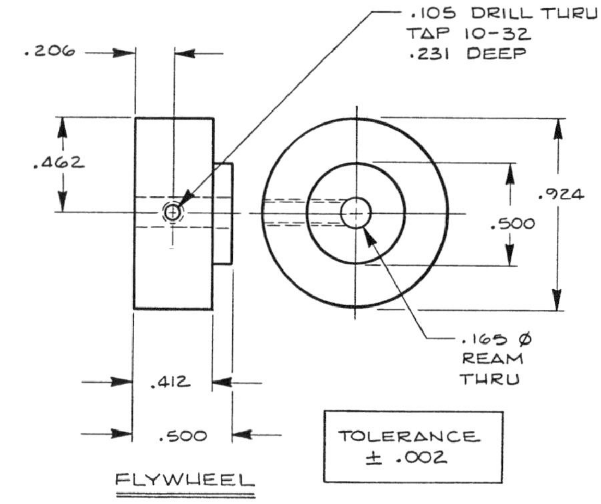

Figure 5

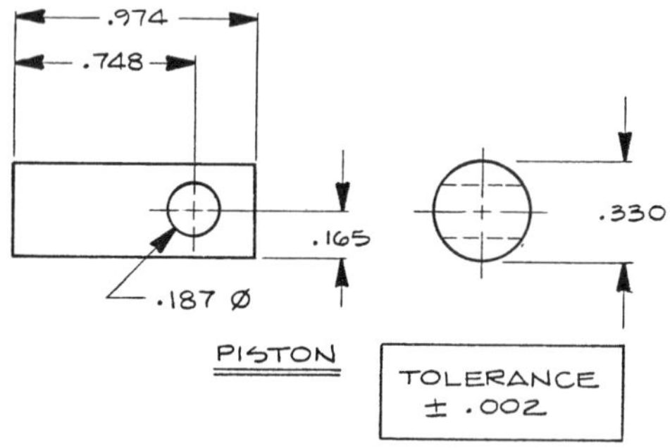

Figure 6

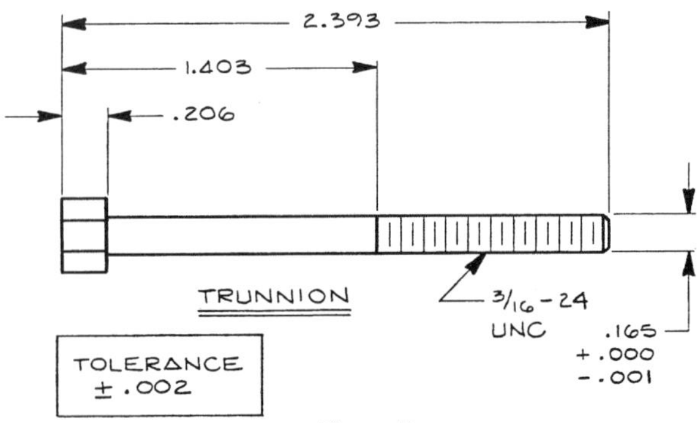

Figure 7

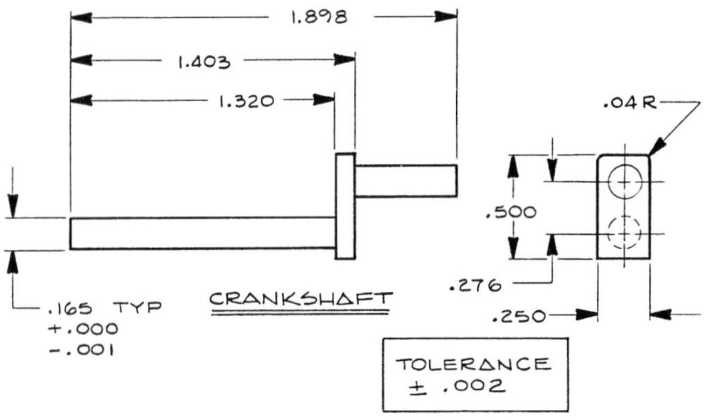

Figure 8

THE FLYWHEEL

Refer to Figure 5. Rough cut stock to length. Face one side in the lathe using a three-jaw chuck, or whatever method you prefer. After facing one side, turn the piece around, place the faced end on a parallel in the three-jaw chuck and machine the other side, first removing the parallel before starting the lathe. Once both sides are parallel, drill and ream a hole through the center of the flywheel. Now mount the flywheel on a mandrel, and place the mandrel between centers with the lathe in back gear. (A variety of methods may be used to accomplish this task.)

A second method to use if a mandrel is not available would be to use a three-jaw chuck, turning a piece of cold rolled steel to the proper diameter and then cutting it off to length. The remaining side could then be faced.

Drill and tap the hole for the setscrew. Now remove the flywheel and set it aside.

THE PISTON

Rough cut the required round stock to length (Figure 6). Face both ends of the piston and bring it to its correct length. Drill a hole in the side of the piston to allow for the crankshaft. For longevity, drill the hole oversized and press fit a brass bushing into the piston.

THE BOLT

Figure 7 gives the details for the bolt or trunnion. Rough cut a piece of hex CRS to 5/16 × 3". Rough turn to .165" + .001". Cut threads 3/16-24 UNC. Cut the threads on a lathe or use a die. I prefer cutting my own threads on the lathe for practice.

THE CRANKSHAFT

Caution: the layout of the crankshaft is extremely important, because sloppy work can cause a crankshaft that will bind. Cut rough stock to a rough length (Figure 8). Turn both ends to length or mill to length. Cut off excess areas on the band saw or with a hacksaw as shown. Using a four-jaw chuck, turn the short end of the crankshaft. Center your work with a dial indicator. Before cutting any metal on the shaft, support the shaft with the tailstock center. This will keep the crankshaft from bending when its diameter is reduced. Proceed with light cuts to avoid any bending of the crankshaft.

When the short end of the crankshaft is finished, open only two jaws and reverse the crankshaft in the chuck. Tighten the chuck and check the position of the crankshaft with the dial indicator. Make sure it is centered with no taper. Using the tailstock center for support, turn the remainder of the crankshaft to size.

THE NUT

Buy a nut to fit, or if you so desire, machine a nut to size using hex stock. Cut the nut to its desired width. Drill to the major diameter. Using a boring bar, cut 60° internal threads.

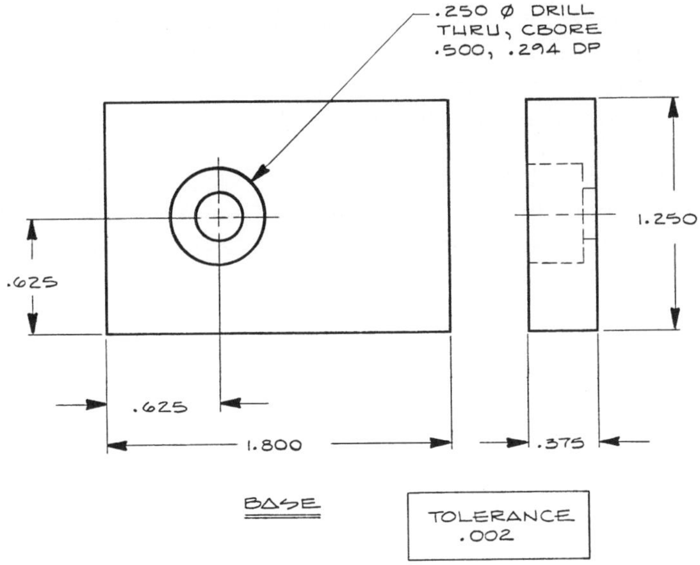

Figure 9

BILL OF MATERIALS

PART NO.	Q'TY	DESCRIPTION	MATERIAL	ROUGH SIZE
1	1	Spring	stainless steel	to suit
2	1	Nut	CRS	3/16-24 UNC
3	1	Cylinder	CRS	1/2 × 1/2 × 2 1/4"
4	1	Piston	aluminum	7/16" dia. × 1 3/16"
5	1	Crankshaft	CRS	1/2 × 1/4 × 1"
6	3	Washer	brass	3/8" OD × 3/16" ID × .60"
7	1	Trunnion	hex CRS	5/16 × 3"
8	1	Standard	CRS	1/2 × 1/2 × 2 13/16"
9	1	Flywheel	CRS	1" dia. × 9/16"
10	2	Stepped washer	brass	to suit
11	1	Base	aluminum	1 7/8 × 1 5/16 × 7/16"

THE BASE

Any interesting shape may be used for a base, as well as any available material as long as it is heavy enough. I've used the cross-section of a railroad tie, which I ground smooth on a surface grinder with excellent results. This part of the construction should bring out your creativity (Figure 9).

WASHERS

Washers may be purchased, but you will probably want to machine your own using a drill in the tailstock and a cutoff tool to cut the washer to length.

Once the parts have been machined to the proper tolerances, a variety of surface finishes can be applied. The following is an example of one. Place a pencil with the eraser down in the drill chuck. Keep the RPM low and bring the eraser in contact with the metal surface. Continue this procedure covering the entire surface. The result will be a beautiful, small radius circular finish. The finish is unique and pleasant to the eye.

Once all the pieces have been completed to meet your standards, oil all parts and begin to assemble the steam engine. Test your engine with air pressure to see if any parts are sticking. Sand down any burrs.

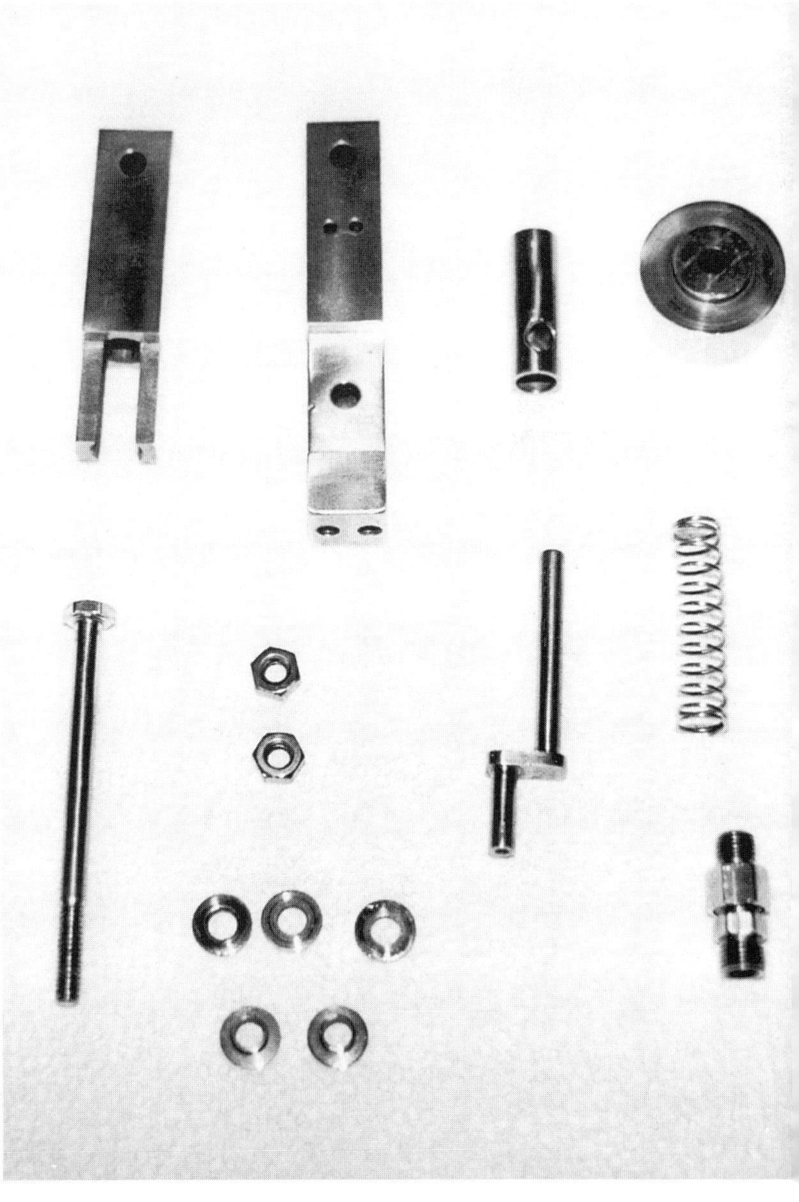

Cylinders and Pistons

for Two-stroke Cycle Model Engines

by George Genevro

Photos by Author

Successful model engines are a combination of parts that are not only well machined and fitted but are also designed to achieve a particular goal. Of this assemblage of parts, the cylinder and piston are most crucial to effective engine operation and provide great possibilities for experimentation. These important components, in rapid succession, must capture the fuel-air charge, compress it, contain it as it burns, help to convert the energy being released into linear and then rotary motion, dispose of waste gases, and reject waste heat to the cooling medium. In the case of some high performance model airplane engines, this process is repeated well over 25,000 times a minute, and in one .40 cubic inch displacement engine, the power output is an astounding 7.9 hp per cubic inch. To achieve this level of performance, however, requires use of processes not generally available to those of us who build engines as a hobby.

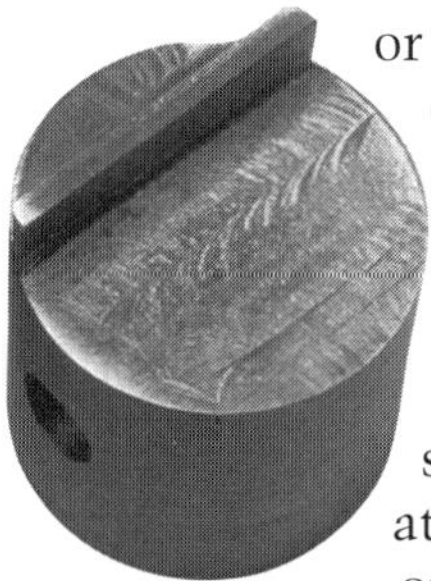

1 ***Cast iron pistons with straight baffles characteristic of cross flow scavenged two-stroke cycle engines.***

Therefore, here we will deal with materials and processes within reach of practically all model engine builders. By making judicious material selections for various components, and choosing appropriate machining processes, we can produce cylinder and piston combinations that are quite efficient as well as durable. In the interest of simplicity, we will deal only with single-cylinder engines, but those who choose to build multi-cylinder engines will find that very similar design and construction decisions and choices must be made.

CYLINDER ARRANGEMENTS

Several different basic cylinder designs have been successfully used in model engines, each of which has some advantages and drawbacks. Whether one is designing an engine or evaluating an existing design, such factors as the overall rigidity of the engine, how the cylinder, sleeve, and piston will be machined, the method of assembly, heat paths, gas passage shapes and sizes, and geometric relationships with the other parts of the engine must be considered. Whether the engine is a one-off type machined from bar stock or will be produced in quantity from sand or die-castings will also be a major consideration, largely because of the cost of the necessary tooling.

Model engines of the pre-World War II era almost all had cylinders that were separate from the crankcase, and were attached either with two or more screws or by means of an external thread on the lower part of the cylinder that screwed into an internal thread in the top of the crankcase. The threaded cylinder and crankcase arrangement was light and worked well on low and medium performance engines, but the threads were susceptible to damage during assembly and disassembly. Higher output engines such as the Hornet generally used four cap screws to secure the cylinder to the crankcase. Of the pre-World War II engines, the Ohlsson .23 and .60 were unique in that the cylinder was held in the crankcase by two spot-welds.

When a steel cylinder with the fins machined from the steel billet was used, the exhaust port, bypass, and, in

2 ***Ports milled in a cast iron sleeve. Note the marks for positioning the ports and the radii at the corners of the ports.***

3 ***The cylinder/crankcase casting in a faceplate fixture for machining the crankcase bore. Note the counterweight on the faceplate.***

4 ***Machining the exhaust opening in the cylinder/crankcase casting of a 1.8 cubic inch displacement engine.***

5 ***Machining the cylinder sleeve bore in a casting for 1.8 cubic inch twin exhaust engine. The lower surfaces of the crankcase lugs are the reference plane for this critical machining operation.***

the case of a piston-ported intake engine, the intake tube were generally attached by brass or silver brazing. The cylinder was then honed to the desired size and to achieve the roundness necessary for a good piston fit. The cylinder could have an integral head or could be open at both ends and have an aluminum head attached with machine screws. Since cylinders of this type are not symmetrical, either in construction or in heat flow patterns, they are sometimes prone to minute changes in shape while the engine is in operation – a condition that can adversely affect both internal friction and the piston/cylinder seal. Therefore, they are generally not suitable for high output engines. Cast iron cylinders with the bypass cover and the exhaust stack attached with cap screws were used in some pre-World War II engines. The rigidity and stability of the iron casting was a major factor in keeping the cylinder bore round during operation.

In the never-ending search for more power, some designers devised cylinder-crankcase combinations of several types that provided greater rigidity and were still quite light. In the case of the McCoy engines, the cylinder and the main part of the crankcase were cast as a unit, and the crankshaft and rear rotary valve were carried in housings attached to the crankcase with machine screws. A removable cylinder sleeve was used. This type of cylinder-crankcase combination could be sand-cast quite easily

6 ***An aluminum piston mounted in a fixture on a rotary table. A ball end mill is used to remove excess material from between the piston pin bosses.***

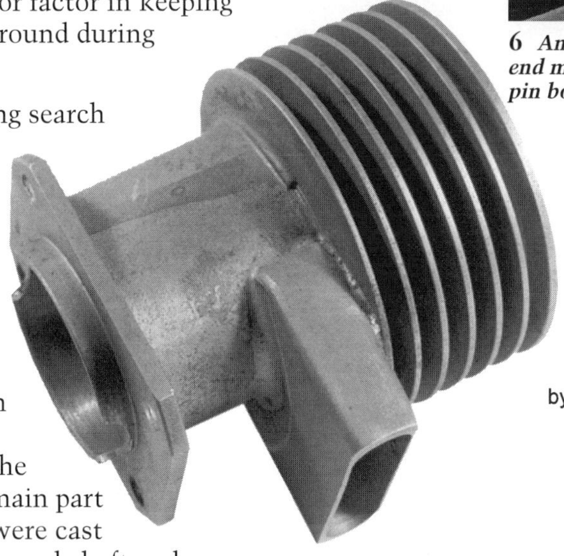

7 ***The cylinder for my Crusader .60 engine machined from leaded steel. The bypass cover and the exhaust port are silver brazed in place.***

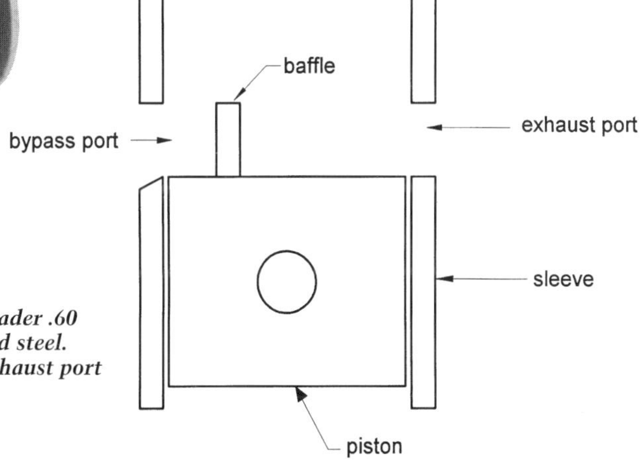

Relationship of piston, exhaust port, and bypass port at bottom dead center.

8 ***The Crusader .60 crankcase, head, and drum rotary valve backplate are machined from aluminum bar stock. The cylinder is of leaded steel with a lapped cast iron piston.***

9 ***The cylinder and crankcase for a 1.8 cubic inch twin exhaust engine mounted on a stub arbor to machine the area that will carry the breaker point assembly.***

and experimentation aimed at finding the ideal bypass shape and size required relatively inexpensive changes in the foundry patterns and coreboxes. Pressure die-casting of this type of cylinder-crankcase combination was considerably more expensive and complicated, particularly when the bypass cavity was complex in shape.

Another arrangement incorporates the crankshaft housing that may also carry the front rotary valve intake and the combined crankcase/cylinder up to the top of the exhaust port. This type of cylinder can be readily sand-cast, and can be die-cast if the designer is willing to accent a less-than-ideal bypass design unless complex tooling is used. My Titan .60 engine and a 1.8 cubic inch engine now being developed use this arrangement with a sand-cast crankcase. Maintaining the proper geometric relationship between the crankshaft and the cylinder bore during machining operations is relatively easy with this type of arrangement, and the assembly-disassembly procedure is quite simple. The cylinder sleeve can be machined from a steel billet with integral fins, or a steel or cast iron sleeve can be used with shrunk-on aluminum fins.

10 ***Aluminum cylinder fins to be shrink-fitted to the cylinder sleeve of a 1.8 cubic inch engine.***

11 ***The cast aluminum cylinder prior to installation of the sleeve. It will be attached to the crankcase with four machine screws or studs and nuts.***

CYLINDER AND SLEEVE MATERIALS

Over the years, there has been much disagreement about the best material to use for cylinder and sleeves. In automotive engines, cast iron has been the dominant material for cylinder blocks, sleeves, and cylinders for air-cooled engines. On the other hand, in aircraft engines ranging from about 65 horsepower to well over 3,000 horsepower, steel, in either heat-treated or nitrided form, has been used very successfully in both air-cooled and liquid-cooled power plants. Regardless of which material the model engine builder uses, compatibility with the piston or ring materials is of paramount importance. In older model engines using lapped pistons, the most popular combination was a cast iron piston fitted to a cast iron cylinder or sleeve. This provided a good seal and good durability when the engine was well made and properly broken in. Steel cylinders were used with aluminum pistons and cast iron rings in some versions of the Brown Junior and the Mighty Midget in the late 1930 period. Early Ohlsson .60 engines had hardened steel lapped

12 ***A leaded steel cylinder sleeve with ports machined and a piston with a single rectangular ring.***

pistons running in a steel cylinder, but cast iron pistons were used in later engines.

For lapped piston engines, probably the most favored combination is a fine-grain cast iron piston machined from Grade 40 (about 40,000 lbs. tensile strength) bar stock running in a leaded steel cylinder. Several leaded steels are available, the most popular of which is AISI 12L14. This is a low carbon (.14%) screw stock containing about .06% phosphorous, .30% sulfur, and about 1.0% manganese. The lead that is added, usually about .25%, does not alloy with the steel but is dispersed in very fine particles and greatly enhances machinability, while other characteristics such as weldability and tensile strength are somewhat impaired. For those who desire greater strength and hardenability, leaded chrome-molybdenum steels, such as AISI 41L40 and AISI 41L50, are available along with some trade-named steels of about .40 to .50% carbon content. All of these steels can be readily silver-brazed by both torch and furnace processes.

13 ***A fully machined cast aluminum crankcase/cylinder for a 1.8 cubic inch engine.***

CYLINDER CONSTRUCTION

If you are designing an engine, you probably have chosen a particular arrangement and combination of materials, and now is the time to determine how the cylinder will be made. Pistons will not be discussed at this point, but the piston material selected and whether rings will be used are major factors affecting the choice of porting procedures.

If you have chosen a design that involves machining threads at the bottom of the cylinder that will screw into the crankcase, steel – preferably a free-cutting leaded material such as AISI 12L14 – will be the only really practical cylinder material. Also, thin copper gaskets of exactly the correct thickness will be necessary for proper cylinder alignment unless a threaded locking ring is used. Making the cylinder with an integral head reduces the number of fasteners and gaskets, but honing and lapping operations cannot be done as easily.

A cylinder attached to the crankcase with two or four cap screws will be a bit heavier, as a rule, but offers a much broader choice of materials and machining procedures. If leaded steel is used, as in my Crusader .60 bar stock engine, it is relatively easy to cut the fins in the steel and, if care is taken, there is no weight penalty. The exhaust port, bypass, and the intake tube (if a piston ported inlet is used) must be fabricated, fitted and silver-brazed in place. Silver brazing is the preferred process, because an alloy with about 45% silver content will flow readily at about 1,250-1,275° F on a properly cleaned and fluxed steel surface and make a strong, neat joint.

When a cast aluminum cylinder is used, as in the case of the Hornet .60 and some modern engines, the head is generally removable and a cast iron, steel, or chromed brass sleeve is used. The sleeve man be a shrink fit, or may be retained by a flange at the top when the head is installed. There is ready access to the bypass, a feature of interest to those experimenting with the share of bypass cavities in an effort to improve gas flow.

When a crankcase/cylinder arrangement is chosen that incorporates the crankshaft housing and the cylinder up to the top of the exhaust port in one casting, the cylinder sleeve and fins can be made in one of several ways. The sleeve with integral fins may be machined from leaded steel bar stock – a procedure popular with builders of commercially-built engines of about .40 cubic inch displacement or less. When six screws are used to attach the cylinder head, two or three of the screws will extend through the fins and engage threads in the castings, as in the case of the Titan .60 and some commercially made engines. All combustion loads are carried by the screws that secure the cylinder sleeve to the casting.

14 ***An aluminum two-ring piston and a cast iron piston with curved baffle.***

The cylinder fins may also be machined from aluminum bar stock and shrunk onto the sleeve with about a .002" interference fit or with a close slip fit. In either case, there is a barrier to heat transfer, with the shrunk-on fins probably carrying heat away more effectively. The ports in the sleeve must be machined before the fins are installed. Then, great care must be taken to align the exhaust and bypass ports with the corresponding openings in the casting before the

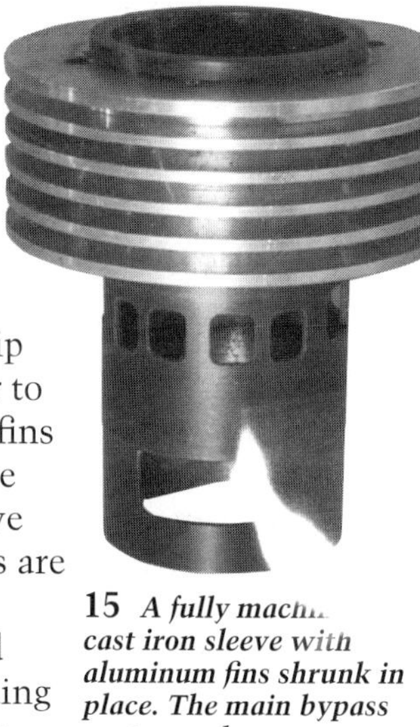

15 ***A fully machined cast iron sleeve with aluminum fins shrunk in place. The main bypass ports are shown.***

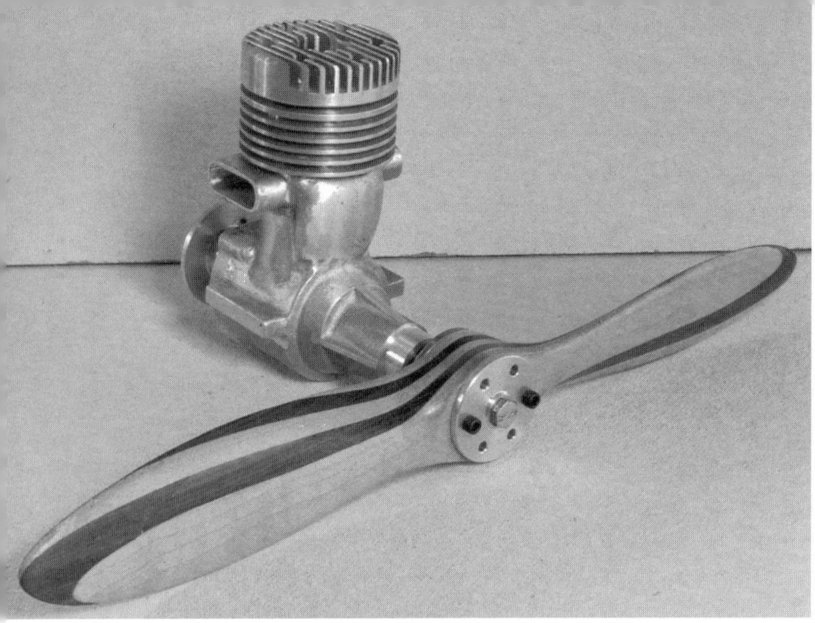

16 *Trial assembly of the 1.8 cubic inch Schnuerle ported engine.*

17 *The Ohlsson .60 engine of the 1940s era. The closed end steel cylinder with machined head fins is held in the crankcase by two spot-welds, one of which is visible at the front of the die-cast aluminum cylinder/crankcase.*

18 *The Hornet .60 of the 1940s era shows the cylinder mounted to the crankcase with four machine screws.*

19 *My Titan .60 cylinder/crankcase and steel sleeve with aluminum fins shrunk on.*

holes are drilled for attachment of the head to the cylinder-fin assembly and the cylinder-fin assembly to the crankcase. Unless a rotary table is available, the use of a drill jig will make this task much easier and will greatly reduce the possibility of error. The sleeve should be honed to size only after the fins are installed and the drilling operations are complete. Ideally, the sleeve should be honed about .0003" per inch of bore larger at the bottom to provide what is referred to as "choke."

While engines with the McCoy-type cylinder/crankcase can be machined from billet, the process can be quite tedious. Also, it is relatively difficult to machine a bypass cavity with a reasonably appropriate shape. Generally speaking, if the means for casting this type of engine are available or if a friendly local foundry will do the castings from loose patterns, it is worthwhile to take the time to make patterns and coreboxes. Regardless of whether the engine is made from castings or billet, a cylinder sleeve must be used, and one must choose whether to press or shrink it in place or machine it to a slip fit with a flange at the top and clamp it in place with the cylinder head. Even with a shrink fit, heat flow through the liner and cylinder interface is less than ideal, but compromises are inevitable in almost any design process.

For a lapped piston engine, the ports in the cylinder sleeve may be cut with an end mill, and the exact machining procedure will vary depending on whether a port with rounded or square ends is desired. When a piston with rings is used, the width of each port should be no more than 8% or 9% of the circumference of the OD of the sleeve. The ports should be machined with a small end mill and a radius should be left in each

20 ***An experimental bar stock engine with rear exhaust, twos bypasses, and curved piston baffle.***

21 *A leaded steel cylinder with partly machined upper bypass port.*

corner. While this may be a small impediment to gas flow, it will extend piston ring life. The lower edge of the bypass ports at the OD of the sleeve may be radiused to improve gas flow.

The vertical placement of the ports in the sleeve must be accurate since this determines the number of degrees of crankshaft rotation during which the exhaust and bypass ports are open. The difference in placement of the top of the exhaust and bypass ports also defines the "blowdown" period, and this can vary considerably depending on whether the engine is designed for a conventional or tuned exhaust system. A blowdown period of about 16° to 18° of crankshaft rotation is a good starting point for an engine not using a tuned exhaust.

In the case of a "three port" engine, the intake tube is attached to the cylinder, and the duration of the intake period is controlled by the placement of the port, its height, and the length of the piston skirt. The intake port is open an equal number of degrees of crankshaft rotation both before and after top dead center, and while this is not an ideal situation in terms of high power output, it is simple and reasonably effective. Since it is almost impossible to alter the intake port in the cylinder once the intake tube is attached, minor changes in intake timing can be made by altering the piston skirt. Remember that, regardless of the type of porting arrangement used, the only really accurate way to measure the duration and placement of these events is with a degree wheel during a trial assembly of the engine.

Honing is probably the most effective method of finishing the inside of the cylinder, regardless of whether the material used is steel or cast iron. The honing machines used to fit automotive piston pins are ideal for finishing cylinders since the operation is done in a flood of oil and a variety of stones ranging from very fine to coarse are usually available. With care, the bore can be slightly tapered (about .0002" to .0003" larger at the bottom) and can be cross-hatched as desired. Your local speed shop owner or community college automotive instructor might be coaxed into performing this task for you.

PISTON MATERIALS AND DESIGN CHARACTERISTICS

Pistons that will be lapped may be made of fine-grained gray cast iron, as previously noted, and every effort should be made to make the piston as light as possible without impairing its structural strength. Milling excess material from the inside of the piston will help, and it is most easily done with a holding fixture mounted on rotary table. Generally speaking, cast iron pistons are very seldom used in engines of over .950" bore since it is difficult to make the piston light enough to avoid severe vibration and roughness in operation.

Normalizing cast iron pistons after all the machining operations have been completed (except turning or grinding to the final OD) is well worth the extra effort. This is done by carefully and evenly heating the piston to about 1,450° F (about a dull red heat in normal light) and covering it so it will cool as slowly as possible. A small thermostatically controlled electric furnace, such as those used by potters, will work very well and can also be used for heat-treating piston rings. It is necessary to make a fixture to hold the piston while the

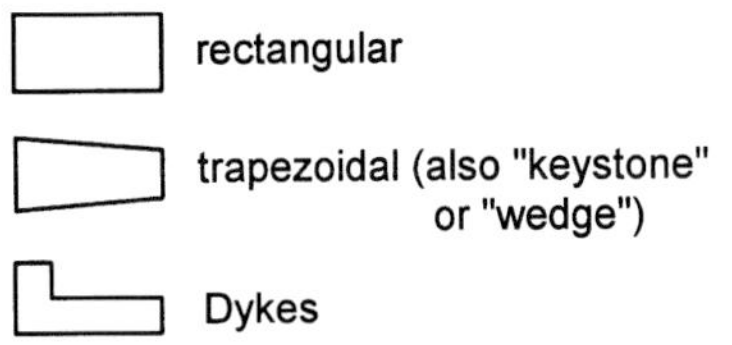

PISTON RING CROSS-SECTIONAL SHAPES

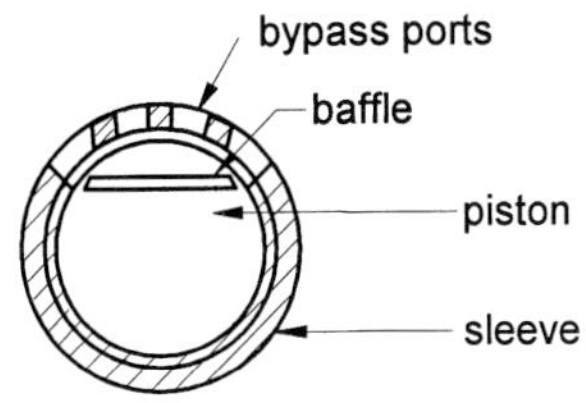

PISTON WITH STRAIGHT BAFFLE

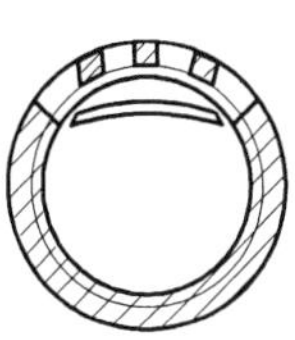

PISTON WITH CURVED BAFFLE

Computers in the Shop

by Roland Friestad

Photos and Graphics by Author

Welcome to a whole new area of machining in the home shop! I mean, of course, the use of computers as a tool in the machine shop and at the drafting table.

Computers have made a tremendous difference in the way we all live. They have an impact so far reaching and pervasive that it appears the Computer Age will join the Bronze Age, Iron Age, and Industrial Age in history books to be written in the future.

As a working toolmaker and engineer for about 30 years, I have had the opportunity to earn my living running machines operated from a ceiling mounted line shaft. I have also programmed and operated computerized machine tools with the ability to make cuts with a resolution of .00004". There is a lot to be said in favor of both of these technologies.

"What?," you ask. "A line shaft is not technology; it's a nostalgic remnant of days gone by." Now, don't be too hasty about that. Did you ever go to a museum and look at some of the really old pieces of preindustrial "manufactured" items? Things like tools, furniture, utensils, vases, or jewelry? I have always wondered about the time and skills it must have taken some unknown craftsman to create some of the incredibly complex and decorative items I have seen on exhibit. To that anonymous craftsman, line shaft operated machinery or computerized machine tools would have been equally marvelous. It's all in your point of view, or, as Arthur C. Clarke says, "Any sufficiently advanced technology is indistinguishable from magic."

A note about my qualifications might be in order here. I've been in the engineering "game" for nearly 30 years now and before that had an interest in machining and mechanical things going back as far as I can remember. My background includes an engineering degree and stints at several aircraft companies, one major computer manufacturer, two contract metal-stamping firms, quite a bit of consulting, and teaching at two vocational-technical schools. I am the second of three generations of engineers and machinists in the family, so it is kind of "in the blood." I get a lot of personal satisfaction out of designing something on the drafting board or computer and then being able to go to the shop and build it with my own hands. Computers have allowed me to design and build a lot more projects and to make them in ways that would be very difficult, if not impossible, using traditional manual methods. However, I started all of this B.C. (Before Computers) and, therefore, have done it both ways. Neither way is the answer to all problems, and computers are just another tool in the "kit" to be used where appropriate.

I am now owner of a small company (Cardinal Engineering) that manufactures metal stampings and builds and retrofits computerized machine tools. I also teach evening classes in CAD/CAM for a local school. Over the years, I have been personally involved in the design, manufacture and retrofit of over 100 CNC machines, ranging from the Unimat or Sherline size up to machining centers with table travels of 48" by 96". Much of the information and many sources for components for this column will come from these experiences, and I am glad to share them with you.

With a sense of healthy respect for what has been accomplished by machinists and craftsmen in the past, and with a sense of amazement at what the computer can do now and will do in the future, I would like to begin this series of articles about computers in the machine shop. The emphasis will be in applications for the home shop. If schools or businesses find some information of interest, that's all well and good. However, the main thrust will be toward home shop users. One aim is to try to reduce the attitude shown in Figure 1. I've been on both sides of that question.

Another aim is to show the reader that it is practical to use computers in the home shop. Computers may be used for designing, drawing, machining or other applications. They can also be much less expensive than you might think. I'll tell you how to find equipment, computers, and software. We will discuss how to program and use computers in the shop. Projects suitable for computerized machining will be presented. Building your own CNC machine tools will also be a topic of discussion. And last but not least, we will try to operate a readers' forum for questions, answers and discussions of related subjects. I welcome your input and will answer your questions through this column.

I am convinced that extensive use of computers in the home shop is "just around the corner" and hope that some of the information and experience I have acquired working with these useful and sometimes frustrating machines will be of help to you as fellow home shop enthusiasts.

We'll start with some basic definitions here. For those of you who already know something about CNC and CAD, please bear with us. I will attempt to provide a mix of beginning and advanced topics for discussion.

What is the meaning of CAD, CNC, CAD/CAM, and some of the other initials that are seen in technical publications? These are acronyms standing for various technical processes using computers.

CAD stands for Computer Aided Design or Computer Aided Drafting. This technology is used to create drawings using the computer as a tool to draw lines, arcs, text, cross hatching, etc. The information placed by the operator in the computer is used to create a "database" that is a recording of instructions used by the computer to recreate the drawing on demand. A drawing on paper is really just a bunch of pencil or ink marks on a sheet of material. Similarly, a drawing on a CAD system is really just a bunch of computer instructions. The real significance of both paper drawings and CAD drawings is the ideas and concepts they represent.

A common technique in manual drafting is the construction of large scale layouts of small mechanisms to check for fits and operation or the making of reduced scale drawings of large projects that are impractical to draw full size. CAD drawings are usually made full scale in the computer and then reduced or enlarged when a paper copy or "hardcopy" of the computer information is sent to a plotter or printer. If nothing else, this method sure eliminates a lot of mundane errors in scale factor calculations. For the model builders among us, and I am one of them, it is really a convenience to make a full scale drawing and then have the computer calculate all of the dimensions automatically for any scale of model.

CNC stands for Computer Numerical Control and is the application of computers to run

Figure 1

machine tools. Other special machines are also run by computer, but CNC has become well accepted as pertaining primarily to machine tools. CIM is another "buzz word" that stands for Computer Integrated Manufacturing and is generally understood to include CNC and Robotics linked together using computers to manage "workcells" or even entire factories.

CAD/CAM is another commonly heard term and is generally understood to stand for Computer Aided Design (Drafting)/Computer Aided Machining (Manufacturing). This refers to the use of some kind of software to generate information directly from a CAD drawing to be used in a CNC machine tool. Don't ask me why this isn't called CAD/CNC. Maybe it didn't have the right sound or something. I have always had the feeling the government has an office somewhere in a cubbyhole in Washington (the Pentagon, maybe) staffed with people who dream up this stuff.

More years ago than I care to recall, I had an assignment to machine some contoured die sections used in tools to stamp wing ribs and bulkheads for small aircraft. This was B.C. (Before Computers). Because of the continually changing contours of these kinds of shapes, it was necessary to make enlarged layouts on special Mylar film, measure all of the coordinates and make up tables of X and Y values and then manually crank the X and Y values into the lead screws of a milling machine. This was also before DRO (Digital Readout), so backlash was a problem whenever the direction of cut reversed. The project was completed with many hours of manual filing to remove the cusps left by the milling cutters along the side of the contoured parts.

Since that time, I have had the opportunity to do similar work using an optical line tracer to follow an ink line as well as a CNC mill to

Either there's a bug in our CAD system or we've just designed a 400 metric-ton sportscar.

Figure 2

follow a programmed line. Generally speaking, the CNC machine produces a better part. CNC can also produce multiple identical parts either at the same time or later on, no mean trick with manual methods.

Another neat trick with CNC is the ability to program a shape and then, using "cutter compensation," machine any given amount on either side of the programmed contour. This is a tremendous way to make matched dies or molds. It is quite easy, once a program is written for a particular part, to instruct a CNC machine to make a mirror image or "left-hand" version, using the same program previously used to machine the "right-hand" part.

Yet another useful feature is the ability to machine parts that are scaled up or down from the given programmed size by a certain percentage. This is a great way to make patterns for castings and allow for shrink factors.

Now a few comments about reliability and affordability. Computers have a certain reputation for being expensive and unreliable in operation. The semi-official "technobabble" term for this unreliability is "flaky" or "flakiness." However, this reputation is only somewhat well deserved. It is frustrating to have a motel reservation fouled up or to have a credit card bill that is impossible to straighten out because "the computer did it." This is usually not the fault of a computer, but of the operators and programmers who are using the computer. When you use and program the computer yourself, you will find it to be much like any other tool in your toolbox. Properly applied, computers are a boon to the machinist and designer. Improperly used, they can be just as much of a problem as you make them out to be. Figure 2 illustrates this feature quite well.

One of the first questions that comes up is cost. Most people think computerized machine tools are out of the reach of the home or "hobby" machinist. Cost is a relative thing. There are a lot of non-machinists who feel that having a lathe or a mill at home in the basement or garage is also an extravagance. Most of the readers of this publication would probably not agree. As mentioned previously, I'll show you how to get started using CAD and CNC for a lot less than you might think.

There are several reasons why a home machinist might want to get involved in CAD and CNC. Chief among these is probably the same reason you are reading this article – just an interest in doing certain kinds of technical things as a hobby. Second might be the capability to more easily machine complex parts, particularly if multiples are wanted. One of the nice things about a hobby is the lack of need to justify it (except possibly to the wife). I always tell my wife it "keeps me off the streets and out of the bars." She has been known to respond with an equally ancient saw, "The only difference between men and boys is the price of their toys."

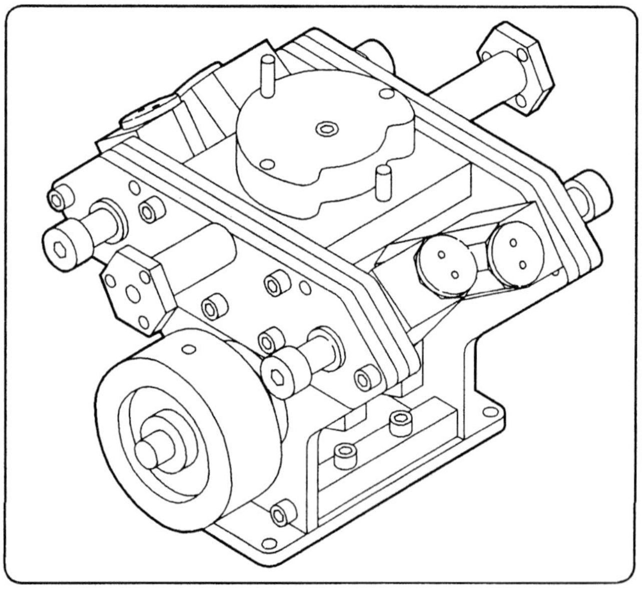

Figure 3
ISOMETRIC DRAWING OF ENGINE ASSEMBLY

One of the implications of using the capabilities of CAD and CNC is the freedom to design and build more complicated parts and projects. I am working on a small four-cylinder steam/compressed air engine that is a modification of the one appearing on page 86 of the book *Steam and Stirling*, edited by William Fitt. An illustration of the completed engine is shown in Figure 3. All drawings were done using CAD and programmed for CNC construction. I am building enough parts for two or three engines and should have them done very soon. You will notice the nice blended corners on all of the parts. Easy to do with CNC. I'll keep you posted if anyone is interested.

The basic principles of CNC (Computer Numerical Control) operations is a natural place to begin our discussions regarding the use of computers in the shop. These fundamental concepts relate to two basic requirements of any control system, whether it is a CNC machine tool, automotive steering system, or the wall thermostat at home. Also covered will be an introduction to the basic programming commands for CNC.

COMMAND AND RESPONSE

The first concept is that of the "command." The command in the case of the thermostat or the automobile is the setting of the pointer or the turning of the steering wheel. The command in the case of the CNC system is a "program." This program normally consists of lines or "blocks" of instructions that are executed one after another by the CNC controller.

The second concept is that of the "response." In the case of the thermostat calling for more heat, the contacts close, sending an electrical signal to the gas valve (assuming a gas forced-air furnace) which opens and sends gas to the burner. The heat generated raises the temperature of the air in the furnace which causes a switch to close and turn on the circulating blower. The resulting circulation of warm air will raise the temperature of the room, causing the heat sensing element of the thermostat to open the contacts and shut off the flow of gas to the furnace. As the temperature then falls, the contacts will again close and the cycle will repeat itself. Most of us are familiar with this type of control and do not normally give it much thought.

The automobile steering system is similar in many respects. The driver, wanting to make a turn, rotates the steering wheel. The linkage to the wheels changes the angles, causing the car to turn. When the proper amount of turn has been achieved, the driver returns the wheel to neutral, cancelling the turn. When another turn is desired, the cycle repeats.

A CNC machine tool operates by reading a "block" or "instruction" into the computer, calculating the amount of motion and timing required and sending the proper signals to the drive motors which move the machine. As the computer sends the signals to the drive motors, it mathematically keeps track of the signals going out and compares the progress to the original instruction or block. When the original block and the calculated motion match, the computer reads the next block and the cycle repeats itself.

HOW DOES THE MACHINE TOOL MOVE?

Most CNC systems use some form of motor, usually electric, to rotate the lead screws that you are all familiar with. Some larger industrial machines use hydraulic actuators, mostly because of their ability to deliver higher horsepower in a smaller physical package.

The electric motors most commonly used in CNC machines are either "stepper" or "servo" motors. Each type has its own characteristic advantages and disadvantages.

PRINCIPLE OF STEPPER MOTOR OPERATION

Stepper motors are a relatively new development and can be thought of as a sort of "rotary solenoid." A solenoid is a familiar electrical device that converts an electrical signal into a push or pull and has two positions, much like a light switch which is either off or on (Figure 1).

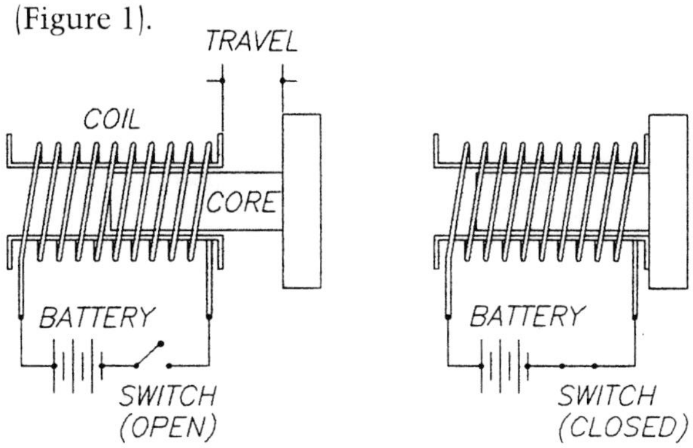

Figure 1 – SOLONOID OPERATION

The solenoid uses the magnetic field generated by the electrical signal to attract a core of steel, producing motion. The force produced is proportional to the strength of the electrical signal applied (volts times amps). The amount of motion occurring depends upon the mechanical design of the solenoid.

A stepper motor can be visualized as an extension of the solenoid as shown in Figure 2.

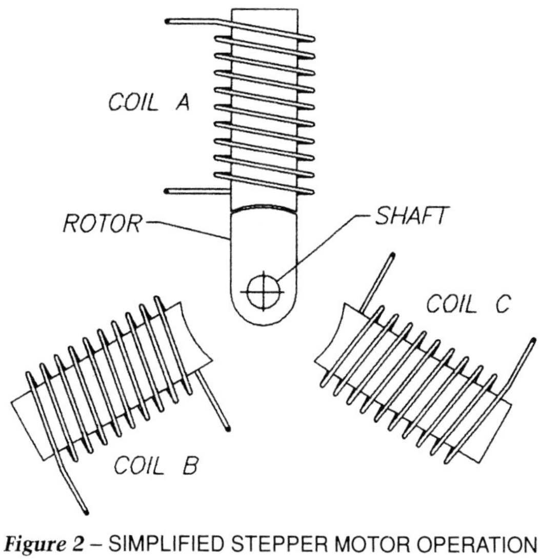

Figure 2 – SIMPLIFIED STEPPER MOTOR OPERATION

The rotor is mounted on a shaft and is free to rotate in either direction. If a signal is applied to coil "A," then the rotor will be attracted to that coil and will align itself as shown in the drawing. If the signal is then applied at coil "B" and removed from coil "A," the rotor turns 120° counterclockwise (a term that precedes digital watches) and aligns itself with coil "B." Similarly, if the electrical signal had been applied to coil "C" instead of "B," the rotor would have turned 120° clockwise.

As in the solenoid example, the torque is proportional to the strength of the electrical signal, and the rotation depends upon the physical design of the rotor and coils.

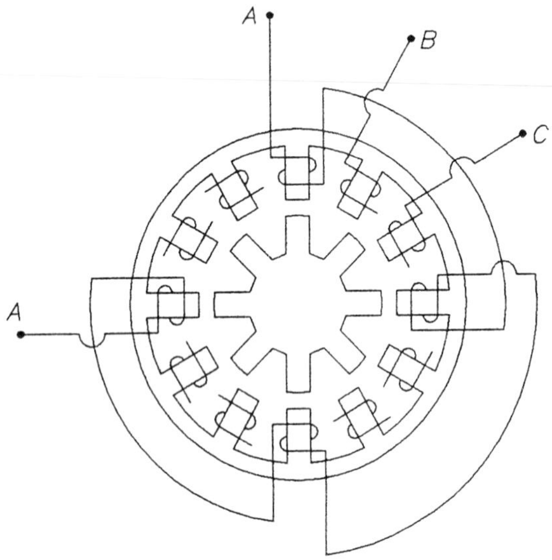

Figure 3 – TYPICAL STEPPER CONSTRUCTION

It is obvious that a switching pattern of A-B-C-A-B-C...will produce counterclockwise rotation and a pattern of A-C-B-A-C-B...will produce clockwise rotation. One of the things that computers are very good at doing is switching things on and off in patterns. Consequently, running a stepper motor is "duck soup" for a computer.

The simplified stepper motor shown has a step angle of 120° while one of the most common step angles found in CNC applications is 1.8° (200 steps per revolution). These stepper motors are more complex but operate on the same principle. A more typical stepper motor rotor and stator design is shown in Figure 3. This motor design would have a step angle of 15°. An extension of this approach is used to produce stepper motors with 1.8° step angles.

Because the computer sends out the proper signals to the stepper motor and presumes that the motor responds correctly, this type of control system is known as an "open loop" system. That is to say, if the motor should skip one or more steps, the computer has no way of knowing about the lost motion.

PRINCIPLE OF SERVO MOTOR OPERATION

By taking the same sketch as our original simplified stepper motor and applying the electrical power in a different manner as shown in Figure 4, a simplified servo motor is obtained.

The principle of operation of a servo motor is not too much different from a stepper motor except in the manner in which power is applied to the motor windings. The electrical connections of Figure 4 can be viewed as a rotary variable resistor with three taps arranged around the periphery. When the wiper shaft is turned through 360°, the electrical signal at each tap varies from zero to its maximum value as the wiper passes over each tap location. If these signals are applied to the corresponding coils on the motor, the magnetic force in each motor coil will vary in the same way. The rotor will be attracted towards the coil with the strongest signal and as the signal in one coil decreases and the adjacent coil increases, the rotor will smoothly follow the position of the wiper.

In reality, the electrical signals applied to the coils are varied electronically by a servo amplifier. This amplifier has the capability to vary the relationship of the electrical signals as commanded by the computer. The servo motor therefore follows the signals sent out by the computer.

One problem with the servo motor as described is that the positional accuracy of the motor is not adequate for controlling a machine tool without further refinements. These refinements are applied by connecting a position encoder to the motor shaft. This encoder sends a signal to the computer, reporting the angular position of the servo motor. The signal sent by the position encoder is used to vary the power level and phase angles of the servo amplifier. The result is that the servo motor can follow the computer signals very accurately. This type of system is known as a "closed loop" system because the computer receives independent confirmation of servo motor motion via the encoder.

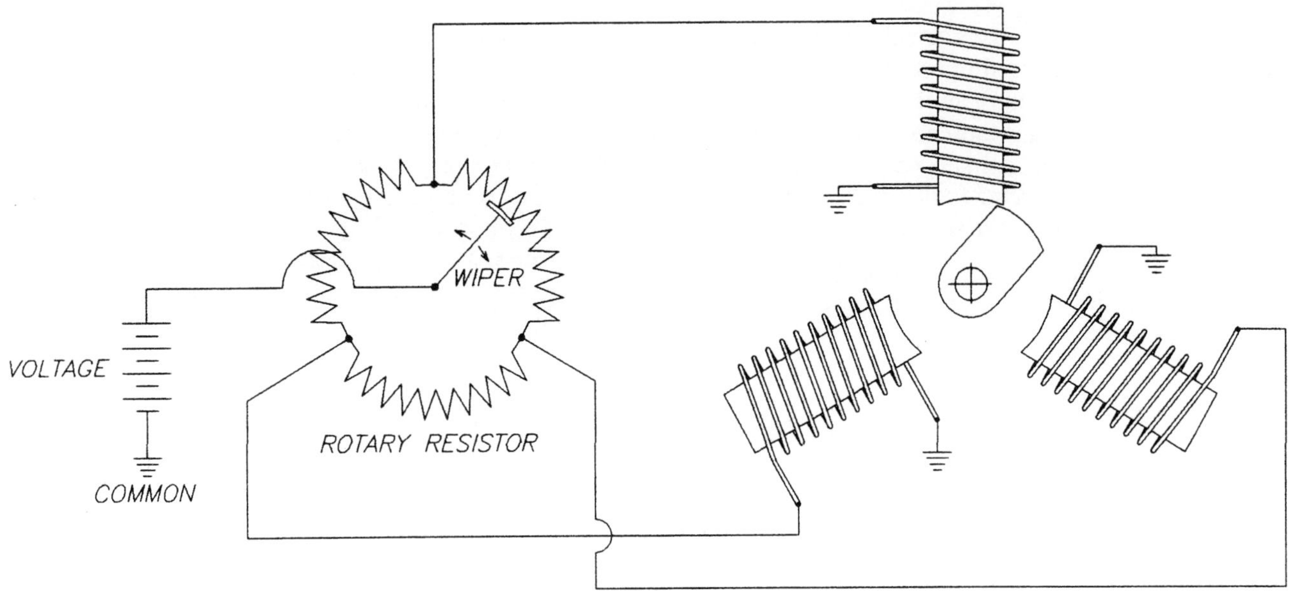

Figure 4 – SIMPLIFIED SERVO MOTOR OPERATION

OPEN LOOP VERSUS CLOSED LOOP CONTROLS

Either open loop or closed loop systems can be used to drive a CNC machine tool. There are a few major differences between the two systems.

A common misconception regarding closed and open loop systems is that the closed loop system is inherently more accurate than the open loop system. This is not true. While the closed loop system does have the capability to feed back position information to the computer via the encoder, this does not automatically assure a more accurate system. There are a great many parameters that enter into the accuracy of a CNC control system, and the fact of being open loop or closed loop is actually only a small aspect of total system accuracy.

A brief comparison of the major features of open loop and closed loop systems is summarized in the accompanying table. This comparison is a guideline only, and many "gray areas" exist where the advantages or features of the two systems are not clear-cut.

FEATURE	OPEN LOOP	CLOSED LOOP
Accuracy	Same	Same
Speed	Slower	Faster
Complexity	Simpler	More complex
Cost	Less expensive	More expensive
Maintenance	Easier to maintain	Harder to maintain
Safety	Runaway impossible	Possible runaway
Service life	Stable over time	Drifts with age

The previous examples and the comparison are necessarily simplified, and some aspects of control theory will be discussed in greater detail later in this series. However, the basic concepts are quite easy to understand if they are reduced to only the essentials.

CNC COMMANDS

Next, how does one talk to the computer and get it to send the correct signals to the stepper or servo motors?

The CNC commands are in the form of a computer language. There are several such languages, but the most common one in use in the USA is known as RS-274-D. Not a very catchy name, but then engineering types are not known for their "Madison Avenue" skills. This language is more commonly known as "G-Codes" or "G and M Codes" and will be referred to as "G-Codes" in these articles. This language is particularly well suited to machining operations because it is closely related to the way technical drawings are made. Much of the information on a drawing is presented in a form that is readily converted into a CNC program using G-Codes.

In the G-Code system there are many commands, but the four most fundamental ones will be discussed briefly here. These fundamental commands are G0, G1, G2, and G3. In some programming manuals, these commands are also called "preparatory codes" or "words."

All programming using G-Codes is based on the assumption that the computer keeps track or "knows" where it is and the G-Codes tell it where to go to get to the next location and what path to follow to get there.

THE G0 COMMAND

The G0 command is known as "Rapid Traverse" and is used when a rapid move from one location to another is desired. This command is not used for actual cutting operations because the motion is normally much too fast for cutters to function. This command is issued followed by X, Y, and Z data and the machine will move to the location specified. A typical command line or "Block" looks like this:

G0X1.0Y2.5Z-0.45

This block will rapidly move to the location X=1.0, Y=2.5, and Z=0.45. For those readers familiar with CNC programming, the assumption is made that execution of the program is done in the absolute coordinate system. Absolute and relative coordinates will be discussed at a later time.

THE G1 COMMAND

The G1 command is very similar at first glance to the G0 command and might appear in a block as shown below.

G1X1.0Y2.5Z-0.45F30

The block is identical except for substituting G1 for G0 and also the addition of F30 at the end of the block. The G1 command is known as "Linear Interpolation." The motion produced by execution of this block would be a straight line from wherever the machine is currently located to the X, Y, and Z coordinates programmed in the block. The "F30" at the end of the block indicates the feed rate or speed of motion along the line. F30 means a feed rate of 3.0 units per minute. If programming is in inches, the feed rate would be three inches per minute. If programming is in metric, the feed rate would be three millimeters per minute. G1 is used when making a cut because the speed of travel can be programmed to match the work and tools. G0 is used for moving rapidly from point to point.

THE G2 AND G3 COMMANDS

The other two basic G-Codes are G2 which is used for clockwise arcs or circles and G3 which is used for counterclockwise arcs or circles. This motion is known as "Clockwise Circular Interpolation" or "Counterclockwise Circular Interpolation" respectively. A typical block would look like this:

G2X1.0Y2.5I1.0J1.25F30

This block would produce a clockwise arc with its end point at the X and Y coordinates shown and its center point at the I and J coordinates shown. I is the X coordinate of the arc center and J is the Y coordinate of the arc center. The feed rate would again be three units per minute.

HOW COMPLEX SHAPES ARE MACHINED

These four commands or G-Codes can be used to machine any finished part desired. There are many more G-Codes that are used for special purposes or to simplify the programming process, but they are all special extensions of these four basic commands. The most complicated parts and shapes can always be broken down to straight lines and circular arcs and the challenge of the programmer is to do this in an orderly and accurate manner. The numerical information included in each block of a program must be calculated from the drawing manually or may be extracted from a CAD drawing semi-automatically.

TWO EXAMPLES OF LOW-COST CNC SYSTEMS

Photo 5 shows a benchtop milling machine that has been retrofitted with stepper motors and ball screws to make a small CNC milling machine. This was originally done to provide a very low cost CNC mill for use in the classroom.

Photo 6 shows a Bridgeport milling machine with a similar retrofit applied. The details of these retrofits and also lathes and special machines will be discussed in subsequent articles, including drawings and sources for components.

These comments about the operation of CNC controllers and fundamentals of programming methods will be expanded upon as we progress. In the next segment, I will present drawings and programs for a simple machine clamping tool that is easy to make with CNC and difficult to make manually. This will get the reader thinking about how his own projects might use CNC. The freedom to design more complex parts, knowing that they are easy to machine using CNC, allows one to make more efficient use of materials and to create a better finished part. Another great advantage of CNC is the ability to make another part just like the first one at any later time. More of this will be covered next time, along with some of the basic mathematics used in manual programming. In addition, there will be a review (the first) of one of several commercially available low-cost CNC systems. Questions submitted by readers will be answered as the topics are introduced. Suggested areas of special interest are welcomed.

MORE ON PROGRAMMING PRINCIPLES

This is a good time to continue with further discussion of some of these basic concepts. This discussion will again concentrate on the RS-274-D language also known popularly as "G-Codes." Other languages will be discussed in later articles if there is enough interest.

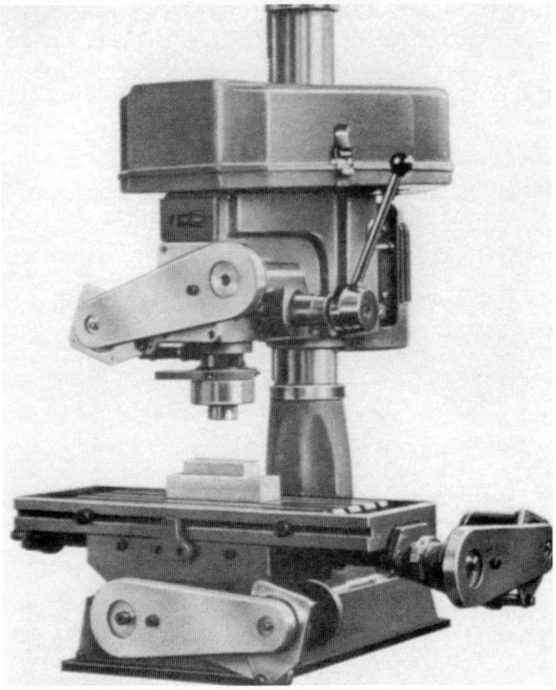

5 ***Benchtop milling machine with CNC retrofit.***

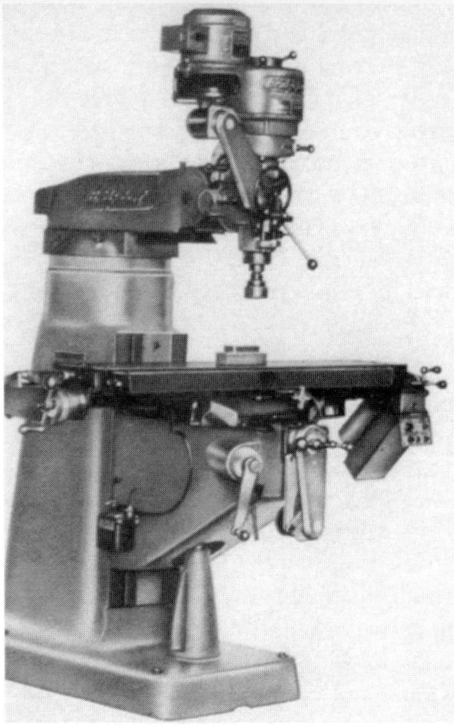

6 ***Bridgeport milling machine with CNC retrofit.***

THE CARTESIAN COORDINATE SYSTEM

A clear understanding of CNC programming requires agreement on how the numerical information will be presented to the computer. All of us who have an interest in machine shop operations already have an intuitive understanding of what is meant by coordinates. We may not use the same terminology, but a good grasp of the basic principles of a coordinate system is necessary to reading any kind of drawing, doing layout work, and making measurements.

The coordinate system we all use without really thinking about it is known as the Cartesian coordinate system and was invented by the French mathematician Rene DesCartes. This occurred about the time Isaac Newton developed the basic principles that would eventually become many of the fundamental rules of physics, mathematics and engineering. It is appropriate to note here that there are many different coordinate systems used in mathematics, physics, and engineering, but for our purposes, the only one discussed at this time will be the Cartesian rectangular coordinate system.

The two-dimensional Cartesian coordinate system we are most familiar with consists of two lines, or axes, which are at right angles (90°) to each other and meet at a common point known as the origin. Refer to Figure 1, and note the direction of increasing X and Y.

Note the points labeled "P1" and "P2" in Figure 1. Looking first at point P1, we see that if we measure the distance from point P1 to the Y axis, parallel to the X axis, we obtain a value of 4.0 units. This value is known in mathematics as the abcissa of the point. In shop or common engineering terminology, this value would be denoted as X=4.0.

Measuring the distance of P1 to the X axis, parallel to the Y axis, we get a value of 2.5 units. This value is known mathematically as the ordinate and, in shop terms, Y=2.5. The two measurements, abcissa and ordinate, taken together are known as the coordinates of the point.

In a similar way, the coordinates of point P2 are X=6.0 and Y=5.0. It is common usage to specify these points in a shorthand way as (6.0, 5.0) where it is understood that the X value or abcissa is always first and the Y value or ordinate is always second. When several points are specified, it is customary to precede the coordinates with the identification of the

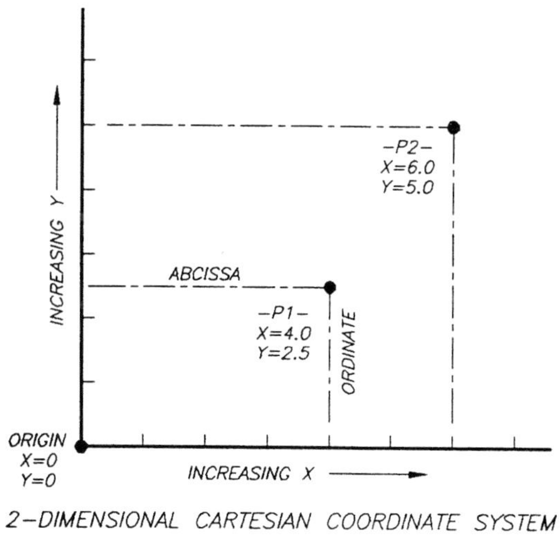

Figure 1

point to avoid confusion. In this way, the coordinates for points P1 and P2 would be written as P1(4.0, 2.5) and P2(6.0, 5.0).

This explanation may seem to be somewhat unnecessary because, as mentioned earlier, we all have an intuitive understanding of this system from reading blueprints and sketches. However, when we intend to tell a machine how to move automatically, it is important to be very particular to define these terms in a specific and agreed-upon way and to use a common terminology.

The advantage of this system is that any point on a given plane may be uniquely specified by the coordinates of that point. In other words, given the coordinates of the point, we know exactly where that point and only that point is in the plane. This is important for programming CNC machines if we want to be certain the machine travels the desired path.

Now, let us extend the two-dimensional Cartesian system to include a third dimension, the Z axis. Refer to Figure 2, and notice that it consists of our original system with an added axis – the Z-axis – which also begins at the origin and extends at right angles to both the X-axis and Y-axis. In common terms, this might be said to be in the "up" direction.

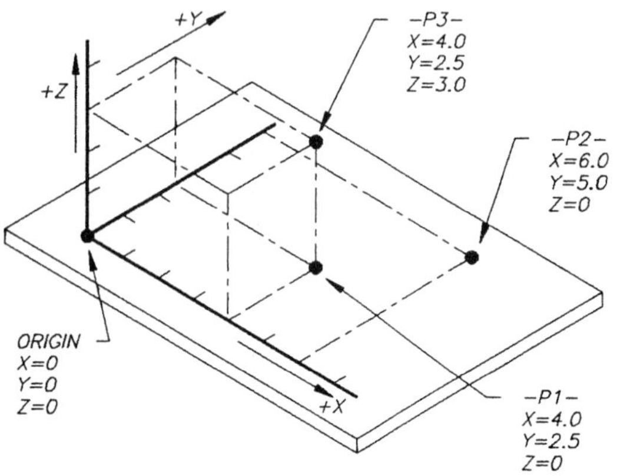

3-DIMENSIONAL CARTESIAN COORDINATE SYSTEM

Figure 2

Note our original points P1 and P2, which now have coordinates of P1(4.0, 2.5, 0) and P2(6.0, 5.0, 0) respectively. As before, the first value is always X, the second always Y and the new third value is always Z. A new point shown in Figure 2 as point P3 would have coordinates of P3(4.0, 2.5, 3.0). In this way, we may now specify a point in space by its coordinates in such a way that there is no mathematical confusion or misunderstanding about exactly where the point is located. This is essential for shopwork with CNC programs and CAD drawings. It is also essential for everyday non-computer shopwork, but because of everyday usage, we normally don't think in this mathematical sort of way. The coordinate system we have just defined is known also as a "right hand" system.

RELATIONSHIP OF COORDINATES TO CNC PROGRAMS

In the last column, four basic G-Codes or commands were introduced. It is essential that the relationship of these commands be understood as they relate to the coordinate system we have discussed above. For those of you who know something about CNC programming, the following examples presume the use of the absolute coordinate system.

THE RAPID TRAVERSE CODE - G0

Let us assume we are looking down at the table of a milling machine, standing in the normal operator's position. The tool, a drill, is located at the point shown as START POINT (0.5, 1.0) and it is desired to move to the location labeled END POINT (1.5, 2.5) as rapidly as possible. Note that we discuss the motion of the tool relative to the workpiece and not the machine slide relative to the machine frame. The "Block" of code that will cause this motion to happen is:

G0X1.5Y2.5

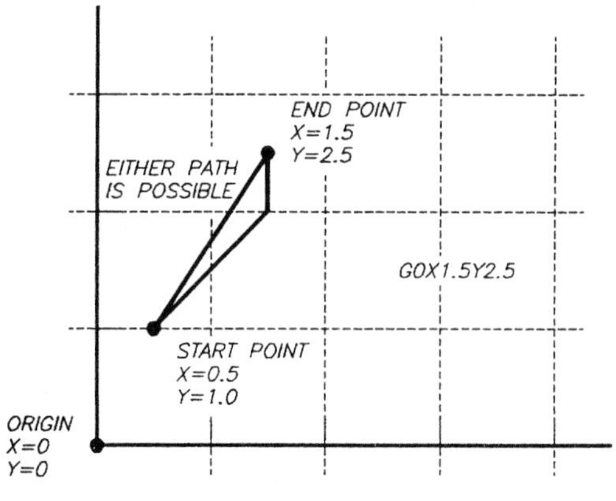

G0 COMMAND – RAPID TRAVERSE

Figure 3

Since we want to move as rapidly as possible, the G0 code is used (Rapid Traverse). Note also in Figure 3 that it doesn't really make any difference what path is taken to get from the start point to the end point since the goal is to move as rapidly as possible. It might be argued, since a straight line is the shortest distance between two points, that a straight line would also take the shortest time to travel. This is theoretically true, but in practice, many other factors come into play, particularly the mass or weight of the machine table, the size and capabilities of tl e motors, and the ability of the computer to process the required information. While most CNC controllers will approximate a straight line between two points in an X-Y rapid traverse, the main guarantee is that the system will get to the end point as rapidly as possible without regard to the path that is taken. It is obvious, therefore, that G0 should be used only when moving from point to point and should never be used to do cutting.

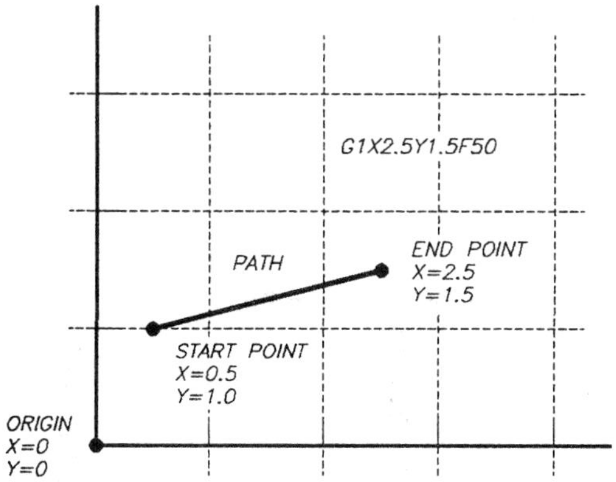

G1 COMMAND – STRAIGHT LINE

Figure 4

THE LINEAR CODE - G1

Now, assuming the same milling machine table, and referring to Figure 4, we are positioned at the same start point (0.5, 1.0), and wish to machine a straight line from the start point to the end point (2.5, 1.5). The block of code required to do this is:

G1X2.5Y1.5F50

The motors of the milling machine table will cause the tool to move in a straight line between the start point and end point at a speed of 5.0 units per minute. The CNC controller will send suitable signals to the motors so the actual motion of the tool will follow the path as closely as possible within the capabilities of the machine tool. Because of the digital nature of computer control systems, the actual path consists of a series of very small steps rather than a true straight line between the two points. These small steps can be seen in parts made on the average milling machine but normally are not detrimental. When necessary, these steps can be made so small as to be undetectable. Since the path is accurately predictable and the speed is closely controlled, it is evident that G1 is well suited for machining operations in a straight line path.

THE CLOCKWISE ARC - G2

When a clockwise arc is desired, it may be programmed using the G2 code in a block that specifies path and the center of the arc (Figure 5).

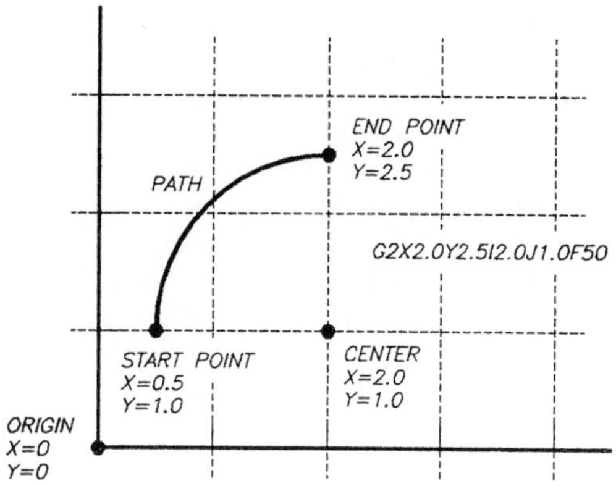

G2 COMMAND – CLOCKWISE ARC

Figure 5

The start point is still at the same location (0.5, 1.0) and it is desired to machine a clockwise arc to the end point (2.0,2.5). The center of the arc is at the point CENTER (2.0, 1.0). The block that will produce this move is:

G2X2.0Y2.5I2.0J1.0F50

The actual path taken by the cutter will again be a series of small steps that will approximate the theoretical path within the capabilities of the machine tool. The letter I is used to designate the X value or abcissa of the center of the arc and the letter J is used to designate the Y value or ordinate of the arc center. Again, since G2 produces an accurately predictable path and speed, it is suitable for cutting.

THE COUNTERCLOCKWISE ARC - G3

When a counterclockwise arc is desired, it may be programmed using the G3 code in a way similar to that used with the G2 code (Figure 6).

The start point is still at the same location (0.5, 1.0) and it is desired to machine a counter-clockwise arc to the same end point (2.0,2.5). The center of the arc is still at the same center point center (2.0, 1.0). The block which will produce this move is:

G3X2.0Y2.5I2.0J1.0F50

Notice that this block is identical to that used in Figure 5 except for the change to G3 from G2. Also compare Figure 5 and Figure 6 to see the difference between the paths the machine tool will travel.

A NOTE ABOUT UNITS OF MEASUREMENT

Note that we have not specified whether we are working in feet, inches, millimeters, meters, or furlongs. Most commonly, the units used in CNC programming are inches and millimeters.

HOW DOES THE COMPUTER UNDERSTAND G-CODES?

The blocks of code given in the examples above

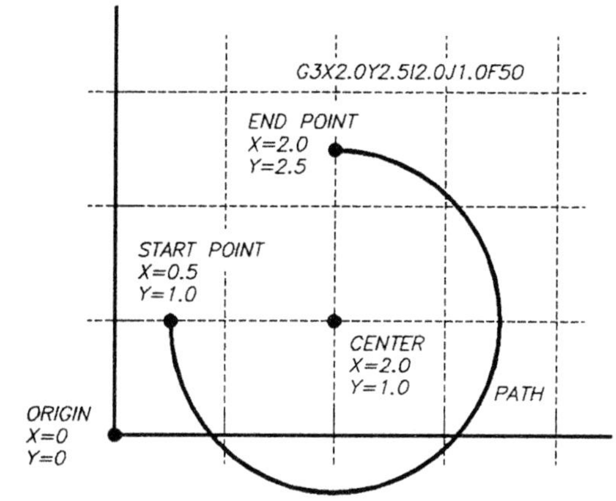

G3 COMMAND – COUNTER-CLOCKWISE ARC

Figure 6

are quite easy for a person to understand, but the computer actually does a great deal of work to convert these blocks into actual motion of the machine table. If we take the examples above, and assign typical values to the table motions, the computer is sending pulses to the motors at a rate of between 100 and 1,000 times per second. That is very fast when we think in terms of mechanical devices, but it is just coasting for electronics. Again, assuming some typical values, even at 1,000 pulses to the motors per second, a typical CNC controller would still have time to do *several thousand* calculations *between each pulse.* This ability to do calculations so rapidly is what makes modern CNC possible.

The CNC controller has the program for the part to be machined in its memory, but most of the memory is used for other purposes. Most of the memory of the typical CNC controller is taken up by another program (or several programs) whose function is to read the blocks of G-code, interpret them, make motion calculations, and send the signals out to the electronics that apply power to the drive motors. This additional program or programs is often referred to as the "Operating System." If you are familiar with personal computers, you will recognize the term as being similar to DOS which is short for "Disk Operating System," a similar program which supervises the functions of personal computers.

These operating systems are very complicated to write and work with, and are usually written by computer experts. They normally use completely different languages that are extremely difficult to understand without a great deal of study and practice. We will not spend a lot of time discussing operating systems in these articles except to recognize them as supervisory programs that allow our computers or CNC controllers to understand what we tell them to do.

SUMMARY OF COORDINATES AND G-CODES

The Cartesian coordinate system allows us a way to specify the location of points in a plane or in space using a series of numbers known as the coordinates of the point. Since the natural language of computers is numbers, this coordinate scheme is easy for a computer to understand. The CNC machining and programming systems that have evolved over the last 35 years use the principles of Cartesian coordinates because they seem to be the easiest to understand by both the human operators and the computers.

The G-code or RS-274-D programming language uses the rectangular Cartesian coordinate system as a means to allow the programmer to communicate information to the CNC controllers. The CNC controllers contain a supervisory program or operating system that interprets the programmer's G-codes and produces the desired movements of the machine tool.

Most CNC controllers use inches or millimeters as units of measurement.

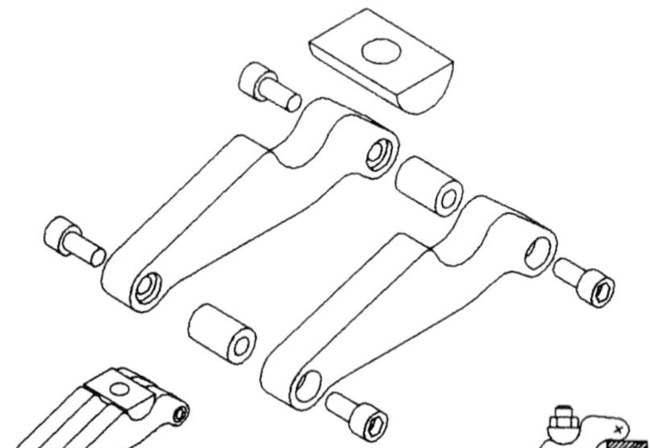

VIEWS OF MULTI-HEIGHT WORK CLAMP

Figure 7

A REAL EXAMPLE OF A CNC PROGRAM

It is about time to introduce a real part to be made using CNC. The drawings shown in Figure 7 are an isometric view, exploded view, and two side views of a clamping device for holding parts down to the table of a machine tool. The nice thing about this design is that it will work with heights from zero to 1.75" and virtually eliminates the searching for shims and loose parts associated with the usual strap-type machine clamp.

Another feature of this design is the use of lines and arcs that blend nicely together. This makes for a good appearance and is much more sound from a structural point of view, but it makes a part that is very hard to make using manual machining methods.

Shown in Figure 8 is the detail drawing of the side plate for this clamp. The spacer bushings are 3/8" OD by 0.525" long and are tapped 10-32 through. The rocker is 3/4" OD stock by 1.150" long, cross drilled in the center to fit a 3/8" stud and milled down to a half round.

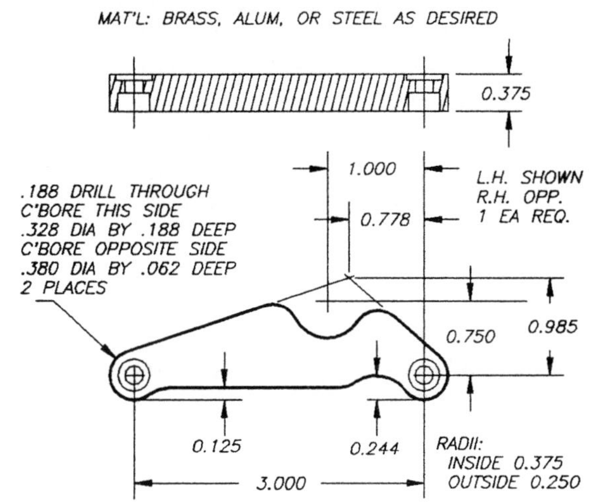

DETAIL OF CLAMP SIDE PLATE

Figure 8

The machining of the clamp side plates would involve first drilling and counterboring a blank of material with the .188" diameter holes and counterbores specified on the drawing. Machine as many blanks as required for the number of clamps desired. Two side plates, two spacer bushings, one rocker and four 10-32 x 3/8" socket head cap screws are required for each clamp assembly.

FIXTURING THE SIDE PLATE FOR MACHINING

One of the requirements for any machining operation, manual or CNC, is holding the workpiece, or fixturing. Fixturing is something like heat treating or plating in that there are many guidelines but no substitute for experience. One very workable method to machine these side plates is to make a simple fixture that can be clamped in the jaws of a vise or directly to the machine table (Figure 9).

The purpose of the two tapped holes is to hold the pre-drilled and counterbored blanks to the fixture for machining. The two holes marked 1/2" DP are reamed to 0.500" for a nice fit to a 1/2" dowel pin. These dowel pin holes are used to realign the fixture if it is necessary to remove it and replace it on the table at a later date. The procedure for this alignment will be discussed in another article. The CNC program is written with the origin of the coordinate system set to the center of the left tapped hole.

Once the fixture has been properly aligned with the machine table, the side plate blanks are clamped in the fixture. It would be a good idea to make a couple of extra spacer bushings and drill them out to fit the body of a 10-32 socket head cap screw. These spacers can be used to block the side plates up from the surface of the fixture so the milling cutter does not cut into the fixture when machining the contour (Figure 10).

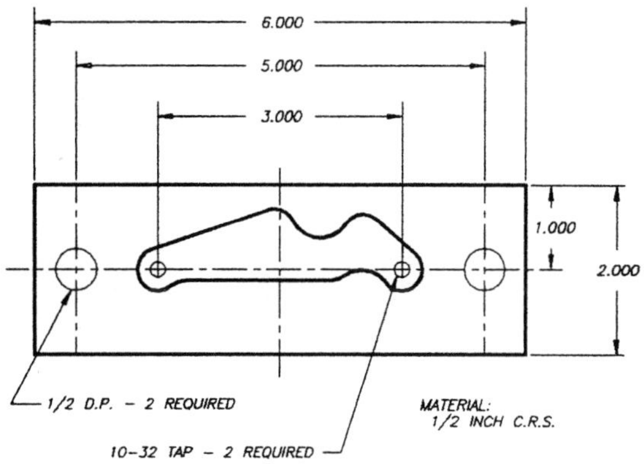

FIXTURE FOR MACHINING CLAMP SIDE PLATE

Figure 9

THE CNC PROGRAM TO MACHINE THE PART

Listed below is the CNC program used to machine the side plates. This program was written using numerical information derived directly from the CAD drawing. The CAD system used was AutoCAD, but many other CAD systems would also work for this. The program could also be written by calculating the required numbers manually, and we will show how to do these calculations in a future column.

```
; PROGRAM TO MACHINE CONTOUR
; OF P/N 2110 CLAMP SIDE PLATE
;
; CUTTER DIAMETER 1/2 INCH
;
; SET Z = 0 PLANE AND HOME TO BE
; 1/2 INCH ABOVE WORK
;
N5G0G90X-1.0Y0.0T1M6
N10G1F50
N15G1F300
N20G42
N25G0X4.0354Y0.1615
```

```
N30G0Z-0.45
N35G1Z-1.0
N105G1X3.6856Y-0.2522
N110G1X2.6825Y0.5958
N115G3X2.3126Y0.5429I2.5211J0.4049
N120G2X1.6652Y0.581I2.0J0.75
N125G3X1.3649Y0.7061I1.4421J0.4683
N130G1X-0.0772Y0.2378
N135G3X0.15Y-0.2I0.0J0.0
N140G2X0.375Y-0.125I0.375J-0.5
N145G1X2.1369
N150G3X2.2839Y-0.0773I2.1369J0.125
N155G2X2.8017Y-0.1523I2.5043J-0.3807
N160G3X3.1614Y0.1909I3.0J0.0
N165G1X2.1582Y1.0389
N170G1X2.5079Y1.4527
N175G0Z0.0
N180G40
N185G0X-1.0Y0.0
; END OF CONTOUR PROGRAM
```

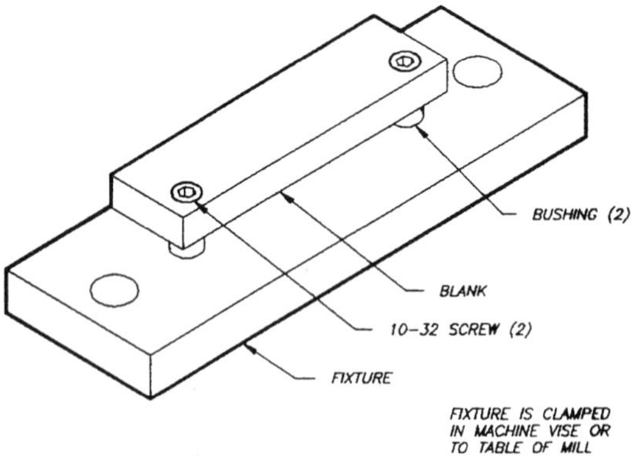

VIEW OF BLANK MOUNTED IN FIXTURE

Figure 10

COMMENTS ON CNC PROGRAM FOR SIDE PLATES

The program listed above has some features we have not discussed yet, and it would be suitable to touch on them at this time.

First, note that some of the blocks of the program are comments for informational purposes. These comments or remarks are preceded by a semicolon (;). The particular controller this program was written for will ignore any information in a block that follows a semicolon. This is a useful way to document your programs for future use. Not all controllers have this feature.

Second, all of the blocks are preceded by the letter "N" followed by a number. This is known as a "Sequence Number" and is used primarily by the programmer to keep track of his place in the program when "debugging" or "smoke testing" a program to see if it will run correctly. The sequence numbers are also used by the CNC controller in more complex programs. Some CNC controllers require sequence numbers, and with others they are optional.

Third, block N5 has a "G90" code and also a "T1M6" code. The G90 is the code that sets a CNC controller to use the absolute Cartesian coordinate system. The T1M6 code tells the controller to use tool number 1 and to load the information regarding that tool (length and diameter) into the controller.

Fourth, block N20 has a "G42" code. This tells the CNC controller to offset the programmed path by an amount equal to one-half of the diameter of the cutter. Reading down to block N180 you will see a "G40" code. This code turns off the automatic offset feature. This ability to perform automatic offset of the tool is known as "Cutter Compensation." Some CNC controllers have this feature and others do not.

RUNNING THE CNC PROGRAM

The CNC program is loaded into the controller and given a "dry run" with no tool in the spindle or with the quill raised clear of the work far enough so cutting does not occur. The program is executed a single block at a time and visually confirmed to produce the desired motions. The tool is replaced or the quill lowered just enough so the cutter will "scribe" the outline of the finished part around the contour. About .010" deep is enough. This will produce a line that can be visually checked for correctness. It will also allow the part to be removed and rough sawn using a band saw to remove most of the material. While such parts can be machined "from the solid," generally band saw blades are much cheaper than milling cutters.

Once the program has been checked out, and the parts have been scribed and sawed, the parts can be replaced, the tool lowered to make a full depth of cut and the parts machined. When they are completed, they are deburred, using your favorite method, and assembled. The resulting product is as shown in Photo 11.

INFORMATION ABOUT THIS PROGRAM AND PART

While writing this article has taken quite a bit of time, the actual drawing of the part, writing the program, and machining the side plates did not take too long to accomplish. The times are listed below. Times to do manual machining operations such as making the bushings, rocker, and fixture and drilling the blanks are not included. The object of this is to give you an idea of how long you can expect to spend to get this kind of part machined.

1. Make drawing in CAD system – 7 minutes

2. Create program from CAD drawing – 20 minutes

3. Machine side plates (each piece) – 11 minutes

The real advantage of this kind of machining is the ability to easily create parts that are more complex and sophisticated than could ever be done manually and to make identical multiples or replacement parts. As a toolmaker, I have spent many happy hours in the long-forgotten past making parts with the kind of tangent intersections on this sample part. Even more hours were then spent making a matching die plate with closely controlled clearances. With the cutter compensation feature noted above, it is only necessary to use the same basic program to tell the controller to machine inside the line instead of outside the line. It is very feasible, for instance, to build a tool steel production die to stamp out parts with a shape like that shown in Figure 8 and to do it in less than one day using CNC.

A REVIEW OF COMMERCIALLY AVAILABLE EQUIPMENT

There are several companies manufacturing CNC retrofit systems that include the brackets, motors, controller, software, etc., required to convert a machine. There are only a few companies manufacturing a CNC controller that is reasonably priced and has the industrial G-code (RS-274-D) language as a part of the controller. Controllers made by Centroid Corporation, State College, Pennsylvania, are among the best available for the money. They currently manufacture their CNC-4 controller which sells for $3,495. Centroid also manufactures a special controller for Cardinal Engineering which sells for $2,500. Information on both of these controllers is available by writing to Cardinal Engineering, Route 1, Box 163-3, Cameron, IL 61423. If you include a check or money order for $15, Cardinal Engineering will send a video tape showing several different milling machines, lathes, and special machines that have been built using the Cardinal/Centroid CNC controllers. Cardinal will also include a floppy disk with the CAD (AutoCAD) drawings for this clamp and the CNC program.

FEATURES OF CARDINAL/CENTROID CONTROLLERS

The Cardinal Engineering and Centroid CNC controllers can be programmed in either inches or millimeters, and in absolute or incremental (relative) coordinates. They have several special G-codes which are known in industry as "canned cycles." Canned cycles greatly simplify the drilling of holes and milling of round and rectangular pockets. Programs can also be written using what are known as "loops" or "macroinstructions" – features that are incredibly useful and allow such things as engraving pre-defined alphabets, machining of multiple parts on a fixture, pattern drilling, and other powerful programming techniques.

These controllers also can be set to any reasonable lead screw and gear ratio combination and can automatically compensate for backlash in the lead screws. While this backlash compensating feature is not a substitute for ball screws or some other zero-backlash system on milling machines, it is very adequate for most lathe operations. There are features in the CNC-4 version that allow the operation of punches, robots, tool changers, and other complex functions in addition to machining. Cardinal Engineering is working on a CNC hydraulic press and a CNC flame/plasma cutter using these controllers. Another project in the works that will use this controller is a combination ram/wire EDM machine with CNC control. Stay tuned!

One of the most popular questions I get is "Can I retrofit my __________?" The answer is generally "YES," but because of the many thousands of different models and variations of machine tools, it is virtually impossible to have your particular machine covered. I have done quite a few retrofits of the small mill-drill tabletop machines, Bridgeports and Bridgeport-clones, and benchtop lathes from the Unimat size to 12 x 36" size. One of the things I have discovered is that even the same models from the same manufacturer are not always identical. This makes retrofits and kits a real challenge. Consequently, one of the major subjects of this column will be the principles involved in doing a retrofit so the reader can adapt these to his own particular machine. The next column will start detailed plans for conversion of a benchtop mill-drill (see Photo 5), and the following will start with plans for a benchtop lathe of the 10 x 24" to 12 x 36" size. The actual machines featured will be from Enco, but the Jet, Grizzly, Wilke, J&L, Rutland, KBC, or other similar machines would work equally well. Plans for a retrofit to the Atlas/Sears type of lathe will also be presented a little further down the line if there is interest.

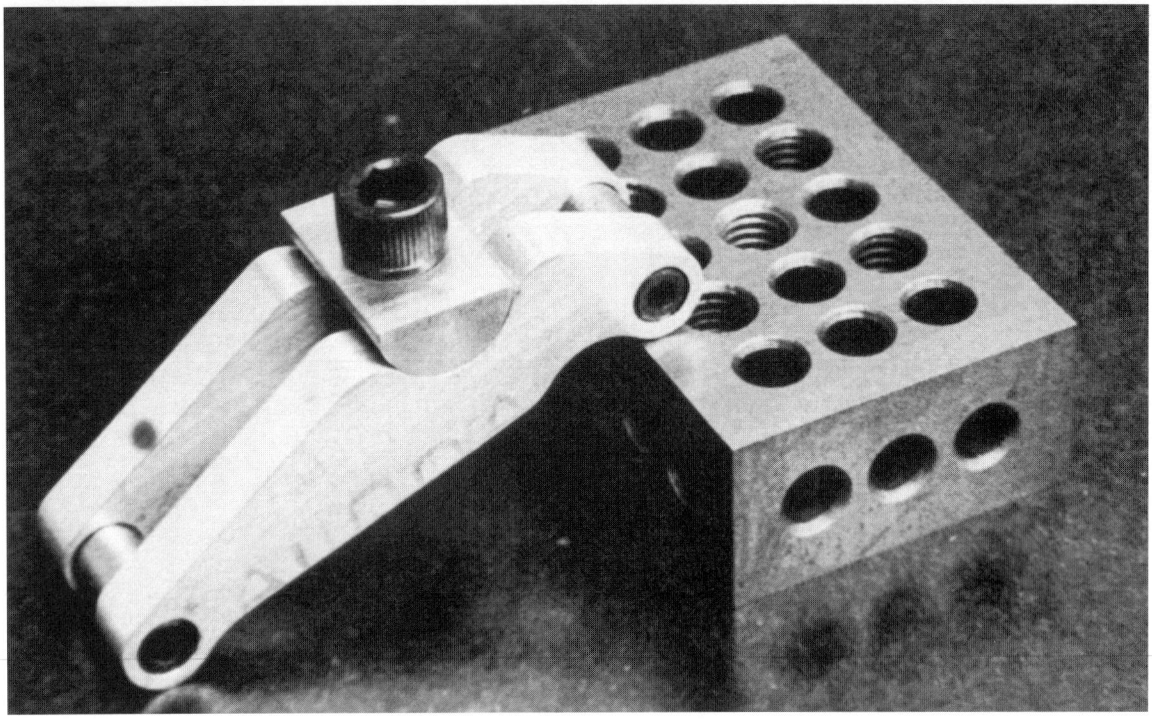

11

Another popular question is "Can I use a CAD program to draw and then machine parts?" This, like CNC, is a very involved topic. The general answer is yes, but there are so many different CAD systems available for several different computers that a specific "here's how you do it for a Whiz-Bang 250-Z with Bozo-CAD" is not possible. I'll start with some of the most popular systems and we'll go from there.

Convert Your Mill-drill to CNC

by Roland Friestad
Photos by Author

Most of what has been discussed in this column up to now has been in the nature of background information. It's about time to begin a first CNC retrofit project. The target machine for this effort is the Enco model 105-1100 mill-drill. A photograph of this machine with the retrofit installed appeared in the November/December '89 issue of this magazine. Many of the inquiries from readers indicated an interest in this class of machine tool as a CNC retrofit, so it was selected to be the first example in this "roll your own" series.

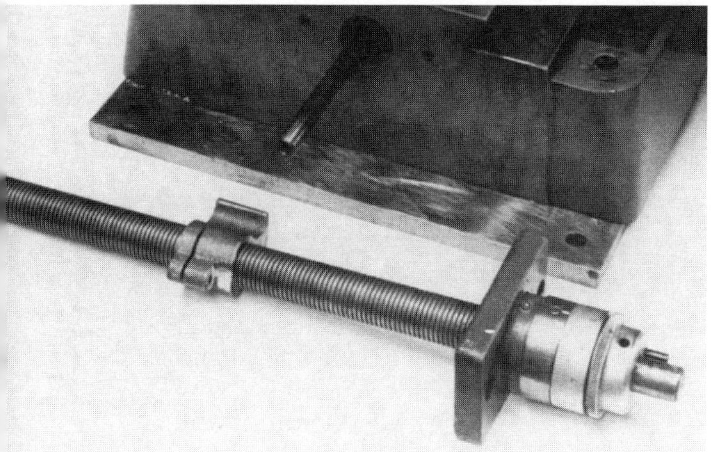

1

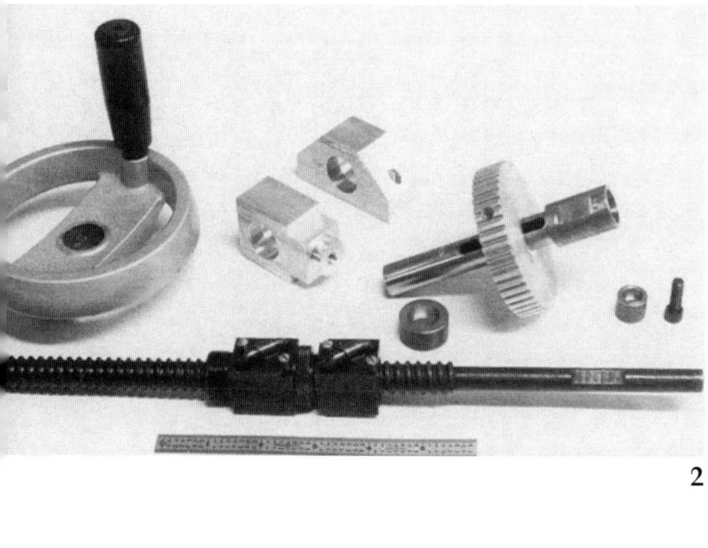

2

The model 105-1100 is the least expensive of six models that Enco carries. The price of this machine in the current catalog is $1,037 but does not include shipping. Shipping would add anywhere from $150 to $300 (guestimated) depending upon where you are located. Since they have 14 locations nationwide, it might be feasible to pick up a mill at one of their warehouses and save shipping charges.

At the time of this writing, we here at Cardinal Engineering have converted 85 of these machines to CNC for various users across the US and Canada. This machine was selected because it is heavy enough and has a large enough motor to make respectable milling cuts in steel using end mills up to 3/4" diameter (although I recommend only 1/2" as a maximum). It also has an R-8 spindle taper which means standard collets and tooling made for the Bridgeport series of milling machines is readily available from hundreds of supply houses. Finally, the machine has a worm gear driven quill feed with a sturdy spring return mechanism which simplifies the application of a motor to the quill.

Many suppliers import these machines, generally from Taiwan. I was told by Enco that there are several manufacturers in Taiwan, all working from the same basic design drawings supplied by a central government engineering group. This would explain the close similarity in appearance between machines advertised by Wilke, Grizzly, J&L, Kabaco, and others. It would also tend to mean that just about any of these machines could be readily reworked to CNC.

THE NEED FOR CONTROLLING BACKLASH

One of the first considerations, after the machine tool has been selected, is whether to replace the standard lead screws with a "zero backlash" screw such as a ball screw. Everyone who has worked with machine tools is familiar with the backlash problem and the need to always approach the work from the same direction if using the graduated dials of the machine. This is normally not a problem when doing manual machining, particularly if a DRO (digital readout) is installed on the machine tool. The familiar admonition against "climb milling," unless it can't be avoided, also derives from this backlash problem.

Most machinists know that, given a rigid machine with little or no backlash, climb milling can produce better finishes. This procedure can provide longer tool life and requires less power than conventional milling. It is much easier to write CNC programs when one does not have to account for backlash. An anti-backlash lead screw is the usual answer to this problem, and the most popular type in CNC applications is the ball screw. Ball screws and other anti-backlash mechanisms will be dis-cussed in detail in another column. The conversion shown in the following photographs uses ball screws for the X and Y axes. The Z axis (quill) has a spring return mechanism that can be set up so a ball screw is not required for most work.

Photo 1 is a view of the front of the milling machine showing the standard lead screw removed from the mill. Note that the casting on the right end of the lead screw contains several internal parts including two sets of ball thrust bearings. These thrust bearings are apparently installed by clamping the lead screw itself into the flanged casting using the calibrated dial mechanism as the other part of the clamp. A hole is apparently drilled through the dial mechanism and the lead screw shaft on assembly, and the whole thing is retained with a roll pin (spring pin). This method is inexpensive, but it has absolutely no means of adjusting any preload or backlash that may occur.

Note also in Photo 1 the device in the center of the threaded portion of the lead screw. This is a cast bronze nut with an integral mount that is retained by a single socket head cap screw. The X axis nut is similar except it is retained by two cap screws. These bronze nuts are replaced by parts machined to accept the nuts of the ball screws.

Shown in Photo 2 are the parts that replace the standard lead screw assembly. The ball thrust bearings and flanged castings are used without modification. The handle shown is bored out to fit the new adjustable thrust bearing assembly. The calibrated dial assembly shown in Photo 1 is used, also, after removing the roll pin and tapping a new setscrew hole. The two parts shown just above the ball screw nut are the new mounting brackets that replace the bronze nuts. These parts are threaded to accept the end of the ball nut.

Shown in Photo 3 are the pulleys, mounting bracket, motor mount, belt, and belt guard as used on both the X and Y axes of the mill. Both axes use the same parts, with the exception of the ball nut mounting brackets and the ball screws themselves. The arm that supports the motor mount and belt guards is designed to clamp over the machined portion of the existing flanged casting that retains the ball thrust

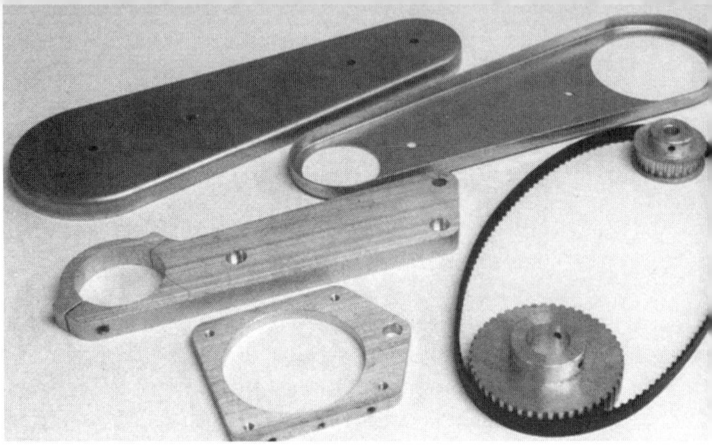

3

4

5

6

bearings. This arm has a cap similar to that on the "big end" of a connecting rod. This allows the arm to be rotated to any location. The pulleys and belt are of the toothed type so there is no lost motion between the drive motor and the lead screws. These toothed or "timing" belts are commonly used on larger industrial CNC machines because of their ability to transmit relatively large amounts of power with no lost motion or backlash. They are also very common as the drive belt for the valve train in some automotive engines for the same reasons.

The parts for the Y axis are shown partially installed in Photo 4. Also shown in the same photograph is the drive motor that will be installed on this particular milling machine. The motor shown has a torque rating of 325 oz-in (ounce-inches) and is adequate for many applications. A higher torque motor of 475 oz-in is also used on this machine when necessary. These motors can achieve a side thrust of 1000 to 2000 pounds using the drive system shown. This is more than adequate to machine even tool steel. These motors can achieve table feed rates of up to 15" per minute using an R-L motor driver and well over 30" per minute with a chopper-type motor driver. Most machining is done in the range of 2 to 5" per minute for steel and 5 to 20" per minute for aluminum and brass. These speeds that can be achieved are more dependent upon the work and the cutting tools than upon the capabilities of the drive motors.

Photo 5 shows the completed installation of the Y axis. Note that the handwheel is fully operative even though the machine can also be driven by a computer. This allows a machine to function as both a manual mill and a fully CNC controlled machine. Photo 6 shows the completed Y axis and a partially installed X axis. As you can see, the installation is quite similar in both cases. Because of the design of the mounting arm and motor mount, they can be installed as either "left-hand" or "right-hand" parts. This allows the motors and mounting arms to be installed in several different locations depending upon your own requirements. For example, the X axis drive system can swing down and allow a workpiece larger than the table surface to be mounted. The X axis motor can be used only on the right end of this particular model of milling machine because the ball thrust bearings are in only the right end of the table. The left end of the X axis is a simple journal bearing with no thrust capabilities.

Moving to the spindle of the machine, Photo 7 shows the head with the handwheels and

7

identification plate removed. The identification plate contains a speed setting chart trimmed to size and re-installed (Photo 10). There are six mounting holes in the front of the casting, and the bottom four are used to retain the motor mounting arm. One thing that is a little hard to get used to is the mixed use of both English and metric fasteners in some of these machines. The holes used to mount the Z axis bracket have been 1/4-20 on most, but not all, of the machines Cardinal has retrofitted.

Before proceeding further with reassembly, you should check the preload settings on the spindle bearings. Most of these small mill-drill machines use tapered roller bearings in the spindle assembly, and if they are set incorrectly, they can have their life considerably reduced. We have noticed the preload settings of the machines as received from Enco range from much too loose to very tight. The design of this bearing arrangement is not the best, but with care, the spindle can be properly set up. The trick is to get the bearing preload set high enough to provide the required stiffness for milling without having it set so high as to cause overheating and burnout of the bearings. Tapered roller bearings are particularly susceptible to excessive preloads and have the unfortunate habit of failing in a sometimes spectacular fashion when this occurs. There is not enough power in the spindle drive motor to make this a matter of danger to the operator, but a seized bearing will rotate in the spindle housing and generally render the casting unusable. A good setting is one that does not cause excessive heating when the spindle is run continuously at its highest speed. This is a cut-and-try operation. When retrofitting these small machines, it is my standard practice to reset the spindle bearing preloads and to test run the spindle for at least two hours at its highest speed. The heat generated by running the spindle should be such that it is possible to hold your hand on the lower end of the spindle assembly. The spindle will normally run quite warm, but you should be able to touch it.

Photo 8 shows the parts that will be added to the Z axis. They are similar in many respects to those used in the X and Y axes, except that, because of the built-in spring mechanism, no special provisions for anti-backlash are provided. The spring return system, which is adjustable, is set high enough to provide a brisk return action to the quill. This is normally enough to provide an essentially no-backlash condition. In those cases where it is required, additional spring return force could be applied with external springs, connecting them to the adjustable depth stop arm. This arm is not shown in the photographs because we normally remove it because it is no longer required.

The mounting arm and a short extension to the quill worm drive are shown mounted in Photo 9. This extension allows the drive pulley to be mounted in place of the original handwheel which is simply moved out to the end of the shaft extension to make room for the pulleys, belt, and belt guard. The knurled shaft, in the right side of the photograph, is the clutch that disconnects the worm feed drive from the quill feed rack and allows the machine to be used as a drill press. The drill press handles are removed from the end of the shaft in the photograph, although they do not interfere with CNC operation and may be retained.

Photo 10 shows the Z axis conversion completed with the exception of the rest of the belt guard and the handwheel. The final view of the assembly is shown in Photo 11. The crank portion of the handwheel has been removed for convenience in operation since it is easier to set up the machine using just the rim of the handwheel rather than the crank.

This completes the mechanical portion of the retrofit project. The resulting machine has the capability of machining nearly anything its larger cousins can with respect to complexity. The only limitations are table size and spindle horsepower. Because the mounting brackets and other parts are specifically made to fit on the Enco mill-drill, some changes in dimensions may have to be made to adapt this design to other brands of machines.

Later, detailed drawings of parts will be published so you can begin to build your own machine. There are quite a few parts involved and it may take more than one issue to get them all in. In the meantime, a list of some of the sources I have used for various parts is given. You can contact any of these companies for information and catalogs.

8

9

10

Milling Machine: Enco Manufacturing Company, 5000 West Bloomingdale Avenue, Chicago, IL 60639. (Note: As mentioned, other suppliers may work also. I haven't tried any machines except those from Enco.)

Belts and Pulleys: Browning Manufacturing Division Emerson Electric Company, Maysville,

11

12

13

KY 41056. (Note: They sell through local bearing distributors.)

Ball Screws: Rockford Ball Screw Company, 3450 Pyramid Drive, Rockford, IL 61109. (Note: They build to customer drawings and would be a good source if you were designing your own retrofit for a different machine. They make rolled thread ball screws only.)

Stepper Motors: Digital Motor, Inc., 13563 N.W. Cornell Road, Portland, OR 97229-5892. (Note: This company sells a lot of stepper motors, mostly to OEM (Original Equipment Manufacturers). However, they will sell small quantities also.)

Plans, Materials Kits, Finished Parts Kits, Motors, Controllers, Ball Screws, Complete Machines: Cardinal Engineering, Inc., RR 1, Box 163, Cameron, IL 61423. (Note: They stock plans, kits, and parts for this project including ball screws and Centroid controllers. They also stock other related supplies and finished machines.)

Controllers: Centroid Corporation, P.O. Box 739, State College, PA 16804. (Note: Manufacturer of CNC controllers.)

Complete Machines: Matrix Machine Tool Modification Corp., 100-Box ART Street, Rochester, NY 14612. (Note: They build and sell a complete machine tool similar in size and using a Centroid controller.

Again, I appreciate your letters and comments regarding this series of articles. Judging by mail I've received, there are a lot of people with an interest in computer control. I've had several letters and phone conversations with Lonnie Morosic of McCook, NE, regarding computer control. He has developed a computer controlled flame cutting machine, written his own software, and even rewinds stepper motors using a computerized machine he built himself. Lonnie sent along a photograph of one of his recent projects, a horse that was cut out using his computer controlled flame cutter (Photo 12).

The last picture (Photo 13) of this series is of a part from Don Creger of Western Illinois University in Macomb, IL. Most of us have seen the woodcarving project that consists of a wooden ball captured in a cubical cage, all carved from a single piece of wood. This part is a variation on that same idea since the central part is separated from the cage by machining a separate project into each face of the cube. Don uses this as a final examination for his students and it is done on a small benchtop mill just like the one described above.

I have been reviewing several CAD systems, and will have a roundup soon of what is available and how they can be used for drawing, design, and CNC programming. CAD capabilities have reached the point where software is available for less than $500 that exceeds the capabilities of systems costing $100,000 or more only ten years ago. Watch for this also in the near future.

As promised before, most of this column is devoted to detail drawings of the various parts required to convert an Enco model 105-1100 mill-drill to CNC operation.

I'll take these parts in order according to their drawing number, and the first is 1727, the Belt Guard Stamping. I made the first prototypes of these parts using a hardwood male form block. The flanges were rough formed with a leather mallet and "spanked" to eliminate wrinkles with a piece of plumber's solder. This is an old trick that I have used in years past when hand forming aluminum aircraft ribs and formers. The demand for these parts became quite heavy, and after everyone in our small shop became tired of hand forming aluminum flanges, we built a steel form die to stamp them. The 1/4" holes are for mounting the guard to the main motor brackets.

Next in line are the X-axis and Y-axis Ball Screws. These are rolled thread units which may be purchased from Cardinal Engineering or the Rockford Ball Screw Company. Cardinal buys them from Rockford Ball screw in lots of a dozen or two at a time and normally has them in stock. I spoke with Ian McBain of Rockford Ball Screw, and he said that if there is enough demand he will make up some of them and keep them in stock for the readers of this column who want to build their own systems. Cardinal sells them for $360 for a set of two ball screws. Rockford will sell them for about $300 for a set, but you may have to wait for delivery. The ball nuts are not detailed in the drawings, since you would not be likely to make them yourself anyhow. Rockford Ball Screw Company has a short catalog that gives information about dimensions and other sizes available. There are quite a few other good suppliers of ball screws also, but this is the one I have used for nearly all of the mill-drill retrofits I have done.

The Drive Pulley, three of which are required, are Browning 24-tooth by 3/8" wide, 1/5 pitch timing belt pulleys. I purchase them with the minimum bore configuration and bore them out to match the motor shaft diameter.

The X-axis and Y-axis Driven Pulleys are modified from a Browning 48-tooth timing belt pulley, as shown in the drawing. There are some other suppliers of similar timing belt pulleys, but be sure they have a configuration that will accept the large bore in the center.

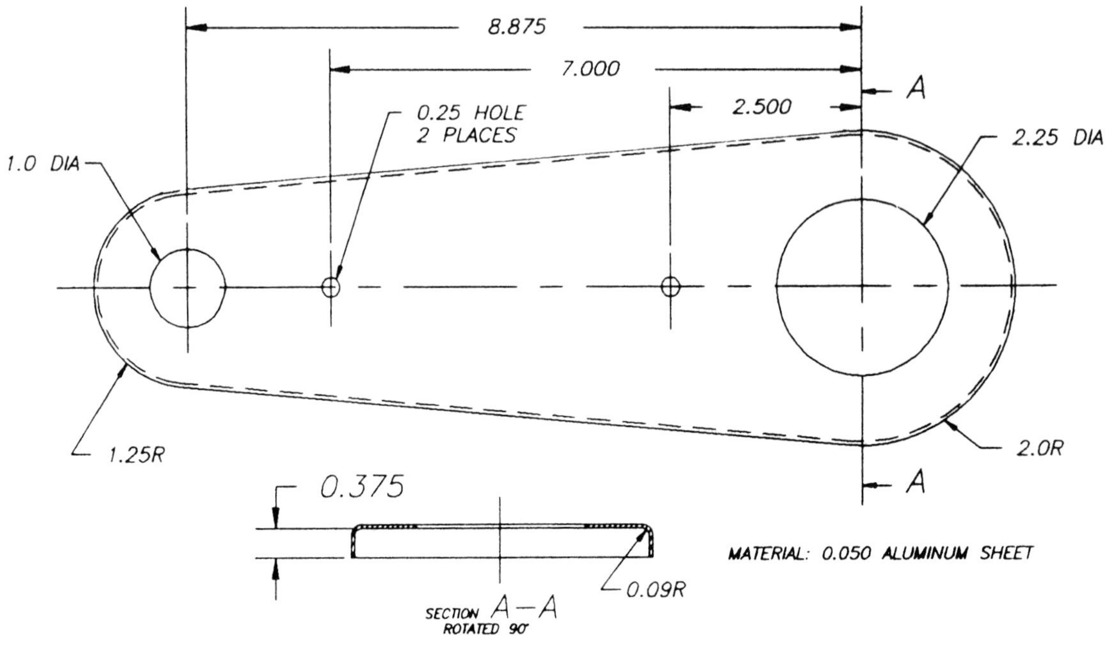

BELT GUARD STAMPING
6 required
Drawing 1727

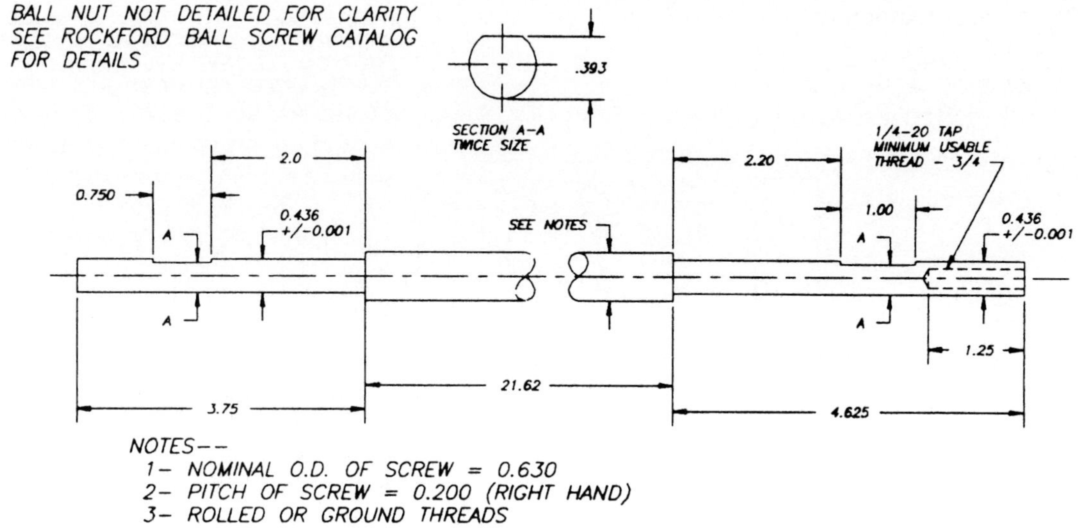

X-AXIS BALL SCREW
1 required
Drawing 1728

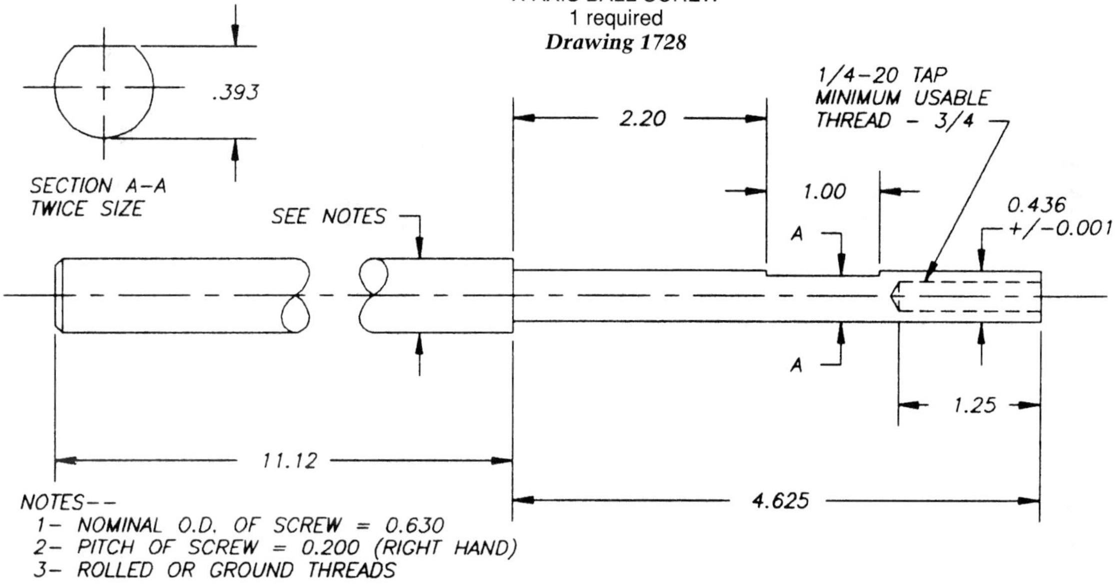

Y-AXIS BALL SCREW
1 required
Drawing 1729

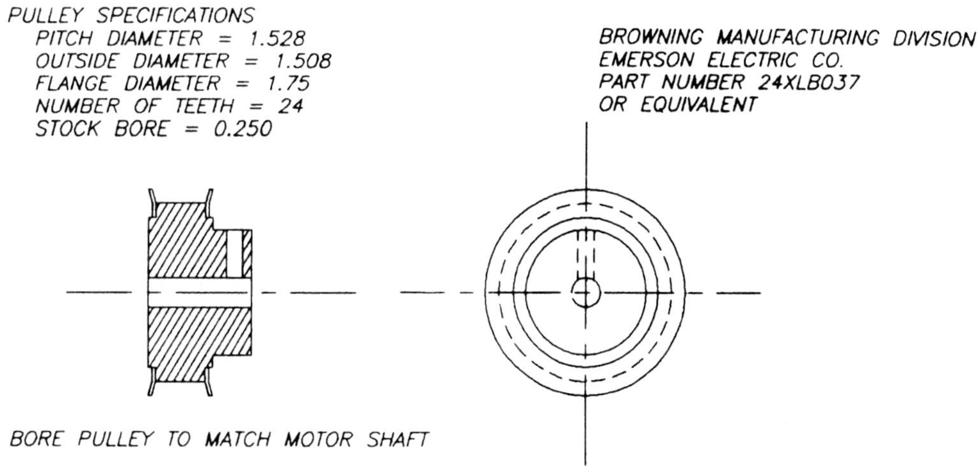

MOTOR PULLEY
3 required
Drawing 1732

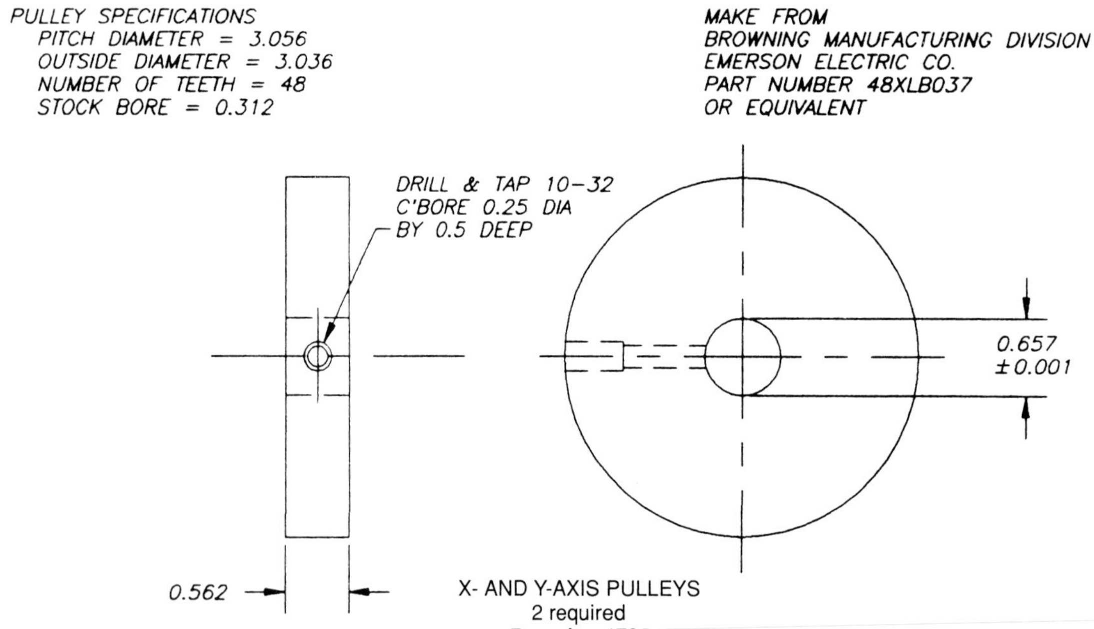

X- AND Y-AXIS PULLEYS
2 required
Drawing 1735

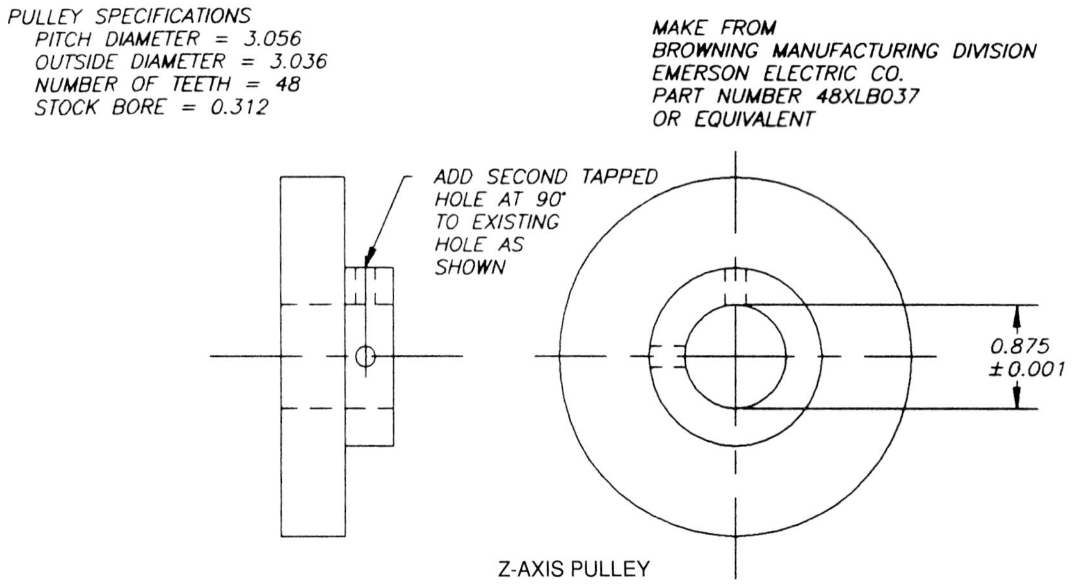

Z-AXIS PULLEY
1 required
Drawing 1736

The Z-axis Driven Pulley starts out the same as the X-axis and Y-axis pulleys, but is modified to different dimensions.

Referring to Photo 9 from the previous issue of HSM, the Z-axis Shaft Extension is shown in place on the head. This extension moves the handwheel forward far enough to allow room for the pulleys, belt, and belt guard behind the handwheel.

The next part shown, the Driver Thimble, is used to make it possible to adjust the preload of the thrust bearings to desired value and replaces the fixed roll pin assembly described previously. Two of these parts are required, one each for the X-axis and Y-axis.

The Thimble Plug is used in the preload mechanism. Two are required, one each for the X- and Y-axes.

Similarly, the Thrust Bushing is also used in the preload mechanism. Two of these are required.

The Idler Thimble is used on the left end of the X-axis and does not carry any of the thrust loads. It functions as a simple journal bearing.

The Z-axis Bracket (1745) is used to mount the drive motor and belt guard and to hold

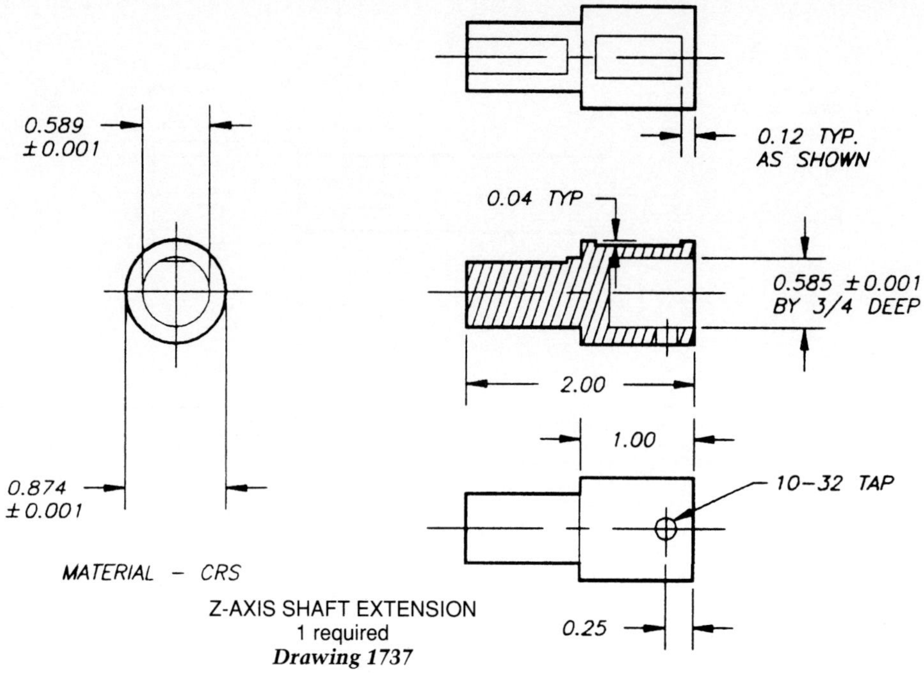

Z-AXIS SHAFT EXTENSION
1 required
Drawing 1737

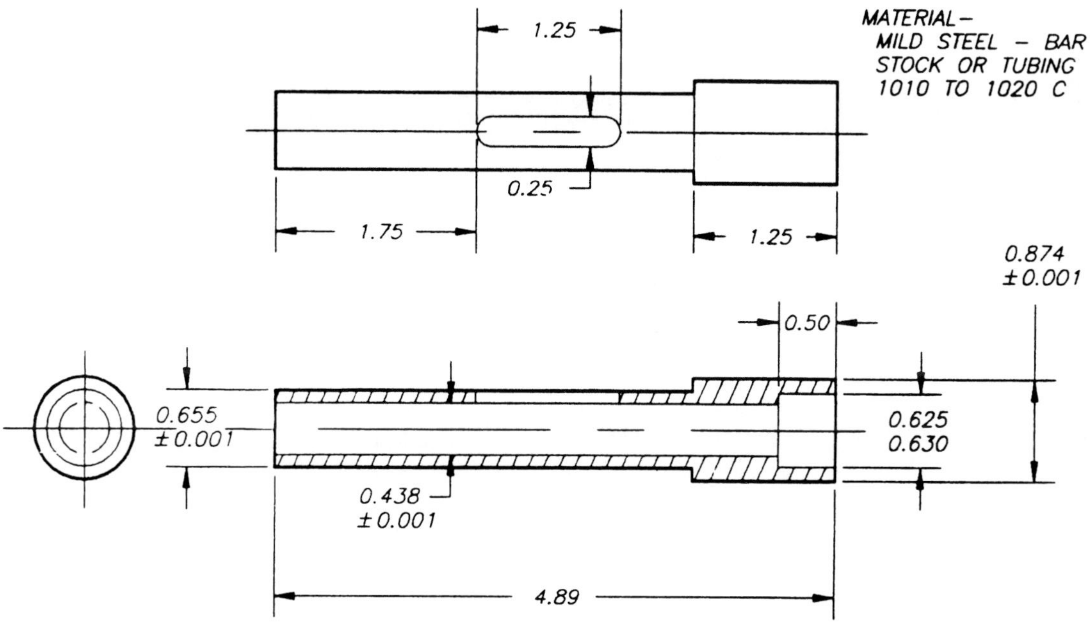

DRIVER THIMBLE
2 required
Drawing 1739

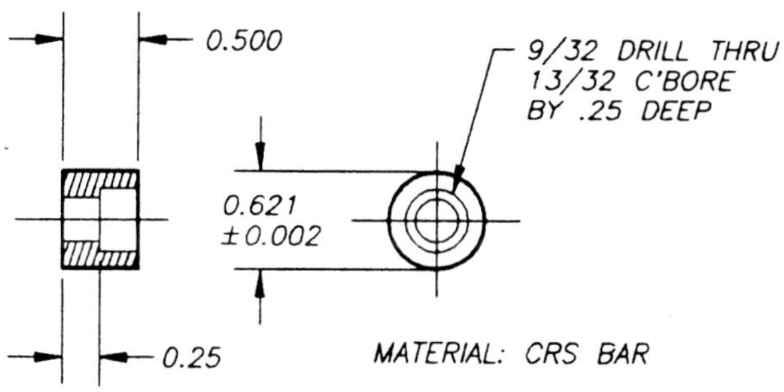

THIMBLE PLUG
2 required
Drawing 1740

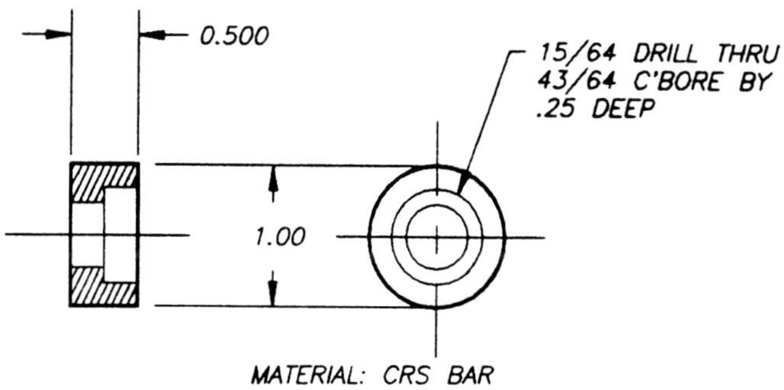

THRUST BUSHING
2 required
Drawing 1741

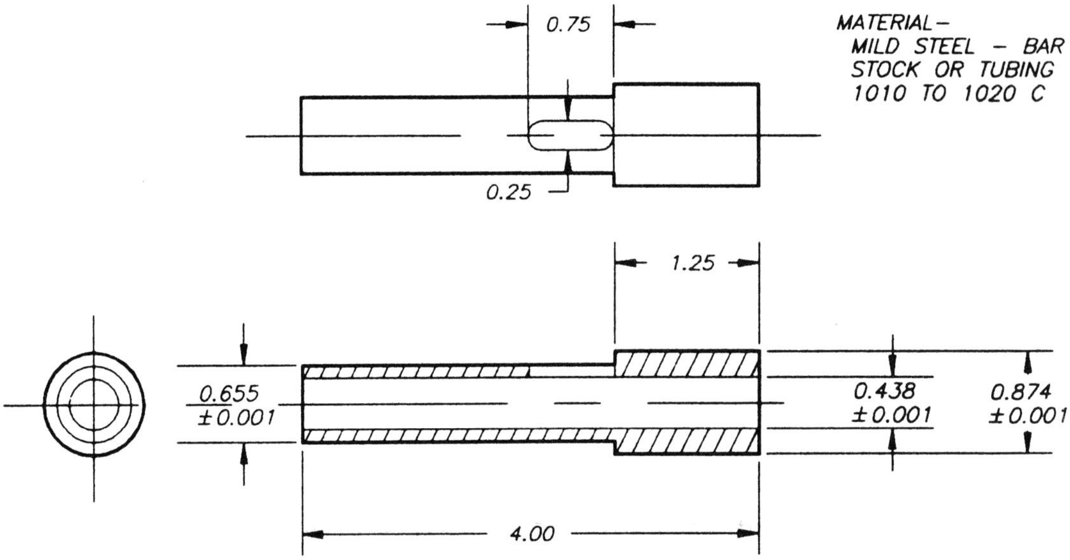

IDLER THIMBLE
1 required
Drawing 1743

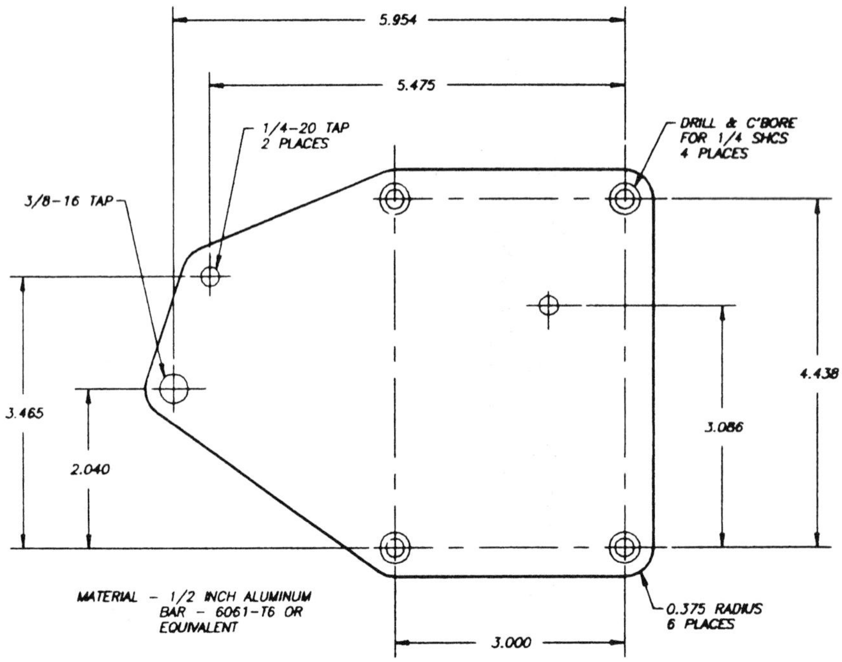

Z-AXIS BRACKET
1 required
Drawing 1745

them in the proper relationship. Some milling machines have the spindle ON-OFF switch located so it interferes with the stepper motor mounted on this bracket. If this is the case, you will have to move the switch to a different location. This bracket is mounted using existing holes formerly used to hold the speed chart. If the speed chart is trimmed down, it can be mounted above the bracket as shown in Photo 11 of the last issue.

The X- and Y-axis Brackets, for mounting the drive motors and belt guards on both of these axes, are identical. They are made to clamp on the machined casting that houses the ball thrust bearings furnished with the milling machine. Check the diameter of the machined surface of your own machine and adjust the part accordingly. A good fit is from "right on" to about 0.010" oversize, which allows a good clamping action and still enables the bracket

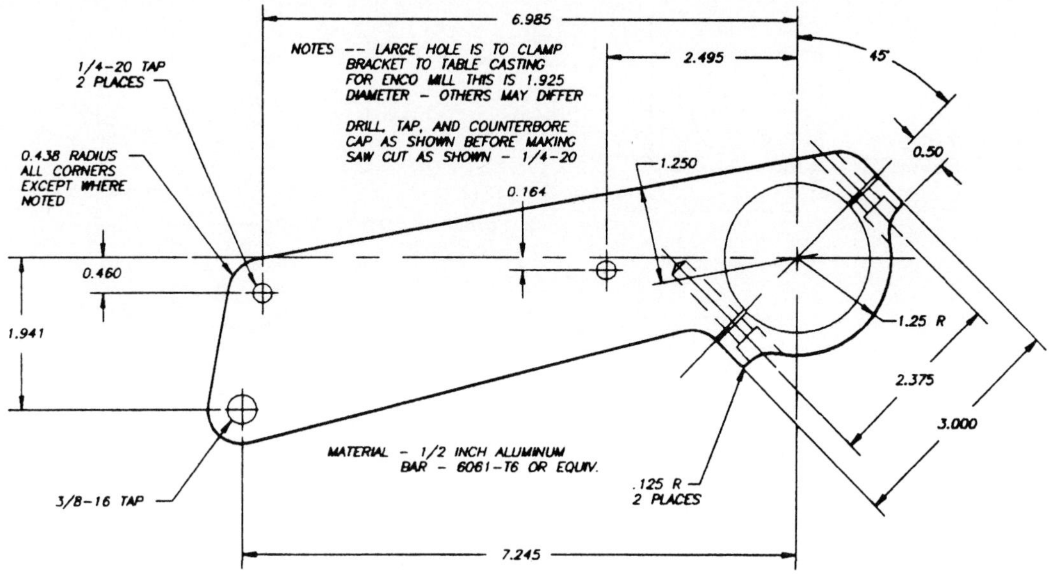

X-AND Y-AXIS BRACKET
2 required
Drawing 1746

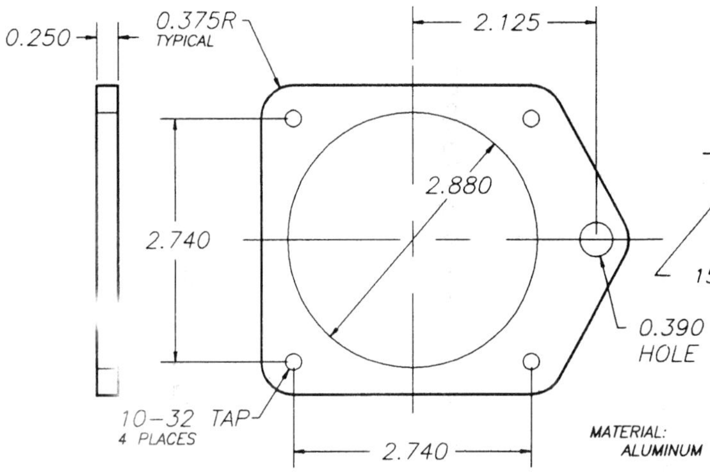

MOTOR MOUNT
3 required
Drawing 2090

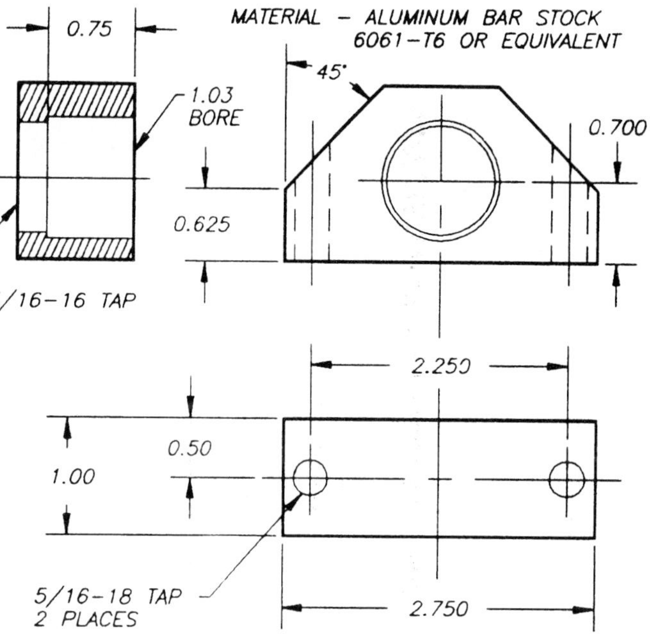

X-AXIS BALL SCREW MOUNT
1 required
Drawing 2217

to be easily rotated by slightly loosening the screws holding the cap in place.

The Drive Belts (1757) are 3/8" wide, 1/5 pitch units with 250 teeth. A typical manufacturer's part number is 250XL037. These are available from most belt manufacturers and are usually stocked, along with the pulleys, by bearing or electric motor supply houses. If the parts are not in stock, they can order them for you and have them in a few days. Cardinal Engineering also keeps these parts in stock.

Just a note about these belts: They are incredibly strong for their relatively small size. I have used them quite successfully on several different projects and still remain surprised by the amount of power they can transmit without slippage.

The Motor Mount (2090) is used to hold the stepper motors and to provide a pivot point for attachment to the bracket of each axis. This allows for ready adjustment of belt tension.

The next two parts (2217 and 2218) replace the bronze parts that thread into the original lead screws. They mount in the same fashion,

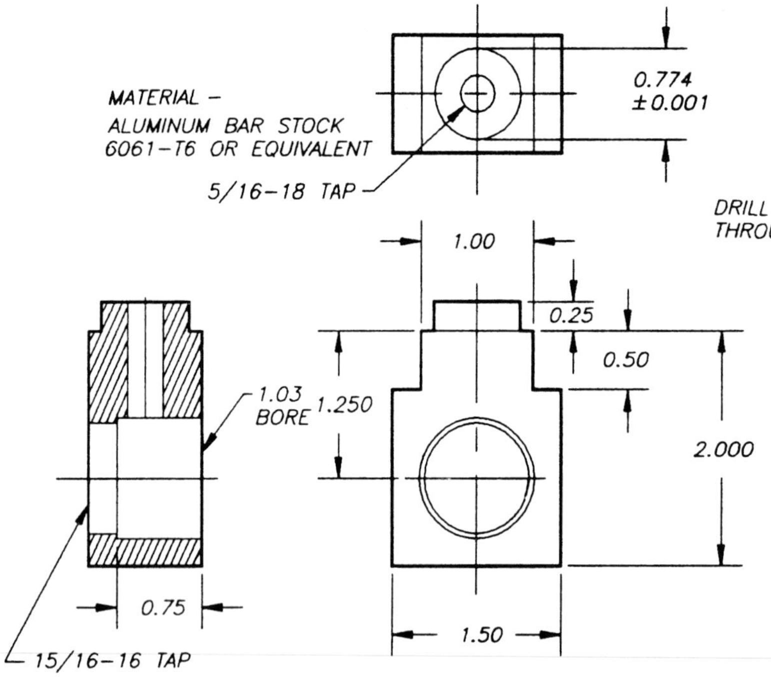

Y-AXIS BALL SCREW MOUNT
1 required
Drawing 2218

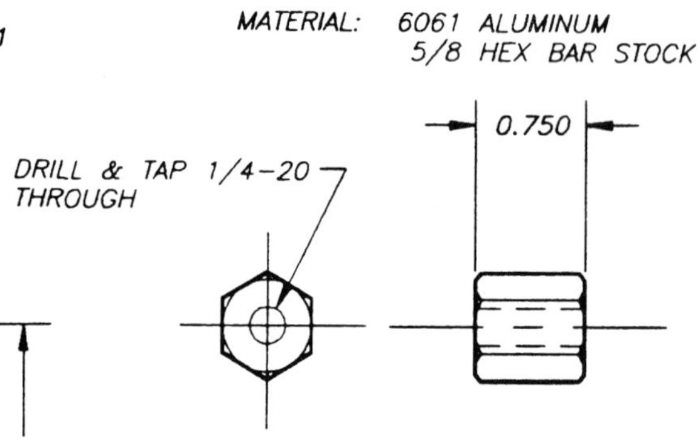

BELT GUARD SPACER BUSHING
6 required
Drawing 2224

but provide a means to install the ball nuts which have a 15/16 by 16 TPI threaded end for mounting. It is very important that these parts are installed so the ball screws are as perfectly aligned as possible. It may be necessary to shim or trim these parts, or to enlarge the mounting holes in the graduated dial castings. Normally this is not necessary since all of the parts that come with the machine have generous allowances for minor adjustments.

The Belt Guards are held in place and properly spaced using two short Bushings on each axis. I used hex aluminum for these parts, but they can be made of any convenient material. If you use round stock, it is handy to machine a couple of wrench flats for installation, thereby eliminating the "vise grip" syndrome.

The handwheels for the table and saddle must be modified by boring them out for a nice fit on the big end of the Thimbles. The graduated dials on the table and saddle sometimes need to be counterbored slightly to fit the thimbles also. This is necessary only when the handwheels have multiple teeth that engage the graduated dials. If the handwheels have a drive pin arrangement that fits into the graduated dial, it is normally not necessary to counterbore them.

Drill and tap a setscrew hole in each graduated dial that will allow the screw to pass through the slot in the Thimble and engage the flats on the Ball Screw. The setscrew shown in the X- and Y-axis Pulley is also intended to pass through the slot in the thimble. When these parts are assembled, the setscrews are left loose until the thrust bearings have been set, then the setscrews are locked in place. This arrangement separates the torque and tension loads so they can be independently set. It also allows a liberal adjustment to allow for differences from one machine to another.

Included in this chapter is general information on choosing the proper controller and motor and some notes about surplus motors.

Two different CNC controllers have been used on the Enco model 105-1100 mill/drill machines discussed in the last issue. The most fundamental difference between the two controllers is the method used to provide power to the motors. Cardinal Engineering P/N 2077 has an R/L (Resistance/Inductance) motor driver system and P/N 2086 has a chopper-type motor driver system. The important differences for our purposes are cost and motor speed. The R/L drive controller is less expensive but cannot achieve motor speeds as high as can the chopper drive controller. The typical maximum feed rate with the R/L drive is 15" per minute. The typical maximum feed rate with the chopper drive is approximately 60" per minute. These speeds are predicated on using 475 oz-in motors. The use of 325 oz-in motors will reduce these speeds somewhat.

For purposes of comparison, a feed rate of 2" to 5" per minute is common for steel and 5" to 20" per minute is commonly used for brass and aluminum. These rates are based on a light chip load and good coolant for the tools. For plunging or heavier cuts in steel the feed rates can be in the 0.5" to 2" per minute range. These ranges are typical and will vary according to your own conditions. This is an area where there is no substitute for practice and experience.

Other factors to be considered in selecting a CNC controller is the intended use. If you are planning on modest, hobby-only machining without heavy usage, the R/L drive controller is probably adequate, particularly if the price difference is critical. If you would like automatic coolant, spindle, vise actuation, or other features associated with an industrial operation, then the chopper drive controller would be your choice. If you can foresee the need to run much larger motors (up to 2500 oz-in), then the chopper drive controller would also be required.

Both of the CNC controllers, the R/L drive and the chopper drive are available from Cardinal Engineering. These CNC controllers are manufactured by Centroid Corporation. Centroid also sells the chopper drive controller (their P/N CNC4), but the R/L drive controller is available only from Cardinal. The R/L drive controller sells for $2,500 and the chopper drive controller sells for $3,500 (plus shipping costs). Write to them for information.

Cardinal Engineering adds a plug panel to the Centroid chopper drive controller in order to make switching motors as easy as possible. The R/L drive controller comes with a plug panel as standard equipment. With drive motors wired with plugs, it is easy to move the CNC controller from one machine to another, reducing the cost to have more than one CNC machine tool in your shop.

Both the R/L drive and chopper drive CNC controllers have essentially the same programming capabilities, although the chopper drive has additional memory for longer programs. Both controllers can be programmed using the RS-274-D language, and programs can be interchanged between the two controllers. The only exception to the ability to exchange programs occurs with the use of the auxiliary functions in the chopper drive controller. These auxiliary functions are not recognized by the R/L drive controller and produce an error message.

Both controllers can be programmed in either inches or millimeters, and absolute or incremental modes. Both controllers have cutter compensation, an invaluable feature that allows for both roughing and finishing cuts without rewriting the CNC program. So it comes down to a choice of speed, cost, and capabilities. If you would like to discuss this with someone before making a choice, contact Jerry Berns at Cardinal Engineering.

Once a controller has been selected, you need to select the motors. I suggest a 325 or 475 oz-in motor as being entirely adequate for this application. I have been using motors from Digital Motor, Inc., and Oriental Motor, but there are several sources available. I would caution you, however, about buying a motor on the surplus market and expecting good results. Many of the surplus market motors have windings that were designed for a particular application. They will not necessarily work well with the Cardinal/Centroid controllers. In fact, certain winding patterns will not even run and can damage the motor driver boards in the CNC controllers. The controllers come with fairly complete instructions on how to hook up a wide range of motors, but caution is required here. The safest way is to purchase the motors already wired, or, if you have surplus motors, to send them in to Cardinal for testing and wiring.

It is possible to get lucky and find some surplus motors at very reasonable prices. New stepper motors are not cheap, and if used or surplus ones can be located, it is possible to save several hundred dollars. One of the brands you will most likely find is SLO-SYN, made by Superior Electric. These motors were very popular in the early days of N/C and they are in relatively abundant supply. Those I have salvaged from old industrial N/C equipment are generally usable but do not provide the torque or speed current production motors do. This is most likely due to improvements in motor design and magnetic materials, although it can also be caused by careless disassembly. Do NOT disassemble a stepper motor unless absolutely necessary as they can be quite sensitive to demagnetization, which will drastically reduce the speed and torque they will produce.

Digital Motor, Inc., provides motors "in the raw" and you will have to wire them up yourself. Oriental Motor normally sells only to OEM (Original Equipment Manufacturer) accounts, so I have not included their address. Cardinal Engineering provides motors that have already been wired and tested with ten-foot cables and plugs already installed. Centroid can furnish motors with cables as an extra cost option but does not furnish the plugs for the ends.

In addition to the mechanical parts, CNC controller, and drive motors, you will need some way to write, edit, store, and send programs to the CNC controller for execution. Any computer with an RS-232 serial port and communications software will work with the Cardinal/Centroid controllers. However, Cardinal has developed special software that emulates (works exactly like) many models of the Bridgeport industrial CNC milling machines. This allows the user to write and edit programs and store them on floppy disk. It also allows programs to be sent to the CNC controller (called "downloading") and received from the CNC controller (called "uploading"). This software, known as CNCED, (for CNC EDitor), is included at no charge with controllers from Cardinal Engineering. The CNCED software has also been used with several other brands of CNC controllers and is available separately for a price of $250. Centroid includes a communication program, CENCOMM, with their CNC4 controllers, but I have not had a chance to evaluate it yet.

CNCED requires an MS-DOS computer (IBM or clone) computer with 640K of memory and a serial port to operate. This is the minimum configuration required. A version of CNCED is also available for the Apple II series of computers. These are the only two computers Cardinal has specific software for. As mentioned, other computers will work as long as you have communications software (sometimes called "modem" software) to work with it.

If you want to use a CAD system to generate programs, there are a lot of choices, all of which add to the cost of the system. Both Cardinal and Centroid have software packages that generate CNC programs from CAD drawings, and there are many other commercially available programs to do this. One of the biggest problems is finding a system that does not cost a small fortune to buy. There are some available and I will review some of these choices in this column.

The designs presented in this column are intended for use by hobbyists in their own shops. Commercial use without permission is prohibited. In other words, if you want to go into the business of making CNC retrofits, you will need to get permission before using these designs.

I want to address some of the questions that have been asked about this project. The most common question has to do with retrofitting some other brand or model of machine that "looks just like the one in the pictures."

Generally, all of these machines can be converted, but there are both minor and major differences in dimensions and sizes of the various mill-drill machines sold by all of the suppliers. The basic principles hold for all of these machines, but specific differences must be allowed for in the design of the motor brackets, ball nut mounts, and ball screw end designs. The drawings presented in the previous column only apply to the ENCO Model 105-1100 mill-drill, and even there, I have noticed some variation from one machine to another. One of the big challenges of these retrofits is to try to come up with a design that is as widely applicable as possible.

Consequently, for those who would like to retrofit some other small milling machine, here are some general guidelines so you can adapt the design to your own needs. Later, you'll find some comments regarding the retrofitting of the larger, Bridgeport-type milling machines.

BELT/PULLEY RELATIONSHIPS

I made the belt and pulley choice for this project with a view towards making it as universal as possible for many different machines. This allows standard belts, pulleys, motor mounts, and belt guards to be made for nearly all of the different machines that might be converted. The belts and pulleys chosen have been extensively tested and used successfully with stepper motors ranging from 150 oz-in to 2500 oz-in torques and on machines from the size shown in the last issue to over twice the size of a Bridgeport milling machine.

A 1/5" pitch timing belt was chosen because of its thin cross section, which produces a very small loss of power due to belt flexing. The energy lost because of flexing can be much larger than one might think, and it is desirable to deliver as much of the motor power as possible to the machine tool instead of to the belt. The capability of this small belt to deliver the full capacity of the motors is amazing.

The combination of belt length and pulley ratios was chosen as a good compromise that allows use of a 1300 oz-in motor on a Bridgeport-type mill, as well as the relatively small motors shown in the last issue. A 250-tooth belt, combined with 24- and 48-tooth pulleys, is used. The center line distance is 8.9", which gives an adequate "wrap angle" to prevent tooth slippage, and the pulleys can be reversed to give two driver/driven ratios without changing the belt guard design.

An interesting side note regarding motor speed, motor torque, and final speed of the machine tool should be mentioned here. During the initial design and testing of these retrofit projects, several pulley ratios were tried with different motors with an interesting result. For a given motor, power supply, lead screw, and machine tool, no matter what pulley ratio was used, the maximum achievable table speed was always about the same.

We had expected to discover some optimum combination that would yield best performance, but were somewhat surprised at this result. Upon consideration, it appears that the relationship between the motor torque/speed curve inversely matches the torque/speed curve of the lead screw and table combination. I have not run exhaustive tests on all combinations of motors, but since the characteristic curves for all stepper motors of this type are nearly identical, I wouldn't be surprised if this were a fairly universal result.

MOTOR MOUNT METHODS

The typical stepper and servo motors used in CNC machine tools and other automation projects in industry have standard mounting dimensions and are generally interchangeable from one manufacturer to another. The motors are classified by "frame size" similarly in principle to the methods used for classifying AC motors. Virtually all of these motors have a flange mount on the end of the motor which takes four bolts and a large pilot hole for alignment purposes.

For motors up to about 800 oz-in of torque, it appears to be adequate to mount the motors as shown in Photos 5, 6, and 10 of my March/April column and Drawings 1745, 1746 and 2090 of the May/June column. By attaching the motor to the motor mount (Drawing 2090) and then attaching the motor mount to the mounting bracket (Drawing 1745 or 1746) using a single bolt, a pivot arrangement is created which makes it easy to adjust the belt tension. Also, with proper selection of dimensions, the pivot point can be selected to fall outside of the area covered by the belt guard, simplifying assembly and access to the pivot bolt. This arrangement

On the other hand, if you think there might be a budget and time squeeze, I suggest we go with this design.

also allows the motor to be mounted on either side of the belt guard assembly, increasing the flexibility of the system.

When motors in the over 1000 oz-in range are used, it is often more desirable to mount the motor flanges directly to the mounting bracket and to achieve the belt tensioning by means of slotted bolt holes rather than a single pivot bolt. This produces a more securely mounted motor and reduces deflections associated with the pivoting mount system. However, this method is not as flexible as the pivoting motor mount, and if enough room exists on the machine you are using, the pivot arrangement can be beefed up with heavier materials, and will be very satisfactory.

BALL NUT MOUNTING METHODS

By far, the biggest problem associated with retrofitting manual machine tools to CNC operation is the mounting of the ball screws. Ball nuts are generally quite a bit bulkier than the standard nut they replace. This would not be a problem except that most machine tools were not designed with a lot of excess available space, and fitting the ball nuts into them is sometimes quite a bit of a challenge. The benchtop mill-drill used in the earlier column is quite simple to work with and the ball screws are easily mounted.

However, for those of you with Bridgeport or "clone" milling machines, there is a bit of a problem. Nearly all of the ball screw suppliers have what they call a "Bridgeport Retrofit Kit" or some similar name. This will normally consist of both the X- and Y-axis ball screw and nut assemblies, along with the casting that mounts the nuts. This casting, called a yoke, is intended to replace the existing part in the milling machine. DO NOT EXPECT THIS TO FIT WITHOUT EXTENSIVE ADJUSTMENTS!! I have yet to find a ball screw manufacturer who has a kit that fits correctly!

In all fairness, this is not surprising since there is really quite a bit of difference between milling machines, and the "clone" mills have dimensions that vary up to 1/4" or more from a Bridgeport. The Bridgeport mills are the most uniform, but they are not totally interchangeable, either. The older Bridgeports seem to be better than the ones made in the last 10 or 15 years.

When selecting a ball screw supplier, you should expect to have to do quite a bit of "fitting." I have yet to have a set of ball screws that didn't require some shimming and enlarging of bolt holes to achieve proper alignment. In a couple of extreme cases, it has been necessary to make a complete new yoke out of bar stock in order to mount the ball nut assembly.

Alignment of the ball screws is very critical for free running and long life. You must resist the urge to flex the threaded shaft "just a whisker" in order to get the mounting bolts to fit properly. You will get the system assembled and it will probably work for a while, but the eccentric loading on the ball nut will invite rough operation and damage to a very expensive part. Properly mounted ball screws are very rugged. Improperly mounted ball screws can be very easily damaged by normal operation of the machine tool.

I have been working on different ways to mount the ball nuts on Bridgeport-type milling machines and will publish sketches of several different methods in the next column. In the meantime, be aware that the $1,500 (or more) ball screws that you purchased may require extensive "fitting up" in order to work in your machine. These "adjustments" will also normally void any warranty that might exist on the ball screws.

BALL SCREW SHAFT ENDS

Another aspect of the ball screws that requires some work is the way the shaft ends are machined. You can select an unmachined ball screw, and machine the ends to match your old lead screws, or you can have the ball screw manufacturer do it for you. Again, this is one of those things that is kind of a toss-up.

Machining your own ends is the less expensive way to go and provides some good experience in the bargain. For a Bridgeport or "clone," I have had good luck with pre-machined shafts. Over the many Bridgeport-type mills I have retrofitted, I have kept a list of dimensions from which a set of "one-size-fits-all" drawings has been developed. These drawings have simplified the installation of the ball screws tremendously because we furnish them to the ball screw manufacturer and have them made to our own drawings.

Two areas I have found to be particularly important with the shaft ends of Bridgeport-type ball screws are the bearing diameter and the Woodruff key slots. Nearly all the commercially made "Bridgeport kits" I have installed have required reworking in these areas.

You should carefully chuck the ball screws in your lathe, either between centers or, if the spindle hole is large enough, in a three-jaw chuck. Wrap a rag or plastic bag around the nut to prevent any swarf from entering and turn the lathe slowly enough to prevent whipping of the shaft. Using a fine emery cloth, work the part of the shaft down that slides in the bearings so you have an easy sliding fit with no drag. It is better to be a little loose than too tight in this area. This is necessary only if the shafts are tight. Most of the commercial "Bridgeport Kits" are too tight.

When all of the shaft ends have been worked down to the proper diameter, you should try the Woodruff keys in the slots. Most of the keys I have encountered needed to be removed with a pliers or even a small chisel. Conventional wisdom regarding the use of Woodruff keys tells us they should be a fairly tight fit in the keyways. The purpose of this is to withstand heavy loads without fretting and "wallowing out" the shaft, key, and mating part.

This is not necessary, and is even a problem with the ball screw installation. It is essential to have the Woodruff keys tight enough to have no play at all but loose enough to be removed by the fingers only. I have found commercial ball screws to be quite consistent in size but always a little tight on the fit. Rather than attempting to enlarge the Woodruff keyway in the ball screw, you will find it more satisfactory to just surface grind the keys themselves for the proper fit.

Resist the temptation to "tap" the key in with a piece of brass. You will never get it out without damage, and the retrofit process will normally require several disassembly cycles for alignment of the ball screws. Unless the key fits properly, you will damage the shaft end of the ball screws and introduce another problem into your already difficult alignment process.

MOTOR SIZE SELECTION

Selection of the proper size motor for your retrofit is an important part of the project. while I have done some retrofits of the smaller "Unimat-type" machines, I have a lot more experience with larger machine tools.

A 350 oz-in motor with the Cardinal Engineering Trainer CNC controller will produce a feed rate of between 7" and 15" per minute on the benchtop milling machine or even on a Bridgeport mill. This makes a good, low-powered and safe system. The feed rates are high enough to do real machining, but the motors are low enough in torque to readily be stopped with your bare hand if desired. This is also the lowest cost approach.

A 475 oz-in motor with the Trainer controller has a similarly safe torque value but can achieve feed rates of 30" to 45" per minute. A 475 oz-inch motor with the Cardinal/Centroid Advanced controller will achieve feed rates of up to 120" per minute. This is much faster than most machining operations would normally require. I seldom use a feed rate any higher than 30" per minute and most of the work we do is in the 10" per minute range and less. I recommend this size motor for most "hobby" machine shop and school usage.

For heavier cuts and larger machines, I have used 1350 and 2350 oz-in torque motors. These motors are suitable for use only with the Advanced controllers and will often not produce any higher table speeds than the 475 oz-in motors. The reason for this is related to the inductance of the motor windings and the additional mass of the rotors. What these larger motors do provide is greater torque at the lower speed ranges for taking heavier cuts. This is not normally required in the small "hobby" type of machine shop. If you expect to take the full capacity cuts of a Bridgeport milling machine, you will need such motors, however.

DESIGN PHILOSOPHY OF RETROFIT PROJECTS

A major goal in converting the machines I have retrofitted for CNC operation is retention of the capability to use the machine as a manually-operated machine. This is particularly useful for schools where machines must often serve multiple duty. In fact, as I have mentioned earlier in this series, the first retrofits were intended for classroom use as CNC trainers. This need for maximum flexibility of a machine is also useful in the home shop since few of us have the space or funds to purchase a machine for every application. Consequently, many of the design decisions are made with the intention of making a machine that can function as both a CNC machine and a manual machine.

Convert a Milling Machine to CNC

by Roland Friestad

Drawings by Author

This is the first of a series of columns on converting a full sized, floor-model milling machine to CNC control. This series will be quite lengthy. In order to also keep up with other developments for those who would like news of computer control, CAD, and related topics, other items of general interest will be included as they come to my attention and as space permits.

SELECTED MILLING MACHINE

The machine I have selected to convert is the ENCO model 100-1527 milling machine. It has a 9 × 42" table, and approximately 24" of travel in the X direction, and 12" in the Y direction. The quill has 5" of travel, and uses standard R8 or "Bridgeport" collets. The spindle drive on this model is the fixed speed type with a choice of 16 different speeds between 80 and 4800 rpm. The variable speed head version of this same machine is the ENCO model 100-1529, and it is identical in every way except for the head. The conversion is exactly the same for both machines. These machines and the many other brands that appear identical to them are known as Bridgeport "clone" machines. This is because they are near copies of the very popular Bridgeport brand milling machine that has been the mainstay of small and large shops worldwide for many decades. The appearance of the "clone" milling machines on the market about 25 years ago coincided with the expiration of several of the key patents held by Bridgeport Machines, Inc., of Bridgeport, Connecticut. Bridgeport, incidentally, is a trademark of that company, which is still in business, and still making milling machines as well as other types of machines.

SIMILAR MACHINES

Since not everyone will have the same machine I am using, and there are slight differences between different brands, and even between different machines of the same brand, I will note as we go along where to watch out for problems, and point out areas that might have to be "adjusted" for other models. Also, because this series will be quite long, the complete plans for this project are also available from Cardinal Engineering, an advertiser in this magazine. Write for their catalog.

DESIGN CONSIDERATIONS

First, a few comments on the design philosophy I followed when developing this project. Several things were considered of primary importance. The first of these was the ability to use the mill both as a CNC machine and as a manual machine. Most small shops, schools, or hobbyists cannot afford the luxury of assigning machines for a single purpose. This way the mill will serve both functions. To convert from CNC to manual operation requires only the removal of a single bolt. The second criteria was that a minimum number of changes be made to the milling machine itself; in other words, no extensive machining or modifications to the machine tool frame. The third requirement I feel is important is the use of standard components as much as possible. This makes maintenance much easier. Another element in the

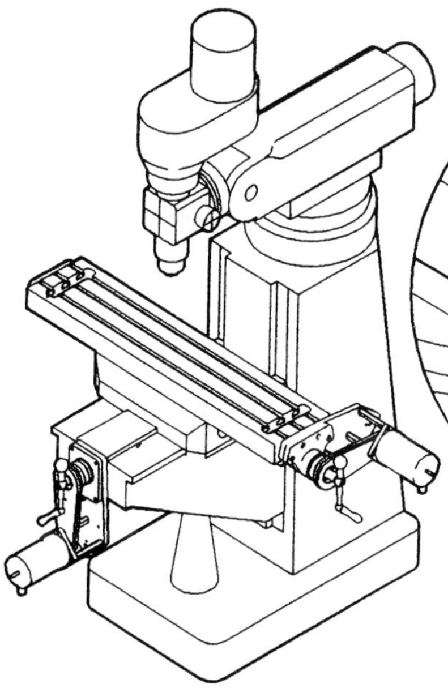

Figure 1
VIEW OF CONVERTED MILL

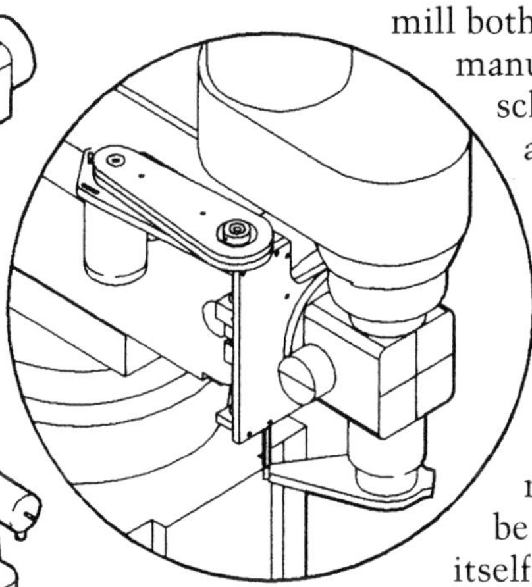

Figure 2
DETAIL OF Z-AXIS CONVERSION

design is simplicity. There is no need to make something complicated if something simple will do the job. The final consideration was ease of construction. The entire project can be constructed using a lathe and the same milling machine you are converting to CNC. Several of the parts used in this project were designed to be machined using CNC equipment, but they can also be made using a manually operated machine.

GENERAL VIEW OF FINISHED PROJECT

For purposes of inspiration, isometric drawings of the finished machine are shown in Figures 1 and 2. The X and Y axes can be driven with any of several different size of motors with only minor revisions. The Z axis can also accept several different sizes of motor, except that a 475 oz-in motor appears to be the best choice for this machine. The selection of motor size for the X and Y axes depends upon the type of work you wish to do but would generally range between 475 oz-in and 2200 oz-in.

THE FIRST CHANGE TO MAKE TO THE MILL

One of the first requirements for ease of use with CNC milling is the elimination of as much backlash as possible from the machine. The X and Y axes (longitudinal and cross-feed) of these milling machines as they are purchased will have a backlash of around 0.010". After they have been used for a while, this figure can increase dramatically, and I have seen machines with 0.050 of backlash in everyday use. For use as a CNC machine, this backlash must be reduced to as low a value as possible. By reducing this backlash, the accurate control of the machine is greatly simplified. Another advantage of reducing the backlash is that climb milling can then be done. Climb milling is not recommended when using machines with significant amounts of backlash. Climb milling, however, has some decided benefits when it can be used. It greatly reduces the power required of the table drive mechanism, and also can produce better finishes and longer tool life, particularly when machining the more difficult, work-hardening materials such as stainless steel.

SOME CAUSES OF BACKLASH

The backlash in the table mechanism of this type of milling machine arises from three primary causes. These three causes – bearing preload, nut/screw backlash, and machine deflections – combine to make a total amount of uncertainty in the actual location of the machine table given a particular reading of the calibrated dials of the lead screw. Of these three, we might as well not worry about machine deflections since for a given design they cannot be changed. The other two causes can be addresed and, with careful work, be reduced to a range of 0.001/0.003". As with manual machining, suitable CNC programming methods can cancel most of the effects of even this small amount of backlash.

BACKLASH DUE TO BEARING PRELOAD

All of the milling machines of this type I have worked with (about 40 at last count) have had the same type of design for the bearings that support the ends of the lead screws. There is a set of two ball bearings mounted in what is called "duplex" mode. This allows them to be preloaded to take up any backlash between the table (or saddle) and the lead screw. Most of these duplex bearing sets are not preloaded properly for use in a CNC installation.

DETAILS OF DUPLEX BEARING MOUNTS

Refer to Figure 3 for a sketch of the internal workings of the duplex mounting arrangement. The bearings are held in a machined bore, and retained by a clamping ring on the end. Two preload rings are usually placed between the bearings. The first thing to do is remove the clamp ring. Then remove the bearings and preload rings from the housing. Clean all of the parts, and then attempt to slide the bearings into the housing bore. They should slide in with finger pressure, and should not be press fit into the housing. If the bearings cannot be installed and removed with finger pressure alone, the housings should be bored out until the appropriate fit is achieved. I would suggest that a clearance of 0.001 to 0.002" be used. However, a clearance of 0.010" is much better

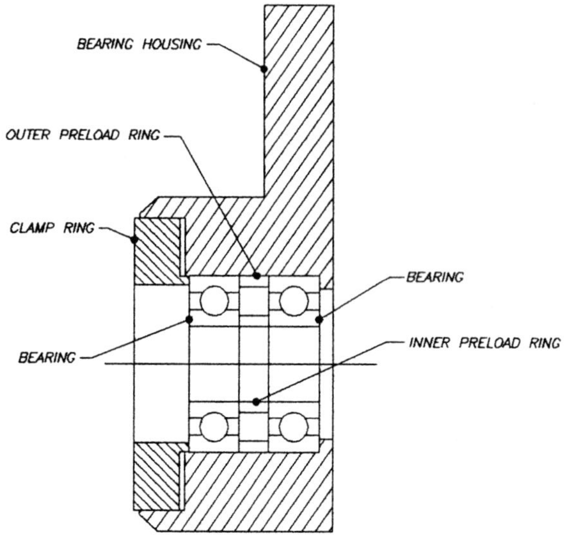

Figure 3
DUPLEX BEARINGS

than a press fit. The Y axis housing and the left end of the X axis are where the duplex bearings are usually mounted. The right end of the X axis has a similar housing but it normally contains only a single bearing. This housing should also be bored out for a sliding fit to the bearing.

MACHINE BEARING HOUSINGS

If required, bore these housings out now to the appropriate clearances. This may be done in a lathe or on another milling machine using a boring head. It is my assumption that the reader is familiar with normal manual machine shop procedures and, therefore, detailed explanations of required manual machining operations will not be provided unless there is something particularly "tricky" about them.

FIT KEYS

While you are at it, it would be a good time to also grind or file the Woodruff or square keys that are used in the lead screws for a suitable fit in their keyways. The keys should fit in the keyways of the lead screws in such a way that there is no free play but they can still be removed by the fingers. Much trial-and-error fitting will be done in this project, and some time spent now in getting this right will be repaid many times over as this project proceeds. Use your own judgment, but don't say I didn't warn you!

MEASURE BEARING BACKLASH

Once the bearing housings have been checked and bored out to the proper clearance as described above, the next thing to do is to measure the backlash in the bearings themselves. Rig up a dial indicator, as shown in Figure 4, to measure the amount of backlash for both the X and Y axes. If it is more than 0.000 to 0.001", you should probably adjust the bearing preload. The procedure to measure this backlash is to lock the table or saddle, and rock the handle back and forth, noting the variation of the dial indicator readings. The difference between the upper and lower dial indicator readings represents the lost motion due primarily to improper bearing preload. The force applied to the handle during this test should be approximately the maximum force you would expect to apply when machining the heaviest job to be done on your machine using manual feeds. Excessive forces applied here will result in preloads that are too high.

DIFFERENT BEARINGS USED

A sketch of two different types of bearing designs is shown in Figure 5. Both types of bearings are used in these milling machines. The first design is quite common, and because of the symmetry of the design, it doesn't make any difference which way it is loaded into the housing. The second design is intended to resist higher axial loads, but it requires that the bearings be mounted in a specific direction in

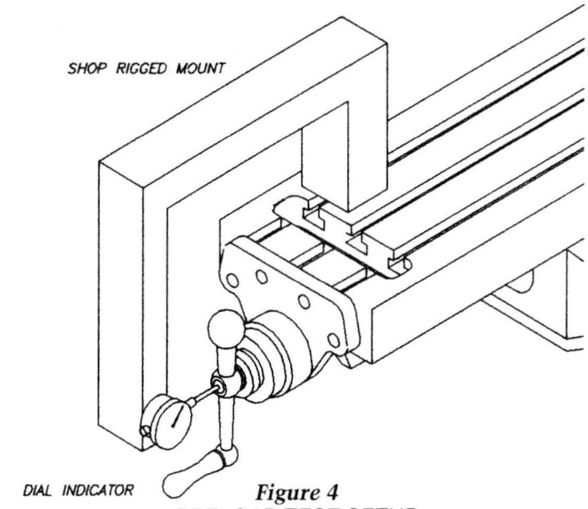

Figure 4
PRELOAD TEST SETUP

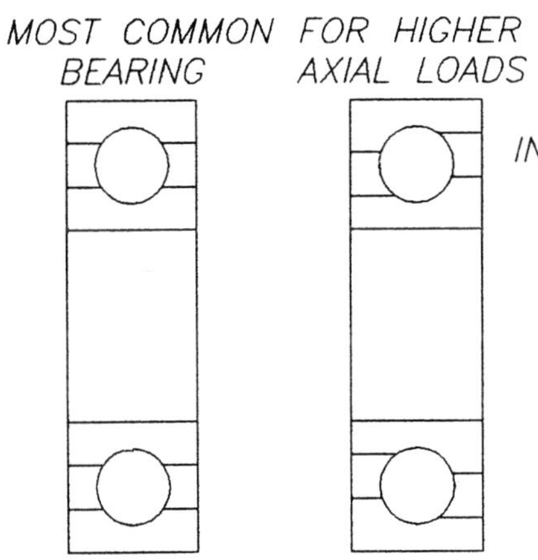

Figure 5
TWO BEARING TYPES

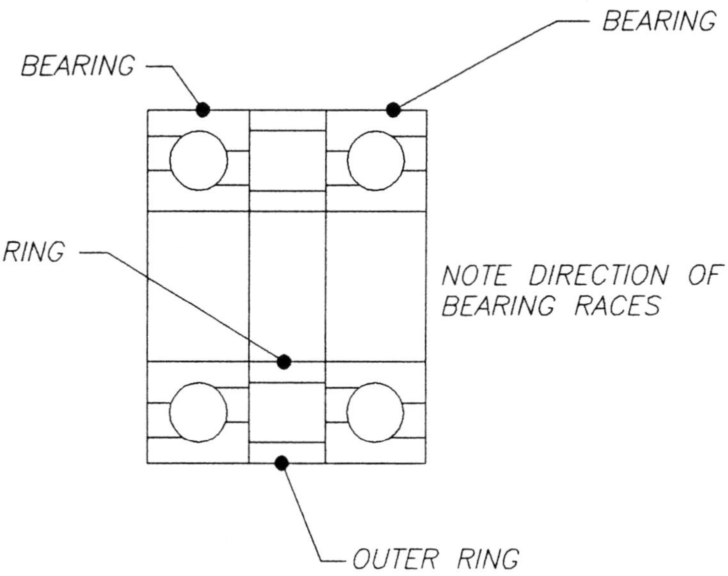

Figure 6
RECOMMENDED MOUNTING

the housing. Figure 6 shows one of the correct methods of mounting the bearings. I suggest this method because the inner races are not clamped until the lead screw is installed. This allows the inner preload ring to float loosely in the assembly, easing the insertion of the lead screw journal. When the graduated dial and handle are assembled and tightened, the bearings become properly preloaded.

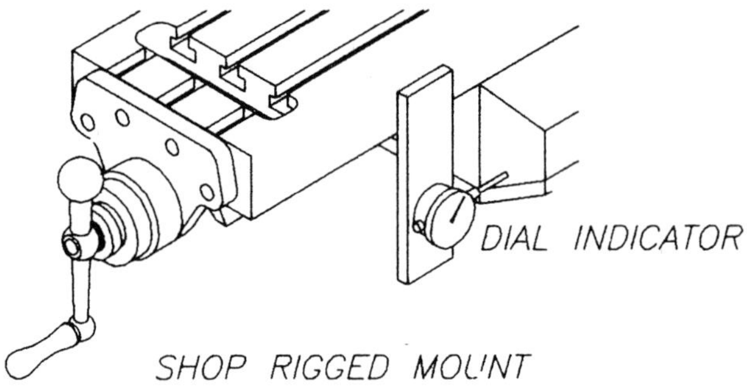

Figure 7
NUT/SCREW BACKLASH TEST

CHANGING BEARING PRELOAD

Note the assembled locations of the Outer Preload Ring and Inner Preload Ring. It is the difference in length between these two rings that determines the preload on the bearings. Now that the bearing housings have the proper clearance, it is a simple matter to disassemble the bearings, and adjust the lengths of the preload rings as required. When the bearings are mounted as shown in Figure 6, the preload will be increased by removing stock from the inner ring, and decreased by removing stock from the outer ring. When increasing the preload, remove an amount equal to the measured backlash plus 0.001 to 0.002". It is important that the length of the preload rings be uniform within a few "tenths" all around. This assures a uniform load on the bearings. A surface grinder is ideal for this work, but a lathe will serve if carefully done.

TEST AND REPEAT

Now that the preload rings have been adjusted, reassemble the table and lead screws, and again test for backlash. Repeat the procedure until the bearings are preloaded to the desired value. When preloaded, the bearings should still feel smooth in their operation. If they feel slightly rough or "bumpy" when the handle is turned, then the preload is too high. Because of the relatively low rpm of a typical lead screw, this method of setting preload is adequate even though it might cause a bearing engineer to wince.

NUT/SCREW BACKLASH

Once the backlash due to the bearing preload has been eliminated, or at least reduced to a minimum, then the backlash due to the nut/screw clearances can be measured. A setup like that shown in Figure 7 will allow this measurement to be taken, but the procedure is different than that used for measuring the bearing preload backlash. Select a setting on the calibrated dial, and approach the setting from both directions with a light drag on the table locks. The difference in the dial indicator reading will be the nut/screw backlash (plus any remaining bearing backlash).

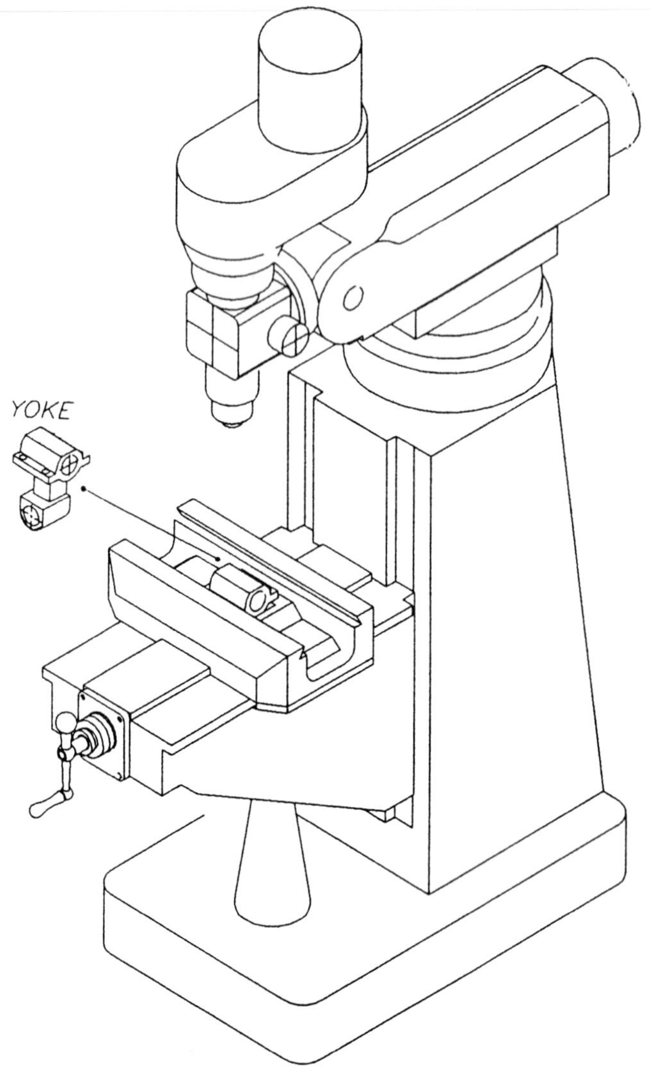

Figure 8
LOCATION OF YOKE

ADJUSTING THE NUT/SCREW BACKLASH

All of the milling machines of this design that I have worked with have had some means of adjusting the nut/screw backlash. This will generally take the form of a one-piece nut that is distorted using a clamp screw to remove the clearance or two separate nuts that have a screw adjustment so they function in the same manner as "jam" nuts. The adjusting screws for both methods are located on the casting, usually called a yoke, that is mounted on the saddle of the machine. Figure 8 shows the yoke and its mounting location. The screws on the X axis can be adjusted without disassembly. The Y axis adjustment will normally require removal of the handle, dial assembly, and bearing assembly for adjustments. The adjustments should be made gradually, reassembling the components where necessary, and retesting for lost motion until the proper settings are obtained. It is much better to "sneak up" on the desired adjustment than to make it too tight, and have to loosen the adjustment because of too much drag on the screws.

FACTORS AFFECTING NUT/SCREW BACKLASH

When a milling machine is new or has been well maintained, the nut/screw backlash can often be adjusted so the combined backlash due to the bearing preload, and the nut/screw will be in the desired range of 0.001 to 0.003". Note that it is exceedingly difficult to get to a figure of less than 0.001". Also note that a value of more than 0.003" should be cause for further work. When the lead screws have been satisfactorily adjusted, they will require periodic checking and adjustment to maintain the desired settings. Even if you don't intend to convert your mill to CNC, the modifications described will result in much improved operation of your machine.

BALL SCREWS

Most industrial CNC machine tools use some form of "zero backlash" lead screw. The most common of these is the ball screw, which is a hardened screw and nut with recirculating balls running between the two in a closed track. This approach allows for easy preloading and very low friction because of the substitution of rolling friction for sliding friction. In addition, the use of hardened screws, balls, and nuts results in a lead screw that will withstand a great deal of usage before adjustments become necessary.

WORN LEAD SCREWS

When a milling machine has had hard use for a period of time, the lead screws will become worn in the areas where the table and saddle are normally positioned. This means setting the nut/screw backlash to the desired value in the area where the lead screws are worn will result in binding when the table or saddle is traversed to the extremes of travel. Conversely, if the nut/screw backlash is adjusted in an unworn area, then the backlash will be excessive when operating in the normally worn areas of the travel.

REPLACEMENT OF WORN LEAD SCREWS

If your milling machine has lead screws worn in the center, then they should be replaced. If you are going to replace the lead screws, then I would strongly suggest you consider ball screws as a replacement. If you do not want to continually have to check, and adjust the nut/screw backlash on your standard lead screws, then I would also recommend that you consider ball screws. In this application they will last forever, require almost no care, and will probably never need adjustments. Ball screws are not inexpensive, but then neither is a new set of standard lead screws. Ball screws for this class of machine can range from $600 to $3,000 or more, depending upon the source and model of your machine. Detailed drawings for ball screws will be presented in a future column so you can purchase them if desired.

ADDRESSES OF SUPPLIERS MENTIONED IN THIS COLUMN:

Bridgeport Machines, Inc., 500 Lindley Street, Bridgeport, CT 06606

Cardinal Engineering, RR 1, Box 163, Cameron, IL 61423

Enco Manufacturing, 5000 West Bloomingdale Avenue, Chicago, IL 60639

FLOOR MODEL MILLING MACHINE

Let's deal with the X- and Y-axis ball screws and parts needed to install them. I've included some detail parts drawings so you can start "making chips." Some news on new software is included at the end of the column.

PREPARATION FOR X AND Y AXIS BALL SCREWS

Should you choose to install ball screws on the X and Y axes (table and saddle motions) of your milling machine, the construction of adapter bushings will be required. Figure 1 shows an exploded sketch of the way the ball screws mount into the yoke (refer to the previous column for location of the yoke). Figure 2 shows the parts assembled.

THE YOKE CASTING

The yoke casting contains the bronze nuts and the adjusting mechanism for taking up backlash in the standard lead screws. Remove the adjusting screws and the bronze nuts, and measure the two bores in which the bronze nuts were mounted. They should be within 0.001/0.002" of each other in diameter. The lengths of these two bores will normally differ with the Y-axis (saddle) being shorter than the X-axis (table). Check the yoke at each end of both bores to be sure all faces are machined square to the bore. If these faces are not square (within 0.001/0.003"), they should be machined square to the bore on each end. Refer to Figure 3 for the location of these faces.

BALL SCREW KITS

When I originally started doing these retrofit projects, I purchased the ball screws as "kits" from several different ball screw manufacturers. They were supposed to be "bolt-in" replacements for existing lead screws. My experience with this was not good, and much rework and adjusting was required to make them fit. One of the biggest problems has to do with the yoke casting. Most of the ball screw kits include a new yoke that is made to mount the ball nuts. This is a good idea, except that the alignment of the ball screws with the bearings discussed in the previous column is critical and is very difficult to maintain with a replacement yoke.

Misalignment does not show up right away because the ball screws run so well, but the side loads on the ball screws due to misalignment will severely shorten the life of the screws. Properly installed ball screws will last as long as the machine tool. Improperly installed screws can be permanently damaged in an incredibly short time. The ball screw kit manufacturers have good products, they just didn't allow for the wide variations in tolerances between milling machines. It's sort of like the "one-size-fits-all" socks available in some stores. Sounds like a great idea, but in reality it doesn't work too well.

After several difficult installations, I came up with the method detailed here which uses the existing yoke that comes with the machine. The keys to this approach are adapter bushings that mount in the existing yoke and receive threaded ends on the ball nuts. By using the existing yoke, the original alignment of the milling machine is preserved. When you disassemble your machine and remove the yoke casting, be sure to check for any shims that might be installed and note exactly where they go. The adapter bushings are detailed in Figures 4 and 5. These are made from 2" diameter steel bar stock.

THE LOCKNUTS

The locknuts shown in Figure 1 are detailed in Figure 6. They can be made to the dimensions shown or purchased from any bearing supply house. They are used to mount ball bearings, and the drawing shown here includes the particular chamfers on these commercial parts. It is not necessary to have the large chamfer on one end of the locknuts if you choose to make them yourself. The slots are for tightening the nuts using a spanner wrench, and should be included because the locknuts are fairly inaccessible when installed in your milling machine (the Y-axis particularly).

BALL SCREW DETAILS

The ball screw assembly for the X-axis (table) is shown in Figure 7. This particular ball screw design can have the drive motor for the CNC system mounted on either end of the table. The ball screw assembly for the Y-axis (saddle) is shown in Figure 8. The dimensions shown are for the lower-cost ball screws and will normally have a lead accuracy of 0.003" per foot. There are a whole series of these drawings to fit different milling machines.

Cardinal Engineering has a full set of drawings for the different variations on all Bridgeport brand mills and most of the "clone" machines. They also stock ball screws made to these drawings.

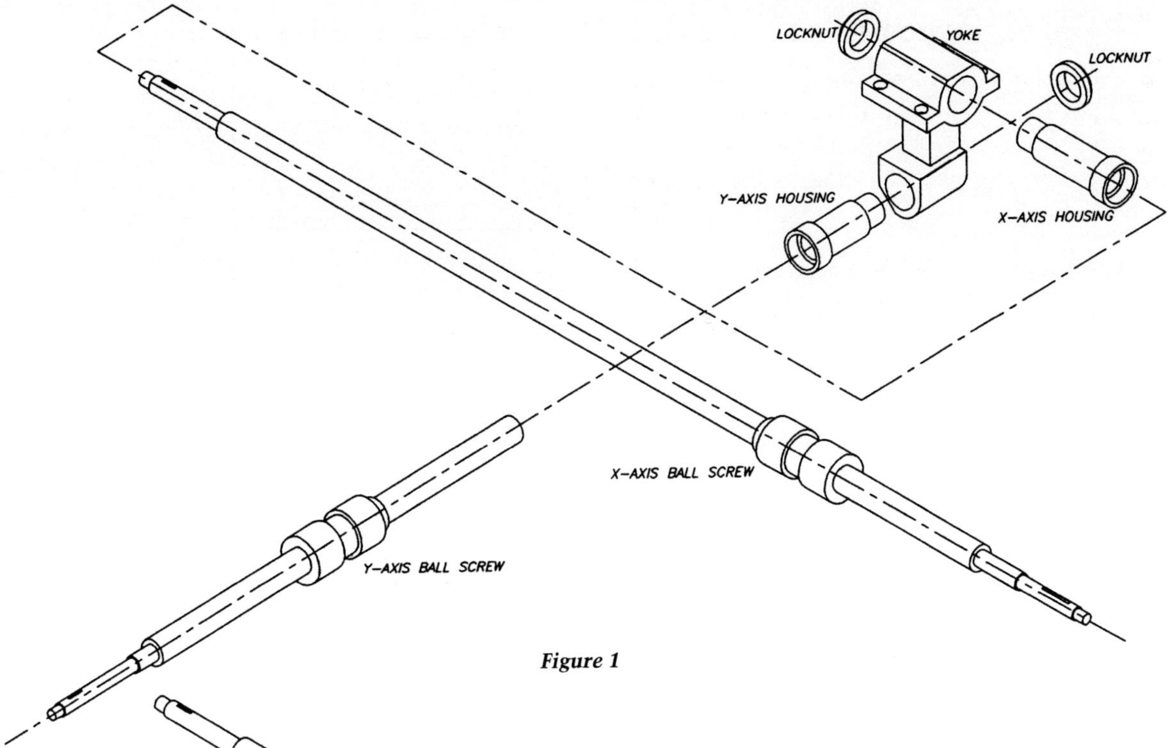

Figure 1

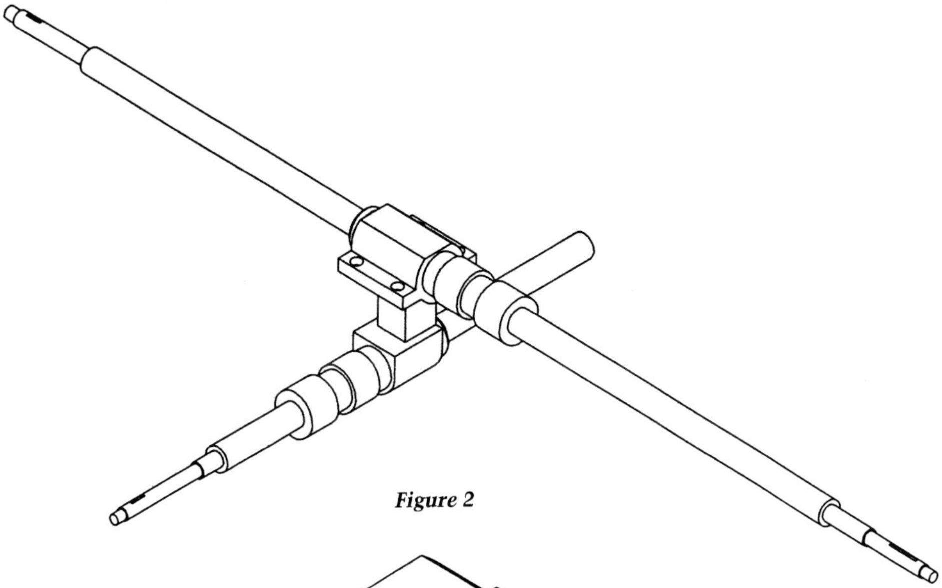

Figure 2

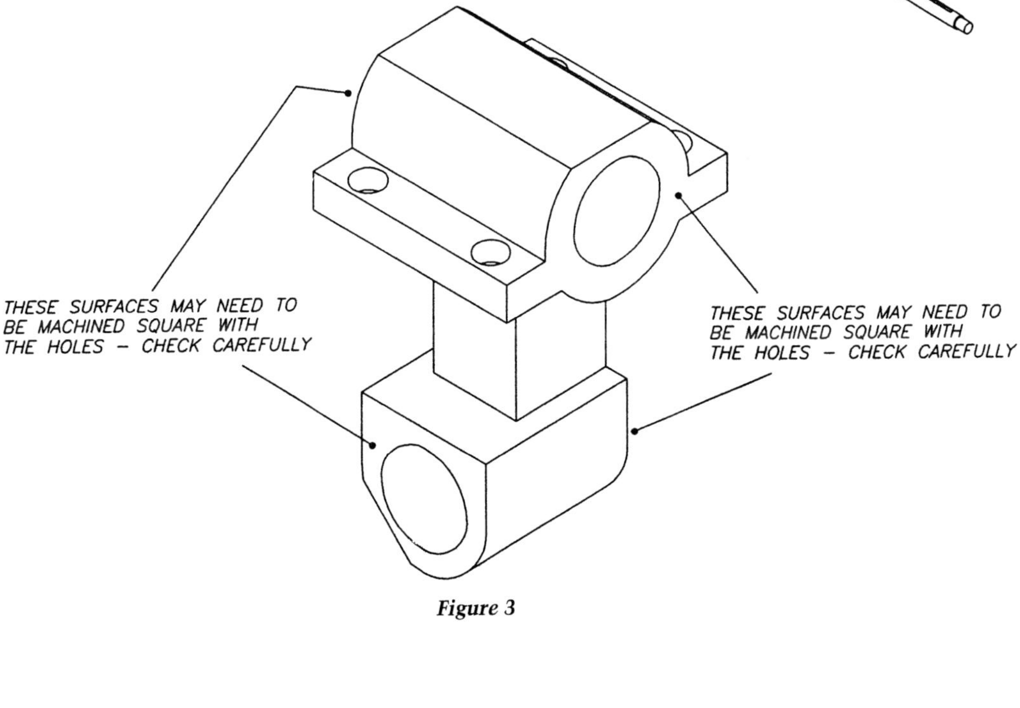

Figure 3

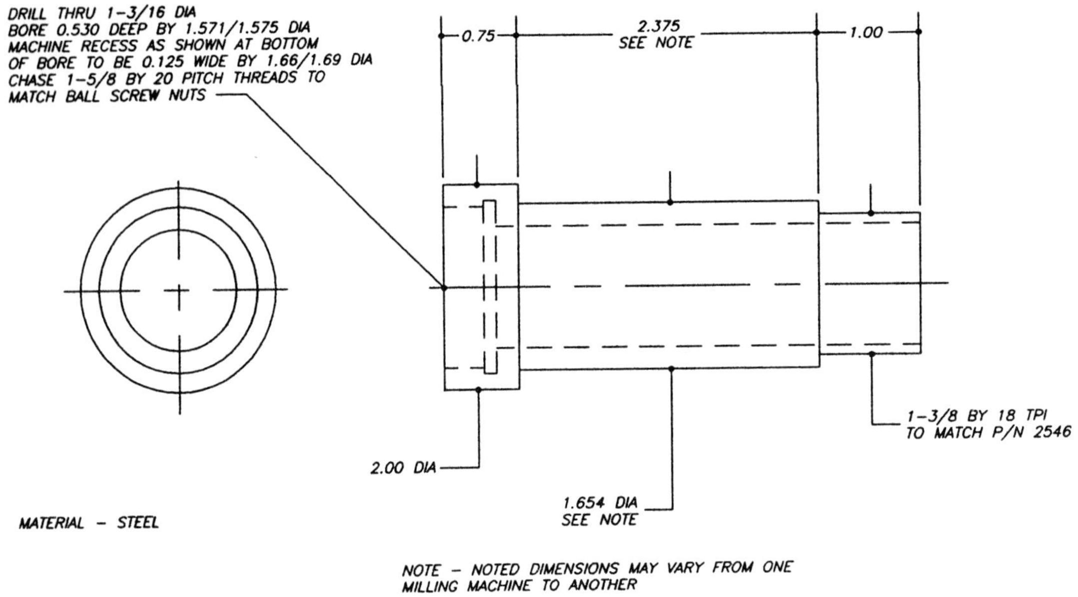

Figure 4

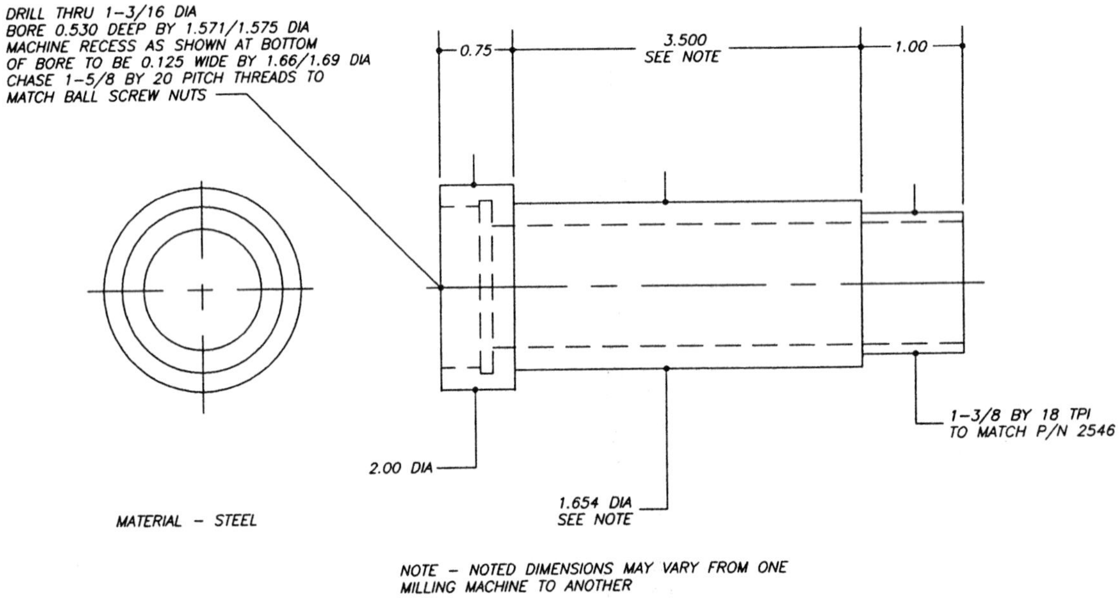

Figure 5

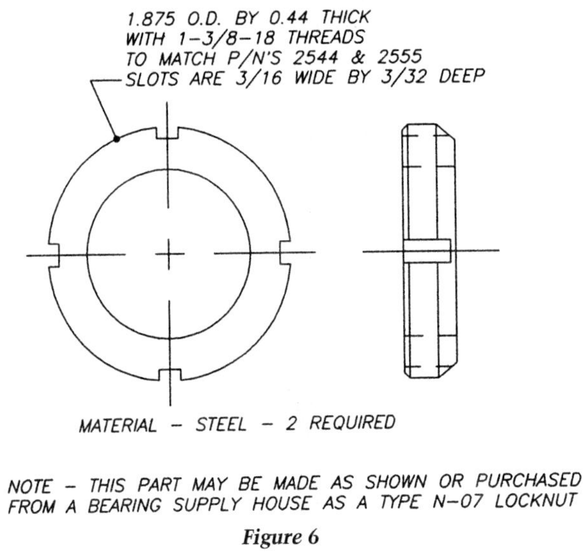

Figure 6

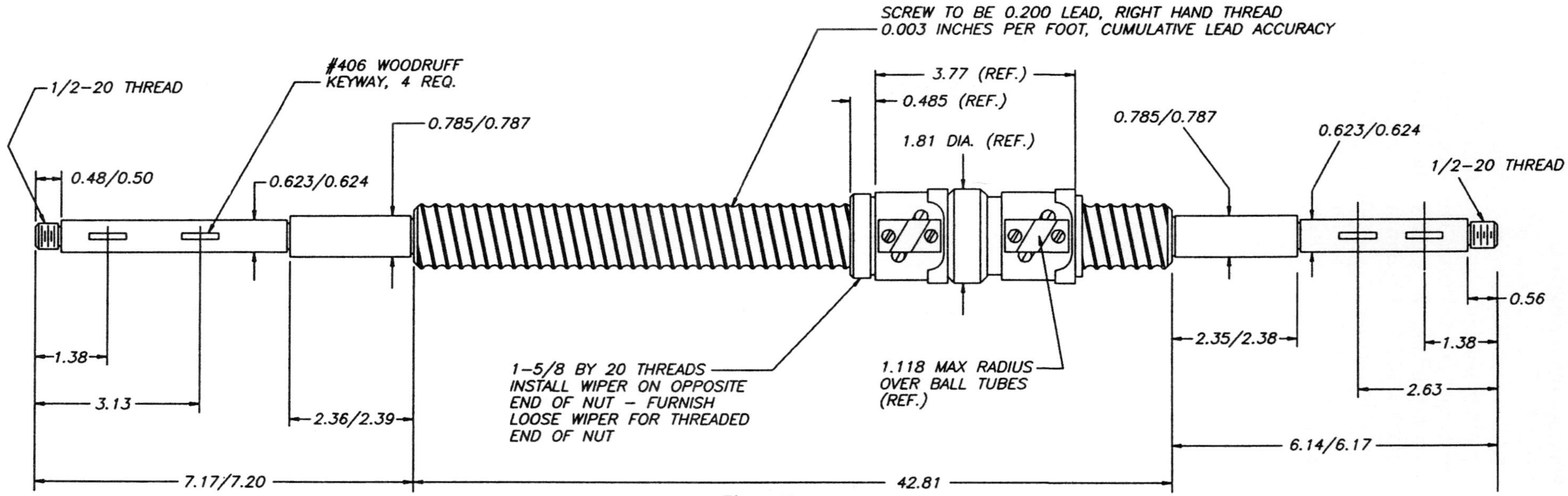

Figure 7

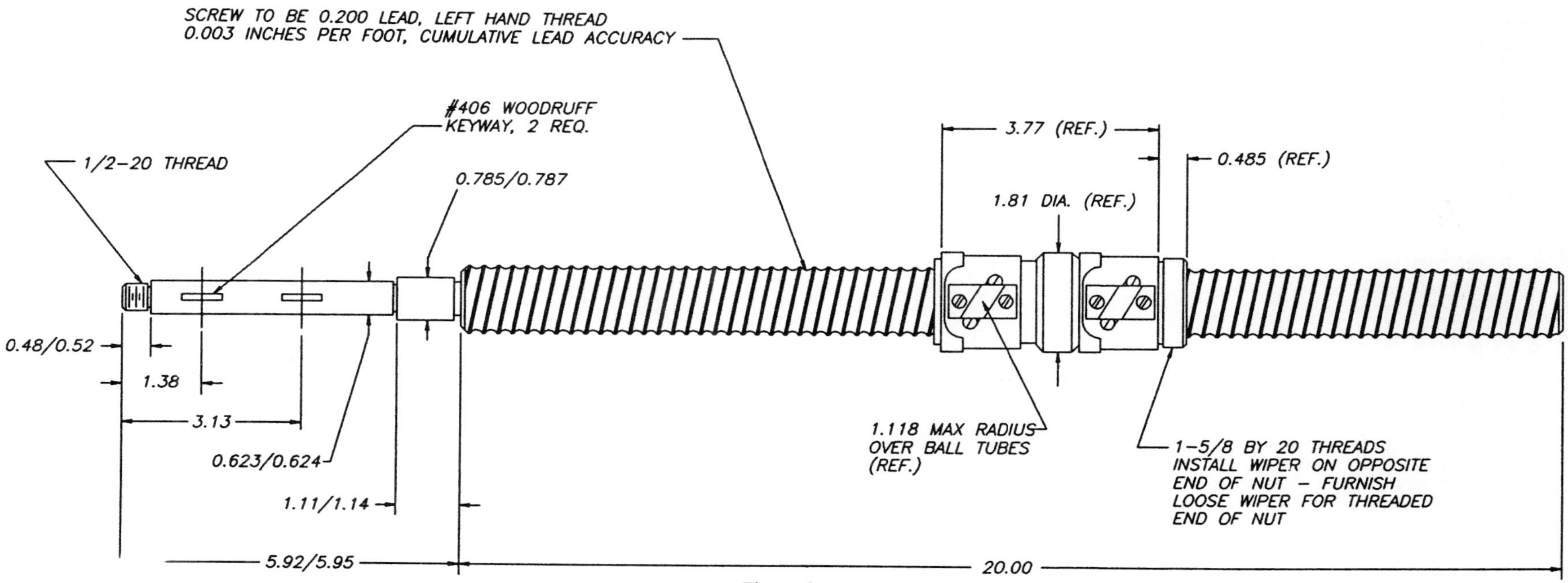

Figure 8

MACHINING CLEARANCE FOR THE ADAPTER HOUSINGS

The yoke, housings, locknuts, and ball screws are shown out of the milling machine for clarity in Figures 1 and 2. When actually assembling the ball screws, the yoke is installed first; then the housings and locknuts are installed and tightened. It will probably be necessary to file or grind some of the saddle casting to clear the way for installation of the X-axis housing (Figure 5). While grinding on your milling machine is not for the "faint of heart," it really isn't that bad once you get started. I use an air powered die grinder with a carbide burr. The resulting gray-iron filings are not particularly harmful to your mill, but it is a good practice to cover exposed working surfaces with shop rags to catch most of the swarf generated by this operation.

INSTALLING THE ADAPTERS AND BALL SCREWS

Once the yoke and adapter bushings are mounted in the saddle, the ball screws can be installed by screwing them into the adapter bushings. If you are using the same ENCO milling machine that I have selected, you can reassemble the mill and continue to use it as a manual machine to make the rest of the parts. You will have to turn one or two bushings for the left end of the X-axis ball screw to position the handle and bearings in the correct location. Those who are retrofitting another brand of milling machine will have to make some bushings and a few other parts. These will be detailed further on in the series.

I have used several different models of CNC controllers over the years but have found the Cardinal/centroid units to be far and away the most "bang for the buck" that is available. However, for those who would like to get into this area with less expenditure, there is a new approach I am working with that will allow those died-in-the-wool tinkerers to build their own CNC systems. A company called Ability Systems Corporation has developed what is known as a *Device Driver* that can be used as one of the key elements of a basic CNC controller.

WHAT'S A DEVICE DRIVER?

Their device driver is a software program that is loaded into the memory of an MS-DOS computer, and will allow the control of up to six stepper motors using inexpensive parallel ports to send out the signals. Their software will allow you to program straight lines and circles (linear and circular interpolation) using their own proprietary language. Their software is known as *Indexer-LPT* and sells for $249. It is available directly from Ability Systems Corporation or from several dealers, including Cardinal Engineering.

USING G-CODES

The normal language used in the shop for CNC programming is the RS-274 (G-Code) language which differs from that used by Ability Systems Corporation. However, Jerry Berns of Cardinal Engineering has written another software package known as CNC-LPT that takes the G-Code language and converts it to the Ability Language. What this means is that you can program in the familiar G-Codes and not have to worry about learning a non-standard language.

BUILD YOUR OWN CNC CONTROLLER

These two software programs, combined with commercially available motor driver cards ($150 to $500 each), power supplies ($150 to $300), and drive motors ($25 to $500), will allow the construction of a CNC controller for as low as $1200 to $1500. I have a prototype unit undergoing test at this time and will include details in the next two columns along with the current retrofit plans.

The Cardinal/Centroid CNC controller can be plugged in and running in a few minutes. However, if you like to tinker, particularly with software and electronics, this new approach will allow you to get started for a few hundred dollars and learn some things about CNC.

NAMES AND ADDRESSES

When contacting these companies, let them know where you heard about them: Ability Systems Corporation, 1422 Arnold Avenue, Roslyn, PA 19001, and Cardinal Engineering, Inc., RR 1, Box 163, Cameron, IL 61423.

FLOOR MODEL MILLING MACHINE

Let's look at details of the parts to make the X- and Y-axis drive assemblies for the milling machine. If you are following along with this series, by now you have modified your milling machine so the bearings and lead screws are properly set up and functioning. If you intend to use ball screws (recommended), then they are installed. Further discussions will proceed under these assumptions.

SELECTING A DRIVE MOTOR FOR THE X AND Y AXES

The selection of an appropriate drive motor for your machine is the next order of business. In my experience, a range of 450 to 1250 oz-inches of rated torque on the stepper motors is adequate for nearly any work that such a machine might be called upon to perform. For a given motor driver package at similar current settings, the motors with larger torque ratings will run slower than those with lower rated torques. This is due to the increased mass of the rotating armature and the higher inductance present in the larger motor. It is not necessarily true that a larger motor (larger torque rating) will produce a more rapid table travel. However, a motor with a larger torque rating will have better capabilities to handle larger cutters at low speeds – more "lugging power," in other words. As mentioned earlier, I favor the motors supplied by Digital Motor in Portland, OR.

The selection of a motor requires matching it closely to the machine and the CNC controller to be used. I have successfully used the Digital Motor P/N HY200-3450-700-B8 (475 oz-in) and the Digital Motor P/N HY200-4270-710-B8 (1350 oz-in) with the Cardinal Engineering P/N 2086 Advanced CNC Controller. I have also used the Cardinal P/N 2077 Trainer CNC Controller on this class of machine, but it doesn't have the ability to deliver enough power to the motors, so I cannot recommend it for this large a machine tool.

THE Y-AXIS DRIVE

Figure 1 shows an exploded view of the Y-axis drive installation. The Y-axis bracket, which is detailed later in this column, is sandwiched between the bearing housing (*HSM*, May/June 1991) and the frame of the machine. The four bolts that hold the bearing housing to the machine frame will need to be exchanged for similar bolts but approximately 3/4" longer to allow for the thickness of the bracket. Figure 2 shows these parts as they are assembled.

Remove the nut that retains the crank on the Y-axis, and slide the crank off the end of the lead screw. Be careful not to lose the Woodruff key which is under the handle. Now remove the calibrated dial assembly from the end of the lead screw. If there are any bushings or shims behind the calibrated dial, be sure to note their location. Remove the bolts that retain the bearing housing and slide the housing off the lead screw. Place the Y-axis bracket over the end of the lead screw, replace the bearing housing and install the new, longer bolts to retain the Y-axis bracket and the bearing housing. Replace any bushings or shims that may have been installed between the calibrated dial assembly and the bearing housing. Replace the calibrated dial assembly, crank, and retaining nut. Tighten the nut and check for free operation.

I have yet to find a milling machine of this general design that doesn't have plenty of room to install the Y-axis bracket without any adverse effects (such as running out of screw travel). However, now would be a good time to check this out. Run the knee over its full travel range to check for operation. Just a note regarding doing this with ball screws. If you run a ball nut off the end of the ball screw, you will have as many as several hundred small balls to attempt to retrieve from the "innards" of your milling machine. This advice is of the "hard-won" variety, so don't say you weren't warned. When you are satisfied the knee has full travel without incident, you are ready to complete the assembly of the Y-axis components.

The motor pulley is a Browning P/N 24XLB037, or equivalent, that has been bored out to fit the selected motor shaft. The driven pulley is a Browning P/N 48XLB037 that has been bored out to fit the extension of the calibrated dial assembly. The belt is a Browning P/N 250XL037. The same pulleys and belts are also used on the X and Z axes, suitably modified in each case.

Install the motor standoffs (Figure 8). Do not install the belt guard now. Install the selected motor on the ends of the motor standoffs using suitable bolts and washers. Slide the calibrated dial assembly onto the end of the lead screw and slide the driven pulley onto the end of the extension of the calibrated dial assembly. Slide the motor pulley on the motor shaft and align the motor pulley and driven pulley. Install the crank and retaining nut and check for proper operation and clearances. The objective is to get the pulleys located so the drive belt will be centered inside the belt guards, still allowing enough room behind the belt guards to operate the locking ring which sets the calibrated dial assembly. Suitable clearance for operation of the crank is also important, but this has never been a problem when using the belts and pulleys specified.

Once you are satisfied with the placement of the motor and driven pulleys, mark the location of the setscrew and disassemble the components so you can machine or file a flat on the motor shaft and the extension of the calibrated dial assembly.

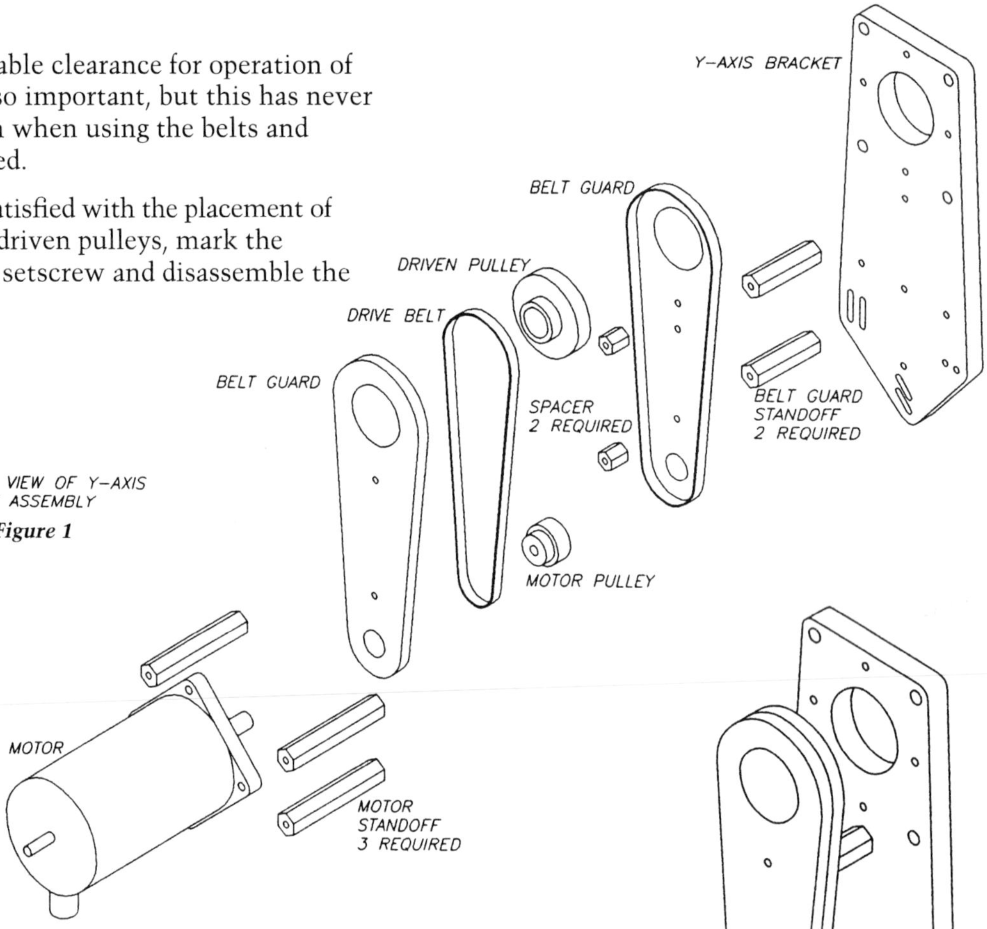

EXPLODED VIEW OF Y-AXIS DRIVE ASSEMBLY

Figure 1

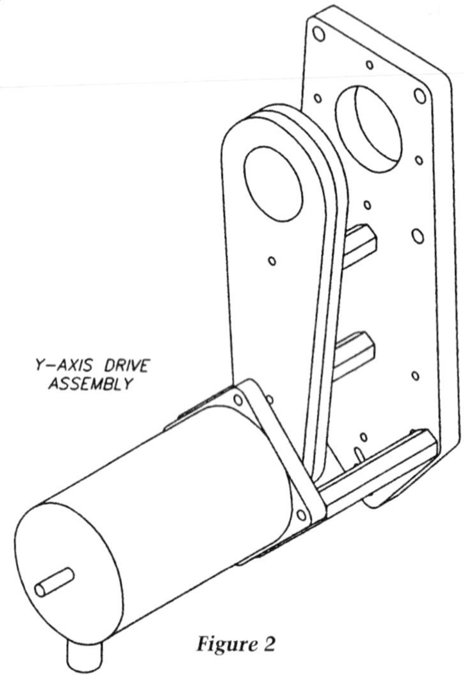

Figure 2

Reinstall the calibrated dial assembly without the driven pulley. Assemble the rear belt guard, P/N 1727, to the belt guard standoffs using the guard spacers, P/N 2224 (*HSM*, May/June 1990, pages 50 and 56). Install the driven pulley, crank, and retaining nut. Lock the driven pulley in place so it clears the rear belt guard, and is centered on the guard spacers. Place the drive belt on the driven pulley, and install the front belt guard. Install the motor pulley on the motor shaft, and attach the motor to the motor standoffs, being sure the motor pulley engages the belt. Note that adjustment of the motor belt tension is done by loosening the motor standoffs where they attach to the Y-axis bracket, rotating the entire assembly, and tightening the motor standoff bolts. These belts do not have to be very tight to transmit the full rated power of the motors. Over-tightening will shorten the life of the motor bearings and the belts. When everything is assembled, turn the crank by hand over the full range of knee travel, checking that the belt tracks properly and does not rub on the inside of the belt guards. Adjust as required. This completes the installation of the Y-axis drive.

THE X-AXIS DRIVE

Figure 3 shows the exploded view of the components for the X-axis drive assembly. The belt guards, belt, motor pulley, motor standoffs, belt guard spacers, and motor are all identical to those used on the Y-axis drive. Figure 4 shows these parts assembled. The driven pulley may require a slightly different bore size to match the extension on the calibrated dial assembly. I have noticed some machines have different diameters on the X and Y axes for this part. Another difference, of course, is in the X-axis bracket which is designed to fit the different bearing housing used on the milling machine table. Finally, the belt guard standoffs for the X-axis drive are a different length than those used on the Y-axis (Figure 8 in the next chapter will illustrate this point).

The X-axis drive is designed to be installed on either the right or left end of the table. If your machine has the power feed unit Enco sells for this model of milling machine, you will be forced to install the X-axis drive on the left end. The ball screws detailed in the previous installment of this series are designed to accept installation of the X-axis drive on either end of the table. If you choose to retain the original lead screws, you will have to do some creative adjusting with bushings and/or make a new inner sleeve for the calibrated dial on the left end of the table. Since this will vary with each case, I will leave this up to you. The following discussion will consider the X-axis drive to be installed on the right end of the table.

The X-axis bracket is installed similarly to the bracket for the Y-axis. The crank, calibrated dial assembly, spacer bushings or shims, and bearing housing are all removed. Longer bolts are procured to replace the originals; the X-axis bracket is placed over the end of the lead screw and retained in place by the bearing housing and its bolts.

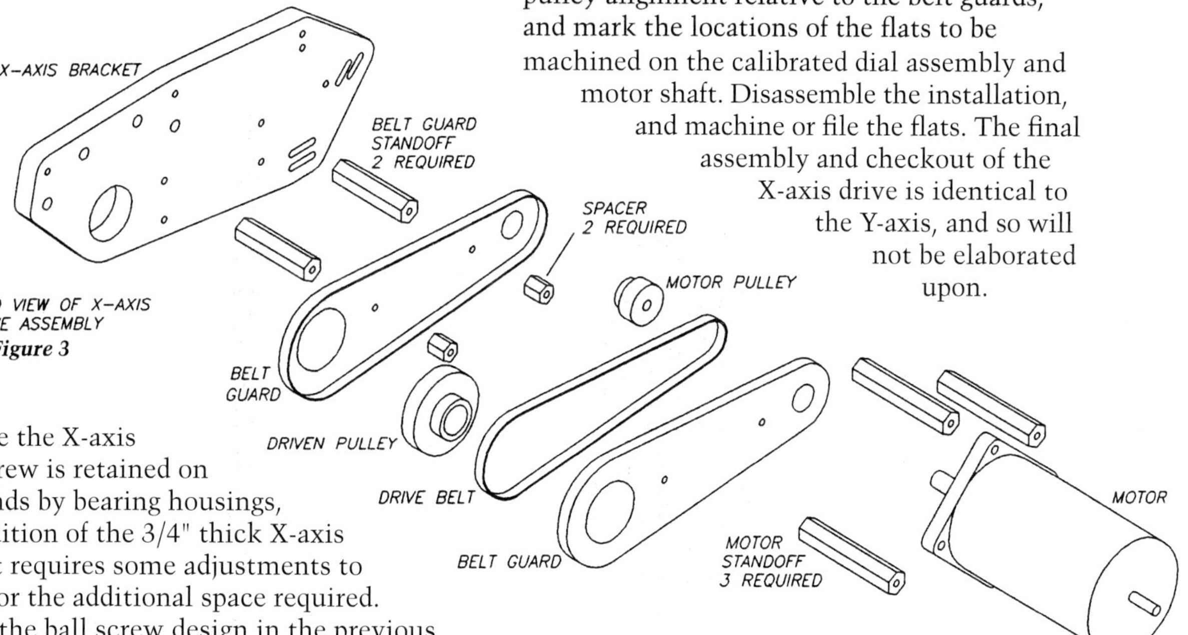

EXPLODED VIEW OF X–AXIS DRIVE ASSEMBLY
Figure 3

Because the X-axis lead screw is retained on both ends by bearing housings, the addition of the 3/4" thick X-axis bracket requires some adjustments to allow for the additional space required. Again, the ball screw design in the previous installment makes allowances for this. If you use the original lead screws, however, it will be necessary to make some adjustments.

ADJUSTMENTS WHEN USING ORIGINAL LEAD SCREWS

Because the Enco milling machine I am using as the basis for this series has a uniform diameter bearing journal for the entire length of the left end of the X-axis lead screw, length adjustments are simplified. A 3/4" long bushing is inserted between the threaded part of the lead screw and the bearings on the left end of the table. This bushing causes the right end of the lead screw to be properly located for installation of the bracket and other parts. However, the left end then does not have enough length to install the calibrated dial assembly and crank.

There are two basic solutions to this problem. The first is to discard the calibrated dial assembly and just install the crank, with suitable bushings to space it properly on the journal. The second is to remove material from the outer face of the left bearing housing, both ends of the calibrated dial assembly, and the center of the crank. By removing a total of 1/2" divided among all three of these items, the crank and calibrated dial can then still be mounted on the left end of a standard lead screw.

These parts are the same as those used on the Y-axis except that, as mentioned, the driven pulley may require a slightly different bore size to match the extension on the calibrated dial assembly.

The installation procedure for the standoffs, pulleys, and belt is identical to that used for the Y-axis. Install the parts, check for belt and pulley alignment relative to the belt guards, and mark the locations of the flats to be machined on the calibrated dial assembly and motor shaft. Disassemble the installation, and machine or file the flats. The final assembly and checkout of the X-axis drive is identical to the Y-axis, and so will not be elaborated upon.

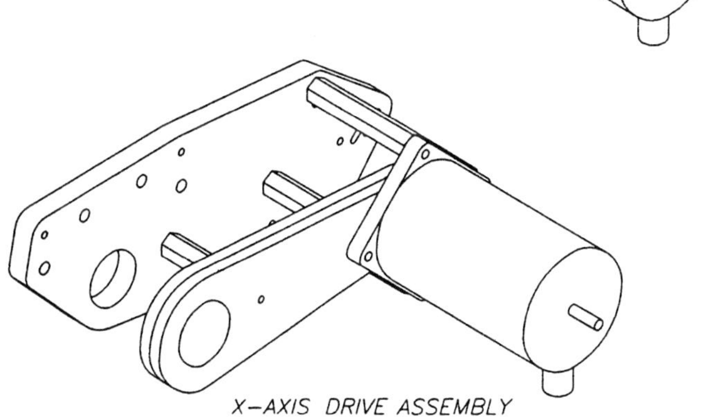

X–AXIS DRIVE ASSEMBLY
Figure 4

COMMENTS ON BRIDGEPORT AND OTHER BRANDS OF MILLING MACHINES

As has been mentioned in this series, most of the parts shown can also be used on other brands of milling machines with little or no change. The most challenging problem is using the existing lead screw and making space for the X-axis bracket. The Y-axis bracket doesn't normally pose a problem because one end of the Y-axis screw is not supported. On milling machines that are not "copies" of a Bridgeport brand mill, you will have to design brackets to suit your milling machine, and also probably come up with your own variation on the ball screws.

When designing your own brackets, if you maintain the relationship between the large hole through which the lead screw passes, relative to the curved slots and pivot holes, then the outside shape is relatively unimportant. This will allow the use of the same pulleys

NOTES:

1. HOLE "A" IS THE PIVOT POINT FOR SMALLER MOTORS
2. SLOT "B" IS 0.281 WIDE AND IS ON A 3.875 RADIUS MEASURED FROM HOLE "A"
3. SLOT "C" IS 0.281 WIDE AND IS ON A 2.740 RADIUS MEASURED FROM HOLE "A" – THE INCLUDED ANGLE OF THE SLOT IS THE SAME AS THAT OF SLOT "B"
4. HOLE "D" IS THE PIVOT POINT FOR LARGER MOTORS
5. SLOT "E" IS 0.281 WIDE AND IS ON A 4.950 RADIUS MEASURED FROM HOLE "D"
6. SLOT "F" IS 0.281 WIDE AND IS ON A 3.50 RADIUS MEASURED FROM HOLE "D" – THE INCLUDED ANGLE OF THE SLOT IS THE SAME AS THAT OF SLOT "E"

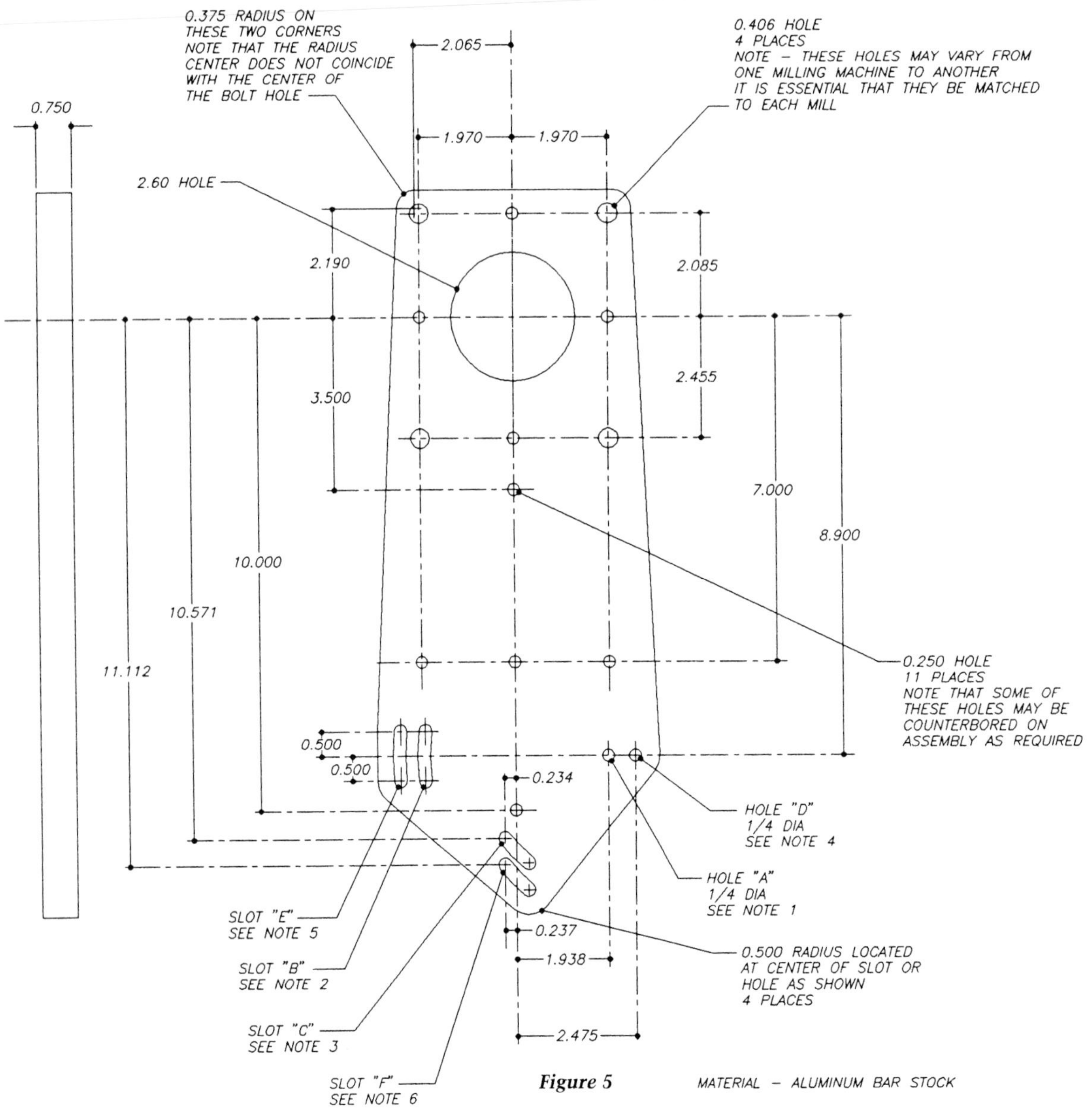

Figure 5

MATERIAL – ALUMINUM BAR STOCK

and belts. If you need to change pulleys and belts, then a whole new layout will be needed, and the belt guards will probably not work. The belt guards are not absolutely necessary, but we use them on most of the retrofits we do in our shop as a matter of good practice. Also, when redesigning belt and pulley relationships, try to keep the recommended number of teeth in engagement or a problem with "skipping" belt teeth can occur. The motors are not powerful enough to strip the teeth from the belts, but if too few teeth are engaged it is easy for the belt to hop out of engagement and "skip" to the next tooth.

Figure 5 is a detail drawing of the Y-axis bracket. It is constructed of 3/4" thick aluminum bar stock. Steel could be used if desired. I normally use 6061-T6 aluminum for these parts, but it is not critical, and substitutions can be made. The softer aluminum alloys will also work but they are "gummier" to machine. If you have access to the harder alloys such as 2024 or 7075, they will also do an excellent job. Some comments are in order on the design of this bracket. Notice the 0.406" diameter holes at the top end of the bracket. These holes are for the bolts that retain the bearing housing and the bracket. The dimensions on these holes will vary depending upon which milling machine you are using. Also notice the four curved slots in the lower end of the bracket. These slots are for adjusting the belt tension, and are sized to accept the two motors referred to earlier in this article. If you elect to use some other motor with a different bolt pattern, enough information is provided on the detail drawing to construct a redesigned bracket, provided you use the same belt and pulley combinations.

Figures 6 and 7 are the final and blank drawings respectively for the X-axis bracket. The blank drawing details the outline of the part and several of the holes used. The final drawing details the motor belt tensioning slots and the 0.406" diameter bearing housing retainer bolt holes in the same manner as described for the Y-axis bracket.

Figure 8 gives details for the standoffs which are used on both the X and Y axes. I normally use 3/4" hex aluminum or steel for these parts

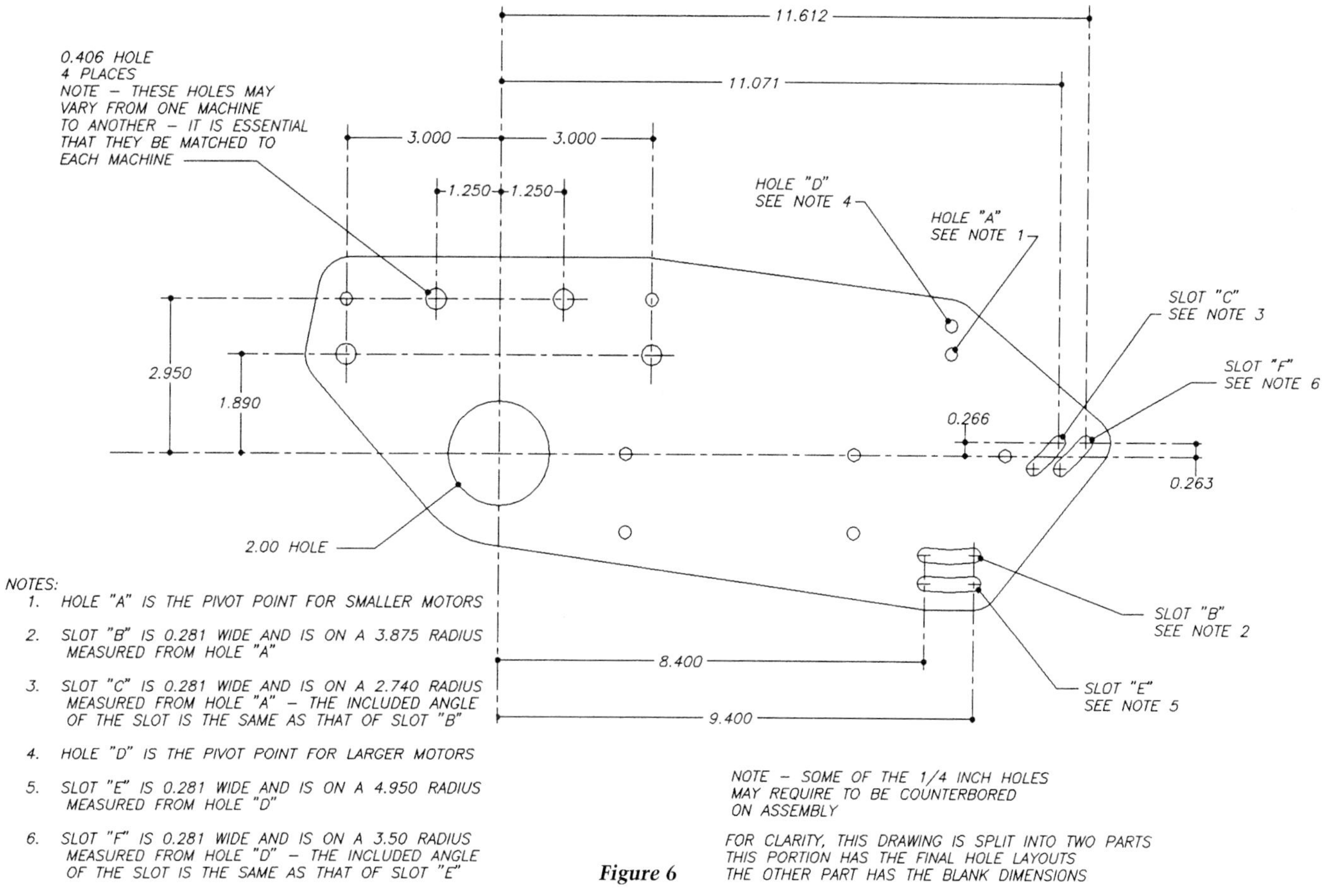

Figure 6

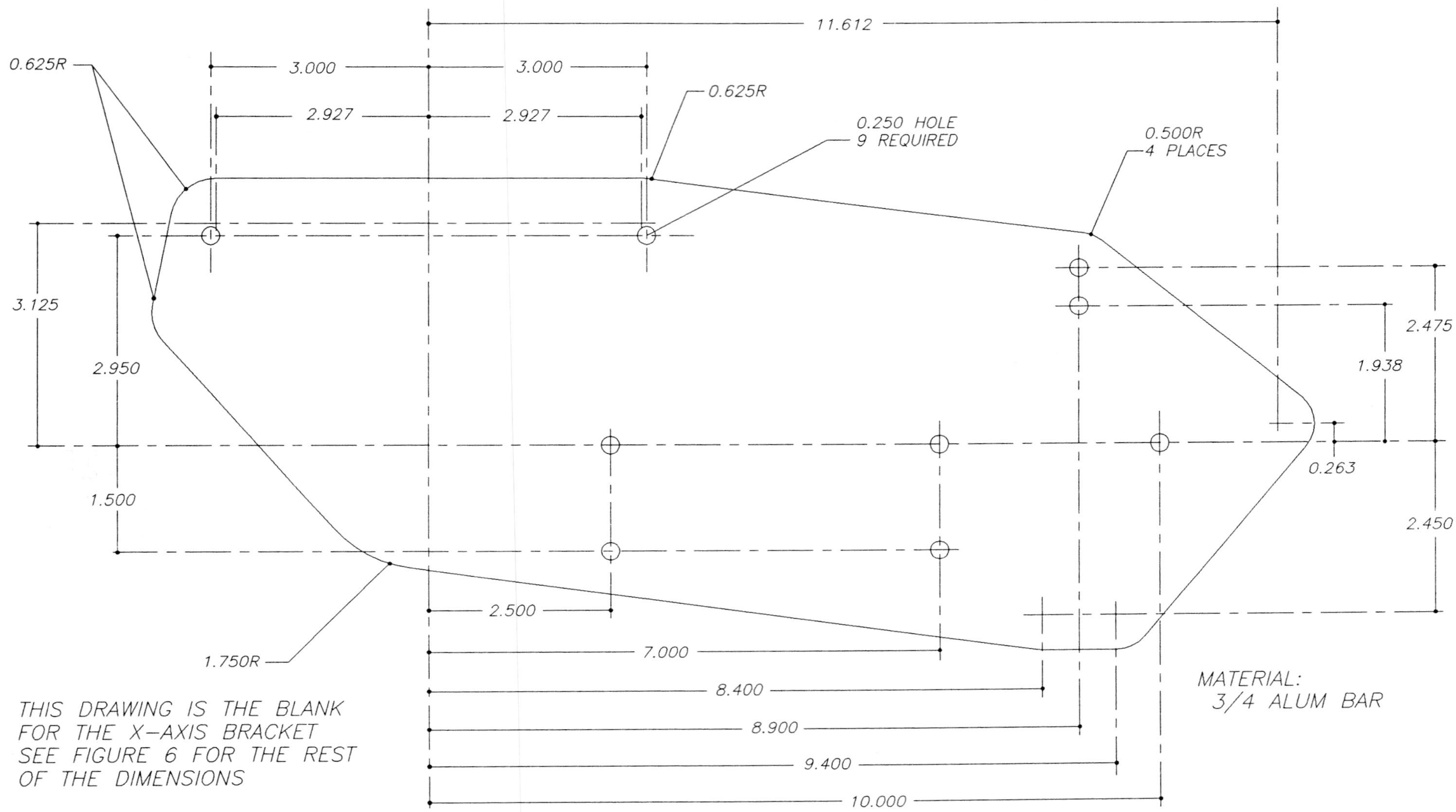
11.612
0.625R
3.000
3.000
0.625R
2.927
2.927
0.250 HOLE
9 REQUIRED
0.500R
4 PLACES
3.125
2.950
2.475
1.938
0.263
1.500
2.450
2.500
7.000
1.750R
8.400
8.900
9.400
10.000
MATERIAL:
3/4 ALUM BAR
THIS DRAWING IS THE BLANK
FOR THE X-AXIS BRACKET
SEE FIGURE 6 FOR THE REST
OF THE DIMENSIONS

Figure 7

so a wrench can be used for installation or adjustment. Round stock would work also if a couple of wrench flats were milled at about the halfway point of each standoff.

THE BUILD-IT-YOURSELF CNC CONTROLLER PROJECT

I promised to have more details about this project in this column and the next. The first prototype system is up and running, and software has been written by Jerry Berns to allow the use of the basic CNC motions, i.e., rapid traverse (G0), linear interpolation (G1) and circular arcs in both directions (G2 and G3). I am investigating several different brands of motor driver electronics packages, and want to test another one or two brands before publishing the results. I am currently using motor drivers from General Controls Electronics, Inc., and Intelligent Motion Systems, Inc. Their addresses are at the end of this article. Specific part numbers and wiring diagrams will be published when tests are completed.

In general, the "Indexer-LPT" software mentioned in the previous article (*HSM*, July/August 1991) can be used to run any motor driver that will accept a "STEP" and "DIRECTION" signal, and requires input voltages compatible with a computer "LPT" port (often referred to as TTL level signals). There are many companies that manufacture boards who will accept these signals and drive stepper motors. While it is not possible for me to test all of those available, a more complete listing of sources will be published in a later column, along with comments regarding their suitability for home-brew applications.

I have been in contact with Mr. Art Volta, author of the "Indexer-LPT" software, and he is continuing to make improvements in his software package. His most recent additions include the ability to use a "joystick" to move the motors without programming. If you have some kind of "wonder machine" you have always wanted to build that requires computer controlled motion, his software would be a good place to start. His manuals are well written and understandable, and he has always been helpful with telephone support for special questions.

REGARDING SUPPLIERS

I read *The Home Shop Machinist* and *Projects in Metal* (as well as several others) with great interest myself, and have been following the "Universal Pillartool" series by Harold Mason. If I can ever get caught up, I would like to build one. However, I would also like to take this opportunity to make a similar statement regarding my recommendations concerning suppliers. With the exception of Cardinal Engineering, my employer, I have no financial interest in any of the companies I recommend. Please mention this article when writing for information. Let these people know you appreciate their cooperation.

"Indexer-LPT" software: Art Volta Ability Systems Corporation, 1422 Arnold Avenue, Roslyn, PA 19001

Timing belts and pulleys: Browning Manufacturing Division, Emerson Electric Co. Call your local power transmission or bearing supplier

CNC controllers and software: Cardinal Engineering. See advertisement in this issue

Stepper motors: Digital Motor, Inc., 13563 N.W. Cornell Road, Suite 255, Portland, OR 97229

Stepper motor drivers: General Controls Electronics, Inc., 2950 Brickvale Drive, Elk Grove Village, IL 60007

Stepper motor drivers: Intelligent Motion Systems, Inc., 511 Norwich Ave., Taftville, CT 06380

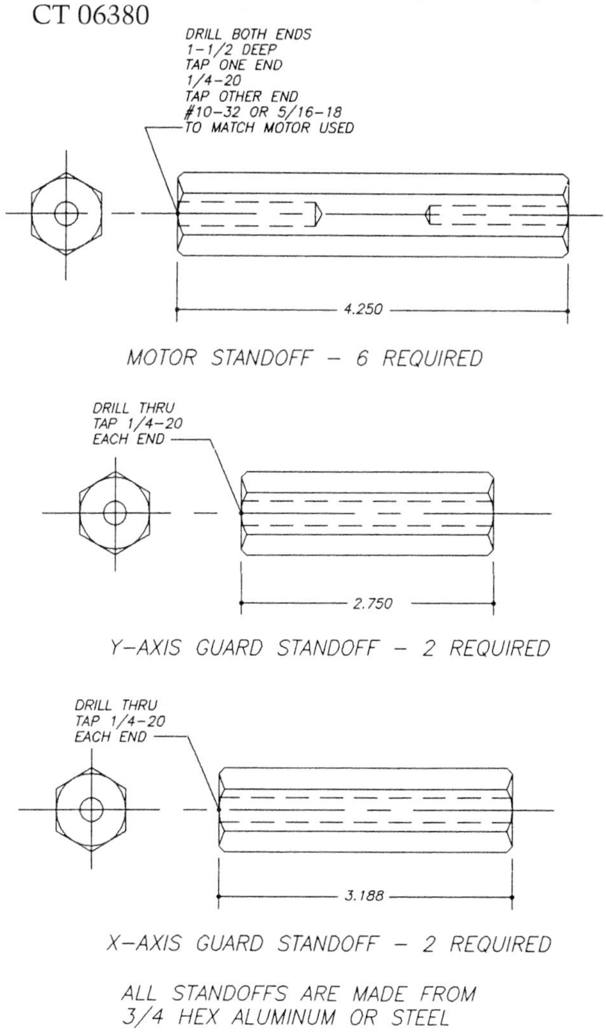

Figure 8

FLOOR MODEL MILLING MACHINE

This time, let's look at the general arrangement and begin the details of the Z-axis drive for the milling machine. I have done about three different versions of Z-axis designs and this is the most successful. It provides for a ball-screw drive to the Z-axis, and also allows for conversion between CNC and manual operation by removal of a single screw.

DRIVE MOTOR SELECTION FOR Z-AXIS

The 475 oz-in drive motor mentioned in the September/October 1991 issue of *HSM* is what I recommend for this installation. Because of the very tight fit, the larger motors are not practical without a redesign of the brackets and additional "beefing up" of the overall design. The 475 oz-in motors seem to do everything required of them, and the use of larger motors for the Z-axis drive would probably not be warranted.

Figure 1 shows an exploded view of the Z-axis drive system. The Adapter Plate and Gear Spacer are mounted between the milling head and the knuckle that allows the milling head to be tilted in both directions. The milling head must be removed to take the required measurements for this phase of the conversion to CNC. Removing the head is not difficult if the proper procedure is followed. When the head has been removed, it will be necessary to have another milling machine available to "dial in" the locations of the clamping studs. As an alternative, you can "jury-rig" a spindle on your existing mill to allow you to pick up the locations of the clamping studs.

Figure 2 shows the sequence to be followed when removing the head (see the note below regarding infinitely variable speed mechanism). Fortunately, all of the "clone" and Bridgeport brand name milling machines I have worked with can have their heads removed by disassembling them in modules. The motor and motor pulley can be removed by taking out the pivot screw and clamping screw used to rotate the motor when changing belt speeds. Some disconnecting of electrical wires may be required, depending upon how your machine is wired. BE SURE THE MACHINE IS DISCONNECTED FROM THE ELECTRICAL SYSTEM BEFORE STARTING ON THIS PROJECT. The drive motor and pulley can then be lifted from the top of the aluminum housing, and set aside. This operation can be done by one person, but two make it a lot safer as the motor can be quite heavy. Now remove the drawbar from the quill if it has not already been done.

I do not recommend removing the motor on the infinitely variable speed head because of the complexity of the mechanism. The motor should be left attached to the large aluminum belt housing and removed with it as an assembly. This will *require* some form of lifting equipment as the assembly is much too heavy and awkward for a single person or even several people to handle manually.

The large aluminum housing holds the main drive belts and also the backgear mechanism. Lower the quill all the way so the splined drive shaft is as far down as possible. This is done so the housing will not have to be lifted as far for removal. The methods used for removing this part are the same for either the step-pulley head or the infinitely variable speed head. There are three studs that clamp the aluminum housing to the main head casting. Remove the nuts from these studs, and the aluminum housing can be lifted straight up until the studs clear their holes and the drive system clears the splined drive shaft. Set the housing aside. (Be careful when doing this operation.)

As I mentioned earlier, this has been done by a single person but it is a whole lot safer and much easier if you have some kind of lifting device and some additional help. Once the housing is unbolted and the lifting process begins, it is very difficult to "back out" of the job gracefully if it proves to be too much for your strength or the available head room.

See Figure 3 for a view of the milling machine with the motor and belt housing removed. The head and quill assembly is held to the ram with four studs. Retract the quill and lock it in place. If possible, it might also be a good idea to remove the quill handle at this time. Many of them are retained only by a spring loaded ball arrangement, and if you grab the handle while removing the head assembly it can come off in your hand, and you can easily drop the head assembly. This can be embarrassing at best, and both expensive and dangerous at worst. If you place a block of wood on the table of your milling machine, and support the end of the quill by raising the table to take the load off of the bolts, it will be possible to remove the four retaining nuts that hold the head assembly to the ram. Carefully remove the head assembly and set it aside. Also remove the studs in the circular T-slot at the end of the ram. If all of the studs or nuts are not the same length, be sure to note where each respective part is installed.

Figure 4 shows the milling machine with the head removed and mounted on the table. At the end of the ram, there is a small gear that meshes with a worm on the head. This gear

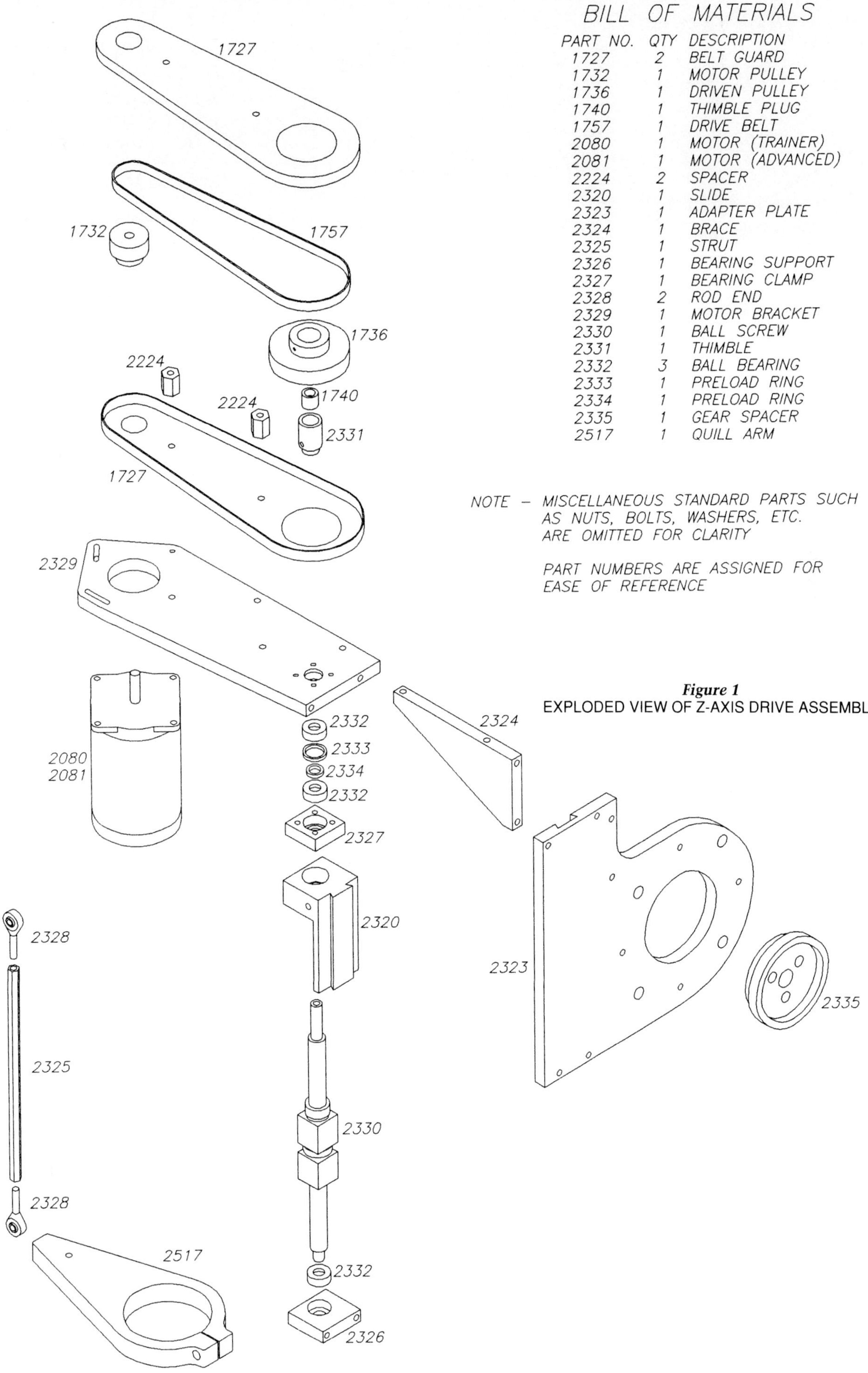

BILL OF MATERIALS

PART NO.	QTY	DESCRIPTION
1727	2	BELT GUARD
1732	1	MOTOR PULLEY
1736	1	DRIVEN PULLEY
1740	1	THIMBLE PLUG
1757	1	DRIVE BELT
2080	1	MOTOR (TRAINER)
2081	1	MOTOR (ADVANCED)
2224	2	SPACER
2320	1	SLIDE
2323	1	ADAPTER PLATE
2324	1	BRACE
2325	1	STRUT
2326	1	BEARING SUPPORT
2327	1	BEARING CLAMP
2328	2	ROD END
2329	1	MOTOR BRACKET
2330	1	BALL SCREW
2331	1	THIMBLE
2332	3	BALL BEARING
2333	1	PRELOAD RING
2334	1	PRELOAD RING
2335	1	GEAR SPACER
2517	1	QUILL ARM

NOTE – MISCELLANEOUS STANDARD PARTS SUCH AS NUTS, BOLTS, WASHERS, ETC. ARE OMITTED FOR CLARITY

PART NUMBERS ARE ASSIGNED FOR EASE OF REFERENCE

Figure 1
EXPLODED VIEW OF Z-AXIS DRIVE ASSEMBLY

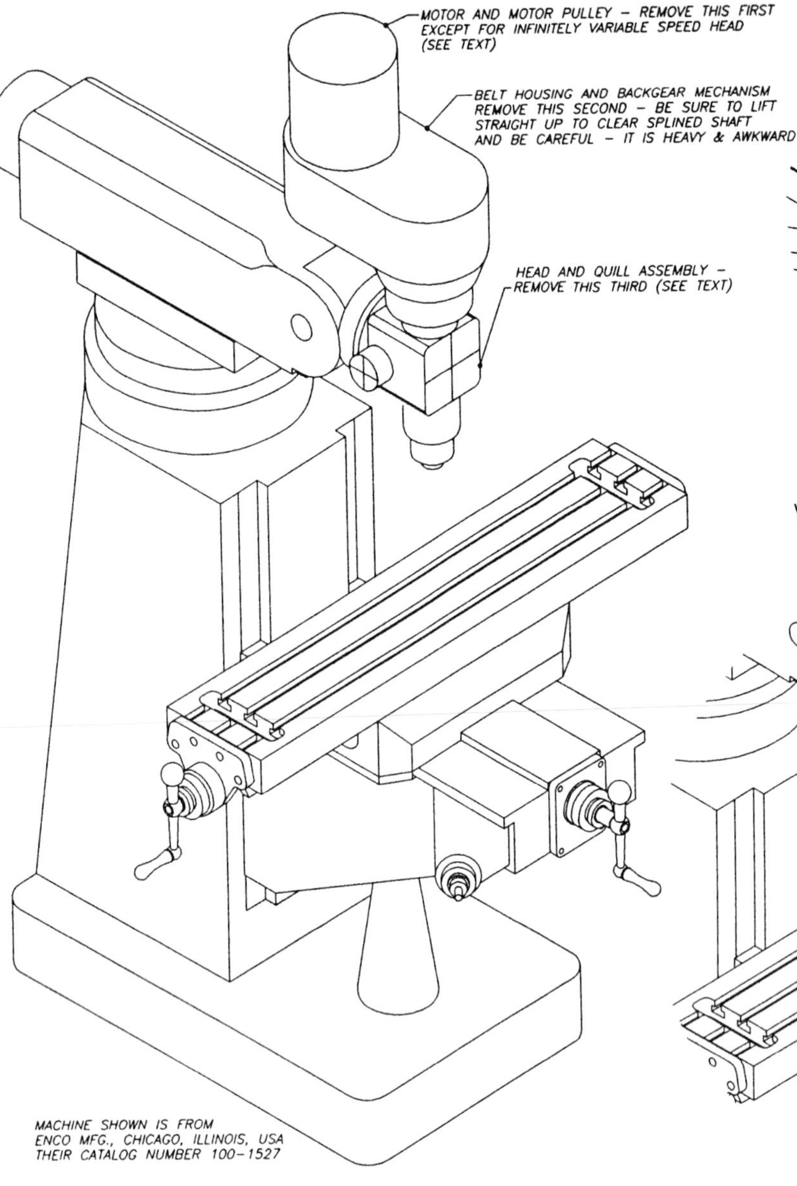

Figure 2
SHOWING SEQUENCE OF HEAD REMOVAL

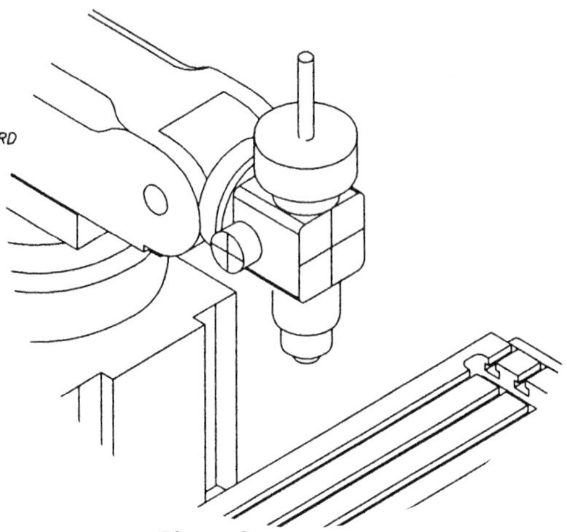

Figure 3
VIEW OF MILL WITH MOTOR AND BELT HOUSING REMOVED

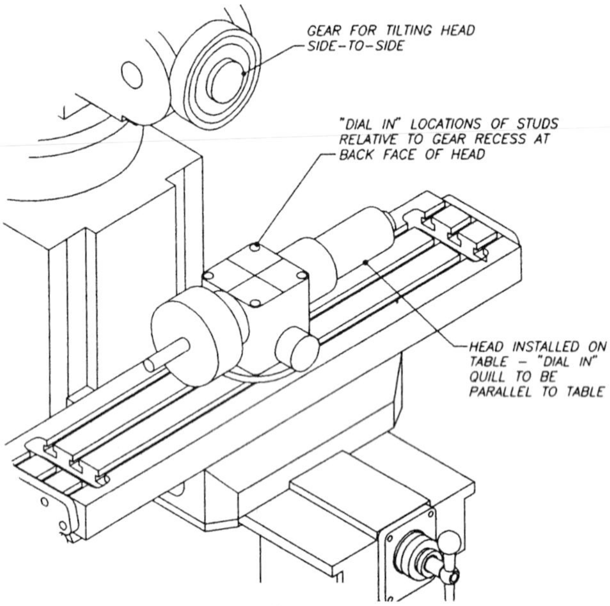

Figure 4
VIEW OF MILL WITH HEAD REMOVED AND INSTALLED ON TABLE

allows the head to be tilted from side to side as required. Because the Z-axis drive mechanism is mounted between the ram and the head assembly that was removed in the previous paragraph, the gear will have to be removed and a spacer (Figure 10, P/N 2335) made to compensate for the addition of the Z-axis drive mechanism. Remove the gear. Most of the gears I have encountered have been retained with three or four cap screws. Some of them have also had a roll pin or a dowel pin installed. The bolts used to retain this gear are normally ⅜" or 10mm. The dowel pins are approximately the same size. A quick calculation shows that a single bolt of that size would be adequate to hold the gear against the rotational shear load placed on the gear by the head in its most critical tilted position. Consequently, it is not likely the dowel pin is required, and it can be eliminated when the gear is reassembled with the spacer. If you are not comfortable with leaving the dowel pin out, I would suggest drilling a new hole through the gear, spacer, and ram casting when you reassemble the gear.

The Adapter Plate (P/N 2323) mounted between the head and the ram must be drilled to match the stud locations in the milling head. I have found these to vary quite a bit between milling machines (both Bridgeport brand and clones), so the pattern should be measured for each head.

Make a fixture to locate the recess in the rear face of the head. This fixture is simply a turned plate with a hole in the center for a 1/4" socket head cap screw. The diameter of the plate is made to be a snug slip fit into the recess of the back of the milling machine head. This recess is where the head side tilt gear mentioned

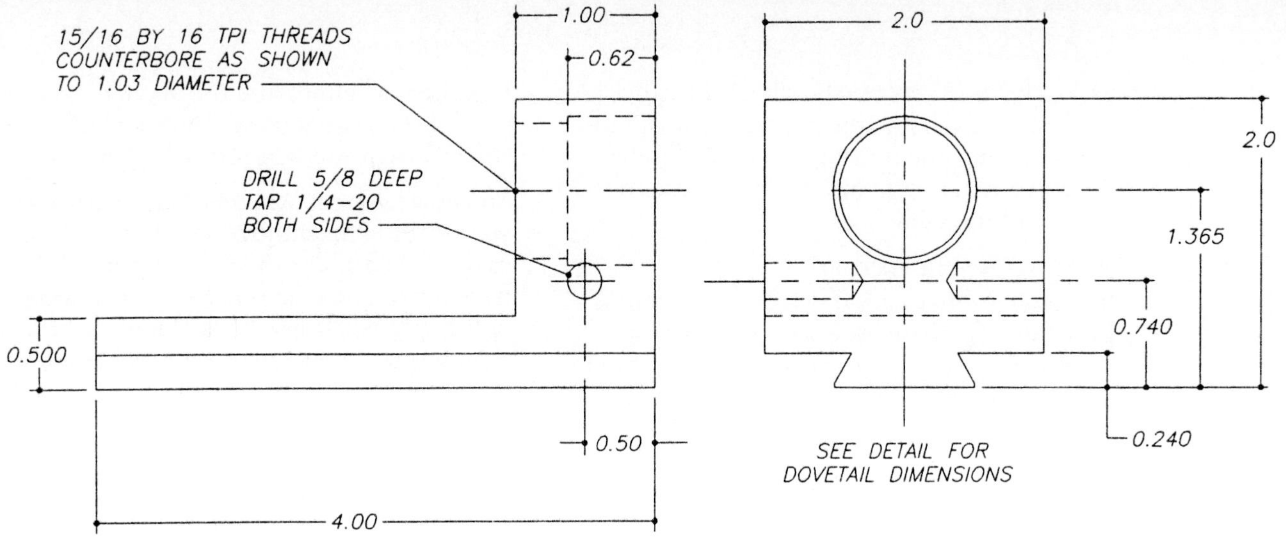

MAKE DOVETAIL TO DIMENSIONS SHOWN
AND FIT TO P/N 2323 WHEN IT IS
MACHINED TO FINAL DIMENSIONS

MATERIAL – BRASS OR BRONZE

Figure 5
SLIDE P/N 2320

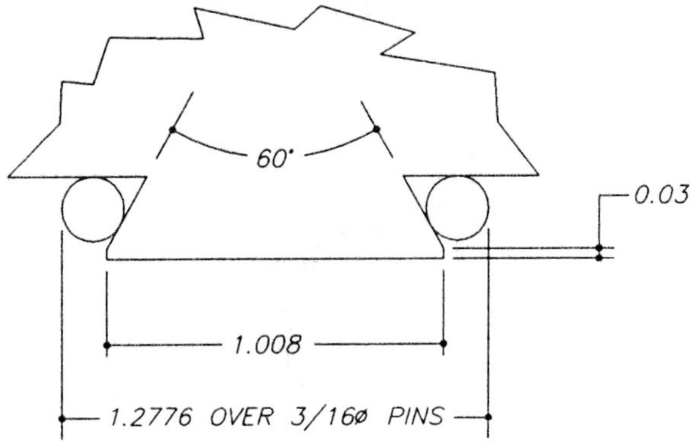

DOVETAIL DETAILS – TWICE SIZE

NOTE:
THIS ADAPTER PLATE WILL FIT BRIDGEPORT BRAND MILLING MACHINES WITH V-RAMS AND MOST OF THE SIMILARLY DESIGNED "CLONE" MILLS.
SUITABLE HOLES FOR THE ATTACH BOLTS THAT HOLD THE HEAD TO THE RAM MUST BE DRILLED IN THIS PART.
THIS HOLE PATTERN WILL VARY FROM ONE MILLING MACHINE TO ANOTHER, EVEN AMONG THE BRIDGEPORT BRAND MILLS. EACH HEAD MUST BE MEASURED TO MAKE THE ADAPTER PLATE MATCH THE MILLING MACHINE ON WHICH IT IS TO BE USED.

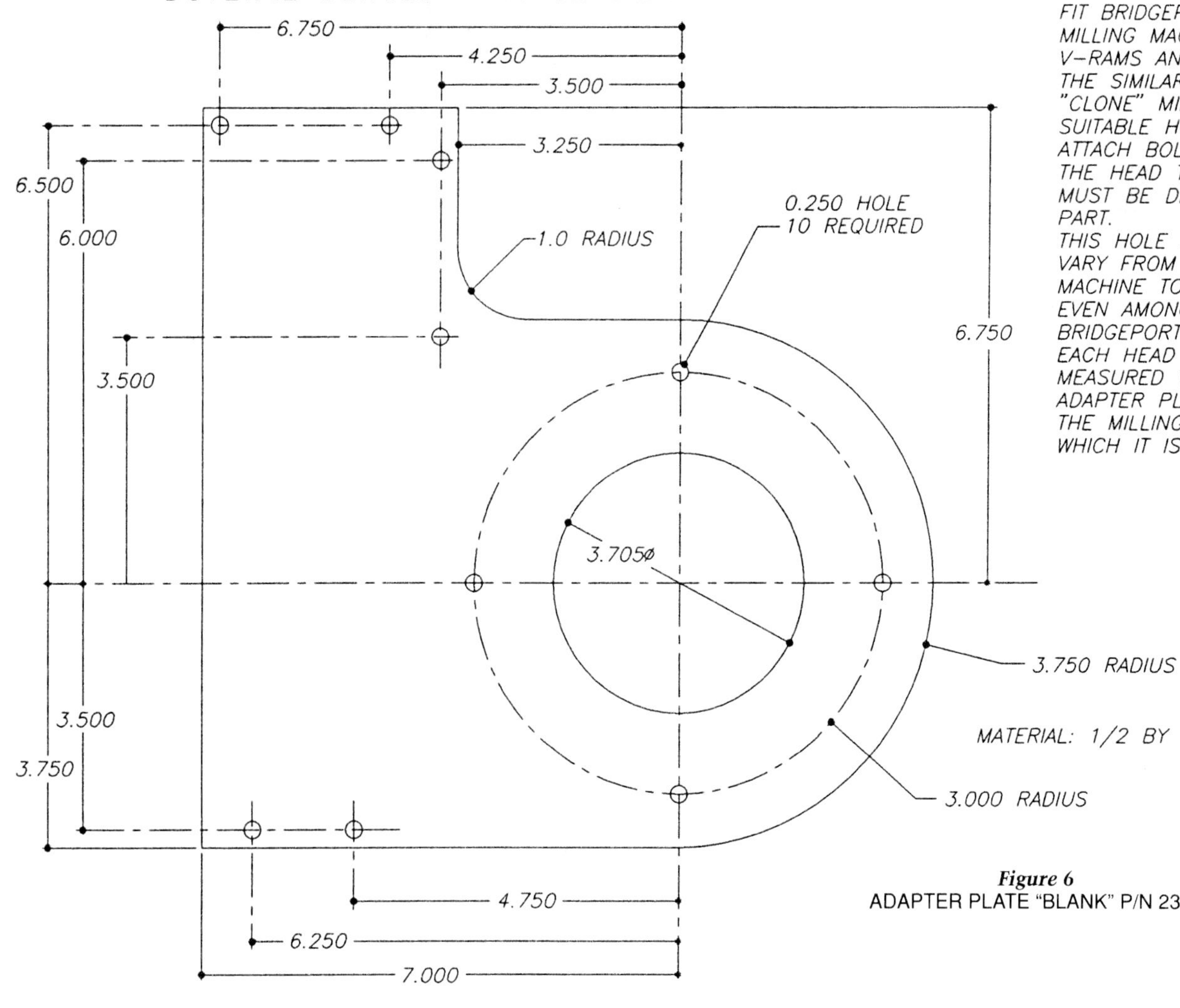

MATERIAL: 1/2 BY 12 CRS BAR

Figure 6
ADAPTER PLATE "BLANK" P/N 2323

earlier fits and acts as a bearing for tilting the head. If necessary, make a T-nut of an appropriate size for the milling machine table you are going to use, and mount the fixture to the surface of the table.

"DIALING IN" THE FIXTURE

If you do not have another milling machine available, it will be necessary to "jury-rig" a temporary spindle that can be used to mount a dial indicator on the ram of your milling machine. If you do have another milling machine, use its table to perform the measurements described. First, dial in the outside of the fixture that has been mounted on the table, and take the center of the fixture as the origin or X=0, Y=0 point. (Note: the "dialing in" of a part is a standard machine shop operation, and I will not describe it in detail here. If enough people would like a detailed description of this, perhaps Joe Rice can prevail upon me or one of the other authors to write it up separately.)

Insure that any burrs or dirt are removed from the mounting flange at the back of the milling head, and place it carefully over the fixture as shown in Figure 4 so the recess in the head is located on the fixture. Then extend the quill to its full length and lock it in position. Again using the dial indicator, dial in the quill itself, and rotate the head until the quill is aligned with the long dimension of the milling machine table. It is not necessary to clamp the head down to the table since all that will be done will be the taking of measurements. When the quill is properly aligned with the table axis and the location of the center of the fixture has been established, dimensions of the stud locations can be taken.

Be careful not to bump the head and disturb the alignment of the quill. Lower the knee of the milling machine until the dial indicator will fit into the four stud holes in the head. This is probably not a good time to suggest it, but the stud holes should be clean of paint or chips to get the best measurements. Now dial in each of the stud holes in turn, writing down the X and Y measurements of each. The measurements do not have to be to particularly close tolerances. Something on the order of ±0.020" is okay, but a gross error will require the remaking of the adapter plate later on.

Carefully measure the length, diameter, and threads used for the bolts that retain the head tilt gear. You will have to procure similar bolts, except ½" longer. Also, measure the location of the bolt holes in the gear itself, and write down the information. This bolt pattern will be machined into the gear spacer when it is made. At this time, also measure the diameter and thickness of the gear and the depth and diameter of the recess in the ram into which the gear fits.

You now have enough information to reassemble your milling machine head if it must be used to make the parts for the retrofit project. If you have the luxury of another mill, then reassembly will not be necessary. If reassembly is necessary, then you will have to again disassemble the machine to install the parts you will make next.

There are three parts you should make at this time: The Slide (Figure 5, P/N 2320) will be fitted to the adapter plate (P/N 2323), and should be made first; it is much easier to fit when done in this order. Then make the adapter plate so the dovetail is a proper fit with the slide. Once the dovetail is properly fit and the adapter plate has been machined or ground to final thickness, then the Gear Spacer (P/N 2335) can be made.

THE ADAPTER PLATE

Figure 6 shows the detail dimension of the basic adapter plate without the dovetail or stud holes. The adapter plate is first machined to these dimensions as a sort of "blank" part. Next, mount the adapter plate on the milling machine table using the circular area only for clamping, as shown in Figure 7. This allows the natural distortion of the material to occur when the dovetail is roughed out. Rough out the dovetail, grind or machine the adapter plate flat, and then finish the dovetail to match the slide, as detailed in Figure 8. A really tight fit is not required on the dovetail as any slight play in the parts will not affect the operation of the finished machine. However, I would suggest you keep the clearance between the slide and adapter plate at a maximum of 0.005". The measurements over the pins for the slide and adapter plate are theoretical dimensions for a tight fit condition.

It is now time to drill the stud holes in the adapter plate. You can see why I cautioned you to be careful with measurements on the head, particularly if you are using the same milling machine to make the parts you will be retrofitting to CNC. Figure 9 details the stud hole locations. The dimensions given are typical of several examples of the ENCO milling machines I have chosen for this project. However, measure your own machine and use your own dimensions. The stud holes should be a fairly tight (0.005") fit in the adapter plate.

THE GEAR SPACER

Figure 10 shows the details of the gear spacer. There are several dimensions adjusted to correspond to the measurements you have previously taken from the milling machine. These variations are described in notes on the drawing.

When the adapter plate and gear spacer are installed on the milling machine, longer bolts and studs will normally be required. The bolts that retain the gear can be purchased commercially, often from a local auto parts house. The studs are a different story, depending upon the way your milling machine is made. Some milling machines have heads that require a stud with a heavy washer and nut for clamping. These machines will require you to manufacture new studs, approximately ½" longer. If you need to make new studs, **do not** use mild steel. They can be made from annealed 01 tool steel, or any other steel with a tensile strength of 100,000 to 125,000 psi. The Enco mill I am using does not require new studs since there are bushings located under the clamping nuts. These bushings should be shortened by the same amount as the thickness of the adapter plate. Then new nuts must be made that are longer by the thickness of the adapter plate. The nuts can be of normal cold rolled steel material.

ASSEMBLY

If necessary, disassemble the milling machine head as was done earlier. Remove the old studs and gear from the end of the ram. Install the gear spacer with new bolts ½" longer than the originals. Be sure these bolts are tight, as they are inaccessible once the head has been assembled. Install the new studs if required. Grease the outside of the gear spacer and the rear face of the adapter plate. Install the adapter plate over the studs, and position the large hole over the outside diameter of the gear spacer. Double check at this time that the adapter plate is thicker than the gear spacer. If the gear spacer protrudes beyond the front face of the adapter plate, the head casting can be damaged when the studs are tightened.

Grease the front face of the adapter plate and the outside diameter of the gear. Carefully reinstall the head and the nuts, washers, and bushings as required on the studs. Snug the nuts down, and then loosen them just enough

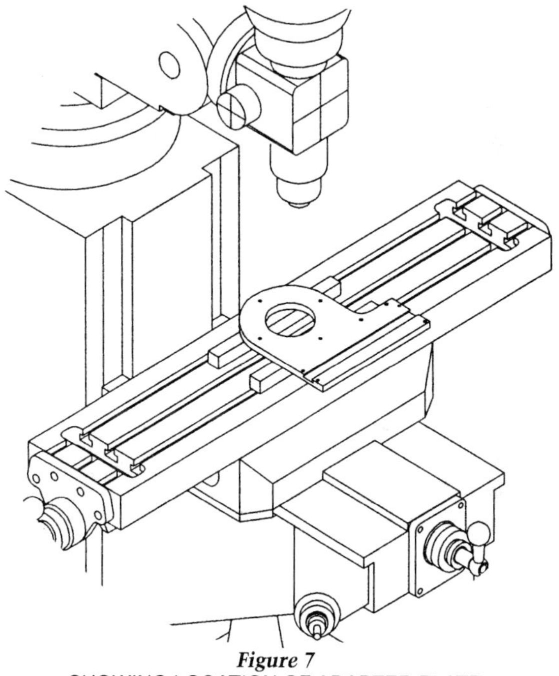

Figure 7
SHOWING LOCATION OF ADAPTER PLATE FOR MILLING DOVETAIL

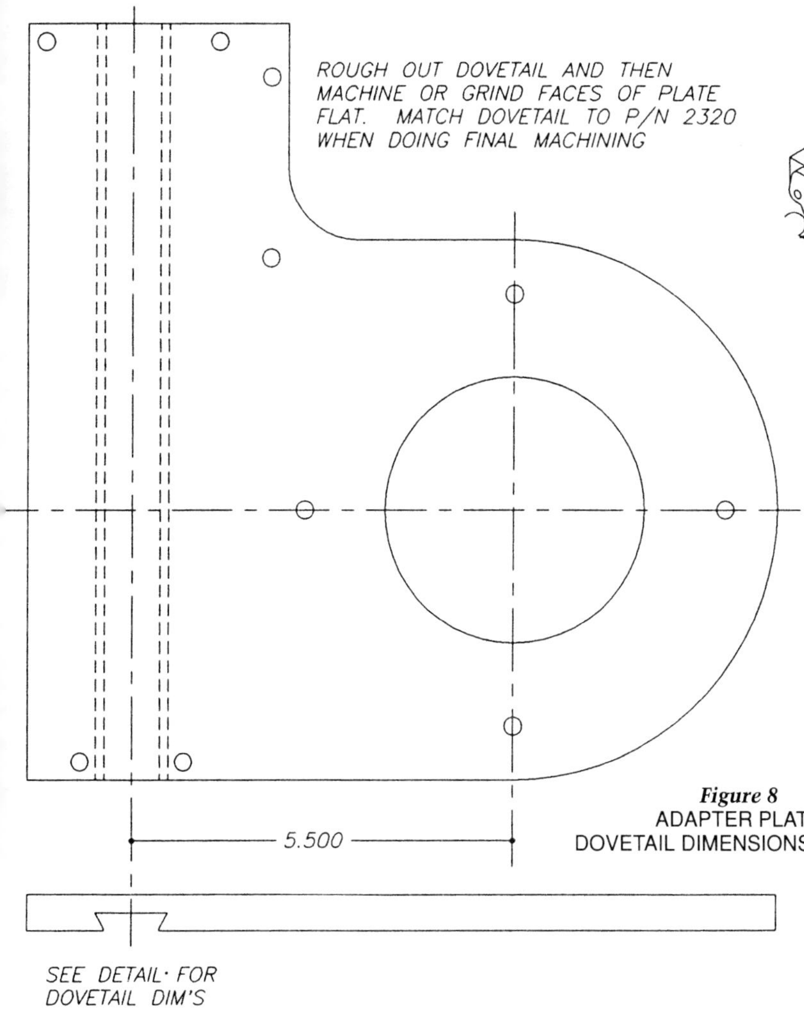

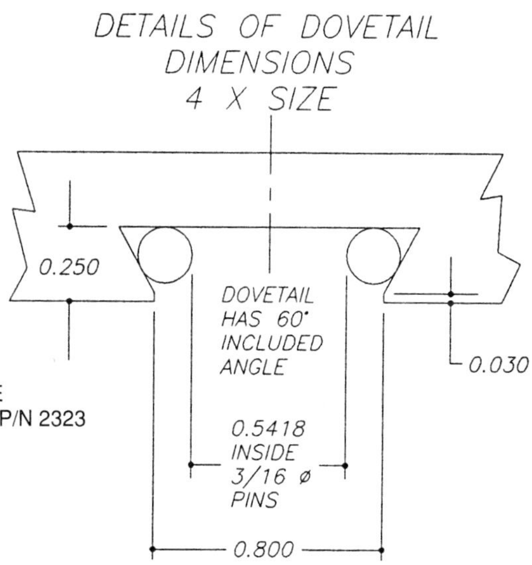

Figure 8
ADAPTER PLATE
DOVETAIL DIMENSIONS P/N 2323

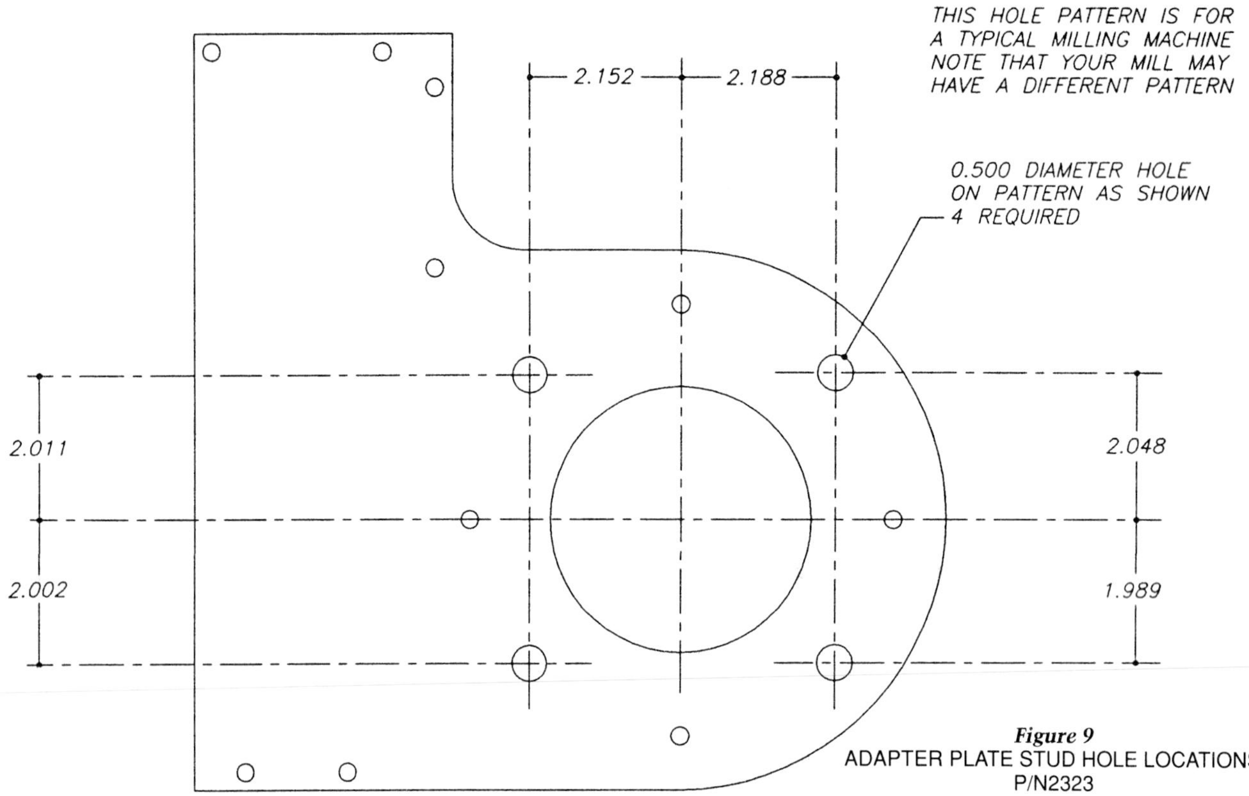

Figure 9
ADAPTER PLATE STUD HOLE LOCATIONS
P/N2323

to check for correct operation of the head tilting mechanism. When you are certain everything is properly installed and functioning, snug the stud nuts down, and reinstall the drive belt housing and drive motor. Reconnect any electrical lines, and check the head for proper function. You may now "tram" the head (align it to the table surface) as it will not be necessary to disassemble it again. Note that the side-to-side tilt and front-to-back tilt mechanisms are still fully functional even with the adapter plate and gear spacer installed.

That's all on the floor model retrofit project for now.

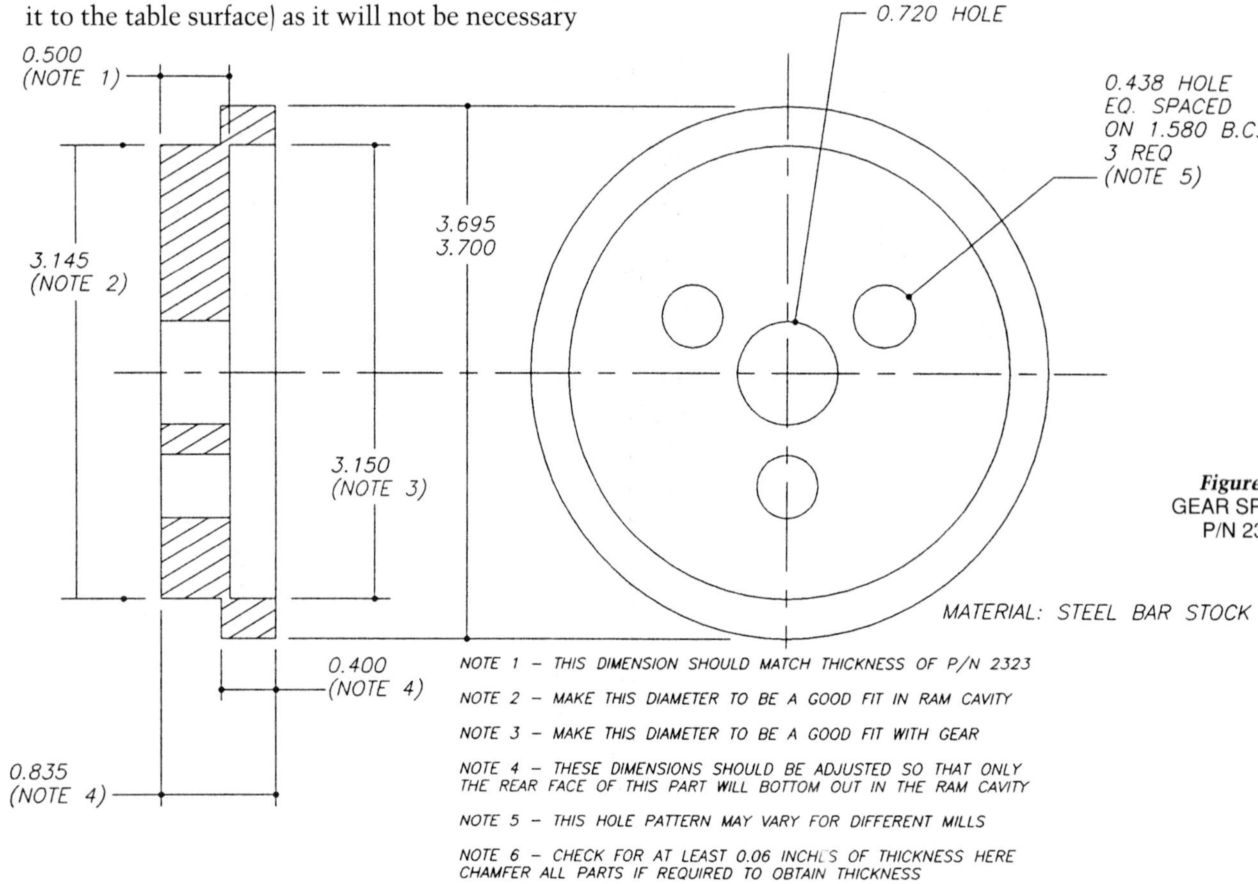

Figure 10
GEAR SPACER
P/N 2335

EXPERIMENTS IN BUILDING YOUR OWN CNC CONTROLLER

Here is the first of several simple projects that will help you to understand how a computer can be used to control motion, and the software/hardware requirements to do so. This is the beginning of our "build-your-own" CNC controller project. Refer to Figure 11 for a drawing showing a typical parallel port card that is installed in your computer. You may have more than one of these cards, but most computers have a maximum of two. It is possible to have more than two parallel ports but we will not go into those details at this time. Two parallel ports will allow you to run three motors using the information presented here.

Again referring to Figure 11, note that the socket located at the rear of your computer is shown in enlarged detail. There may be several sockets but the parallel port sockets are normally of the 25-pin, female variety. The serial port sockets may be similar in shape and pin count but are normally male. If you have one parallel port, it will automatically be assigned by your computer as LPT1. If you have two ports, one will be assigned as LPT1 and the other as LPT2. This is done by DOS (Disk Operating System) and the technician who set up your computer originally.

THE FLASHING LIGHT SHOW

In order to illustrate some of the principles involved with running motors from a computer, we will begin by flashing some lights under computer control. You can plug these lights directly into the parallel port of your computer or you can make an "extension cable" to bring the lights around from the back of the computer to the keyboard area where they are easier to see.

MINIMUM LIST TO PURCHASE

1 package (2 pieces) *Red Light Emitting Diode,* $0.69 each, Radio Shack catalog number 276-026.

1 package (2 pieces) *Green Light Emitting Diode,* $0.89 each, Radio Shack catalog number 276-037.

1 D-Subminiature male 25 position *Solder Connector,* $1.99, Radio Shack catalog number 276-1547.

TO MAKE AN EXTENSION CABLE

1 D-Subminiature flat *Cable Connector,* 25-pin male solderless, $3.99, Radio Shack catalog number 276-1559.

1 D-Subminiature flat *Cable Connector,* 25-pin female solderless, $3.99, Radio Shack catalog number 276-1565.

1 package (5 feet) *Ribbon (Flat) Cable,* 25-conductor, $3.59, Radio Shack catalog number 278-772.

SOFTWARE TO FLASH THE LIGHTS

Figure 12 is a program you can type into your computer, and which will illustrate the use of the computer to control external events (in this case the Light Emitting Diodes). Figure 12

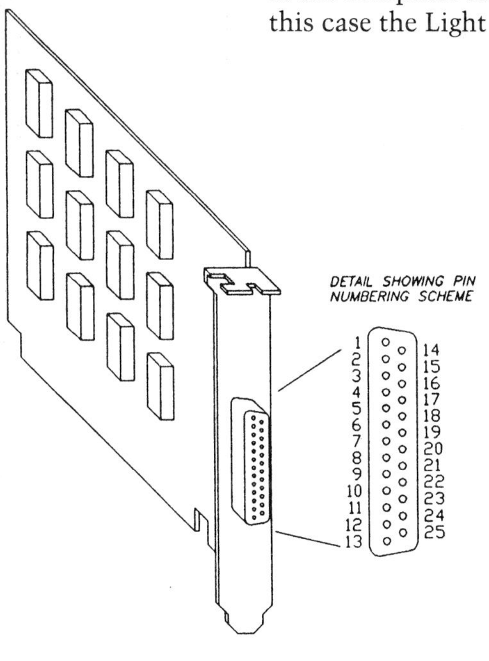

VIEW OF PARALLEL BOARD AS SEEN FROM REAR OF COMPUTER
Figure 11

is a photographic copy of the actual computer printout. Any MS-DOS or PC-DOS computer should have a copy of the BASIC language included. This program should run in any of these BASIC language systems. Refer to the reference manuals that came with your computer for information on BASIC language programming. If you don't want to type in this program, send $10 to Jerry Berns at Cardinal Engineering to get the program on a floppy disk. This program is quite elementary in operation but will demonstrate some of the basics of motion control. In a future column,we will use this same program to run motors. So get busy typing. Save the program using the name "MOTOR1.BAS."

MAKING THE EXTENSION CABLE

If you elect to make the extension cable, assemble it using the solderless connectors and ribbon cable. Most of the Radio Shack parts have some form of instructions on the package. If they do not, ask the clerk or an acquaintance who is involved with computers. There are small numbers molded into the face of all the connectors (I have to use a magnifying glass to read them). Make sure you have the cable and connectors aligned so the pin numbers correspond on each end. Check for continuity and no shorts between lines using an ohmmeter or continuity tester.

TESTING THE LIGHT EMITTING DIODES

Most of the Radio Shack Light Emitting Diodes (LED's) are functional as they come out of the package. However, they are not expensive, and now is the time to check that you have good ones. Using the LPT1 socket at the rear of the computer, plug an LED into pins 16 and 18. The LED should light up. If it doesn't, reverse the pins. If the LED doesn't light in either position, it is bad. Mark the lead that was in pin number 18 when the LED lit up. That lead is the ground lead. Do this with all of the LED's. You will need two red and two green LED's.

INSTALLING THE LED'S

To install the LED's, put the ground lead of a red LED in pin 18 and the other lead in pin 2. Put the ground lead of a green LED in pin 19 and the other lead in pin 3. Put the ground lead of a red LED in pin 20 and the other lead in pin 6. Put the ground lead of a green LED in pin 21 and the other lead in pin 7.

RUNNING THE SOFTWARE PROGRAM

Using BASIC, load and run the program. The delay value is set for 100 but may be changed to suit your requirements. The objective is to have a delay that is long enough so you can see each individual flash of the red LED's. A value of 100 works well with XT type computers. Increase the value to slow the flash rate. If you are plugged into LPT1, type X10. The red LED plugged into pin 2 should flash, and the green LED plugged into pin 3 should be on continuously. Now type X-10. The red LED should flash again but the green LED should not go on at all. Similarly, the LED's plugged into pins 6 and 7 should light up when Y10 and Y-10 are typed. Experiment with different delay values and different numbers. The program will also respond to entry of Z10 or Z-10 but no flashing will occur unless the LED's are plugged into LPT2.

EXPLANATION OF WHAT YOU ARE SEEING

Most of the stepper motor driver systems commercially available use an operating method called "STEP and DIRECTION." In this scheme, the red LED corresponds to the STEP signal, and the green LED corresponds to the DIRECTION signal. When the red LED is flashing, the motor will step once for each flash. When the green LED is on, the motor will turn one direction, and when it is off the motor will turn in the opposite direction. The electronics to accomplish this are contained in a motor driver, which is an electronic circuit that will accept the STEP and DIRECTION signals, and send the correct voltages and currents to the windings of the stepper motors.

These motor drivers can be completely contained on a single integrated circuit chip or can occupy a large circuit board with fan cooling, depending upon the size of the motor they can drive. We will build a small motor driver later on in this series, and use the software you have been working with to drive it.

```
100 ' * * * * * * * * * * * * * * * *
110 '
120 ' "MOTOR1.BAS"
130 ' MOTOR STEP PROGRAM
140 ' COPYRIGHT (C)1991  CARDINAL ENGINEERING, INC.
150 ' ALL RIGHTS RESERVED
160 ' WRITTEN BY JERRY BERNS
170 '         ON 08-15-91
180 '
190 ' * * * * * * * * * * * * * * * *
200 '
210 ' This program was developed for use with
220 ' low cost stepper motor drive systems that
230 ' receive motor control from a parallel port
240 ' of an IBM or compatible computer.
250 '
260 ' * * *  VARIABLE TABLE  * * *
270 '
```

```
280 '  A$           Temporary string to hold user input
290 '  AXIS$        Axis to turn
300 '  BLOCK$       Axis and motor step input
310 '  CLEARPORT    Value to set step data line (2 and 6) low
320 '  DELAYLOOP    Counter for time delay between motor steps
330 '  MOTORDATA    Value sent to PORT to step motor
340 '  MOTORSTEPS   Number of steps to turn motor
350 '  PORT         HEX value of LPT port: LPT1=888 LPT2=632
360 '  RESP         Value of user input
370 '  RESP$        Temporary string to hold user input
380 '  STEPDELAY    Delay time between motor steps
390 '  STEPLOOP     Counter for motor steps
400 '  TMP$         Temporary data string
410 '
420 ' * *  VARIABLE TABLE END  * *
430 '
440 ' * * *  PARALLEL PORT TABLE  * * *
450 '
460 '  LPT1 = 888 (378 HEXADECIMAL)
470 '  LPT2 = 632 (278 HEXADECIMAL)
480 '
490 '  CONNECT MOTOR DRIVER STEP TO DATA LINE (PIN)
500 '  2 OR 6 AS NOTED IN FOLLOWING TABLE.
510 '
520 '  CONNECT MOTOR DRIVER DIRECTION TO DATA LINE (PIN)
530 '  3 OR 7 AS NOTED IN FOLLOWING TABLE.
540 '
550 '  X AXIS: PORT=LPT1 , PIN 2 = STEP , PIN 3 = DIRECTION
560 '  Y AXIS: PORT=LPT1 , PIN 6 = STEP , PIN 7 = DIRECTION
570 '  Z AXIS: PORT=LPT2 , PIN 2 = STEP , PIN 3 = DIRECTION
580 '
590 '  PINS 18 THROUGH 25 ARE SIGNAL GROUND PINS
600 '
610 ' * *  PARALLEL PORT TABLE END  * *
620 '
630 ' * * *  MOTOR DATA TABLE  * * *
640 '
650 '                              ----- PIN NUMBER -----
660 '  ASCII  MOTOR  DIRECTION  2  3  4  5  6  7  8  9  CLEARPORT
670 '
680 '    1     X,Z     CCW      1  0  0  0  0  0  0  0      0
690 '    3     X,Z     CW       1  1  0  0  0  0  0  0      2
700 '   16      Y      CCW      0  0  0  0  1  0  0  0      0
710 '   48      Y      CW       0  0  0  0  1  1  0  0     32
720 '
730 ' * *  MOTOR DATA TABLE END  * *
740 '
750 DEFINT A-Z        'DEFINE VARIABLES AS INTEGERS
760 STEPDELAY = 100 'DEFAULT DELAY BETWEEN MOTOR STEPS
770 BLOCK$ = "X0"    'DEFAULT INPUT VALUE FOR AXIS AND MOTOR STEPS
780 ' * * *  MAIN PROGRAM  * * *
790 CLS
800 PRINT : PRINT
810 RESP = -1
820 WHILE RESP < 0
830   TMP$ = "Enter delay between motor steps <"
840   TMP$ = TMP$ + RIGHT$(STR$(STEPDELAY), LEN(STR$(STEPDELAY))-1)
850   TMP$ = TMP$ + ">: "
860   PRINT
870   PRINT TAB(5); TMP$;
880   INPUT "", RESP$
890   IF LEN(RESP$) THEN RESP = VAL(RESP$) ELSE RESP = STEPDELAY
900 WEND
910 STEPDELAY = RESP
920 PRINT : PRINT
930 PRINT TAB(5); "Enter the axis and motor steps as shown below:"
940 PRINT : PRINT TAB(10); "X100  - X axis move positive 100 steps"
950 PRINT TAB(10); "Y-50  - Y axis move negative 50 steps"
960 PRINT : PRINT TAB(5); "Type DELAY to set new motor step delay."
970 PRINT TAB(5); "Type QUIT to exit program."
980 PRINT
990 ' * * *  DATA ENTRY  * * *
1000 WHILE A$ <> "QUIT" OR A$ <> "quit" 'OR LEN(A$) = 0
1010   PRINT
1020   PRINT TAB(5); "Axis and motor steps <" + BLOCK$ + ">: ";
1030   INPUT "", A$
1040   IF LEN(A$) = 0 THEN A$ = BLOCK$
1050   IF A$ = "QUIT" OR A$ = "quit" THEN END
1060   IF A$ = "DELAY" OR A$ = "delay" THEN GOTO 790
1070   AXIS$ = LEFT$(A$, 1)
1080   MOTORSTEPS = VAL(RIGHT$(A$, (LEN(A$) - 1)))
1090   GOSUB 1130
1100   BLOCK$ = A$
1110 WEND
1120 ' * * *  DETERMINE PORT, MOTORDATA and CLEARPORT VALUES  * * *
1130 IF INSTR("XxZz", AXIS$) > 0 THEN GOTO 1140 ELSE GOTO 1190
1140 IF INSTR("Xx", AXIS$) > 0 THEN PORT = 888 ELSE PORT = 632
1150 IF MOTORSTEPS >= 0 THEN MOTORDATA = 3: CLEARPORT = 2
1160 IF MOTORSTEPS < 0 THEN MOTORDATA = 1: CLEARPORT = 0
1170 GOSUB 1350
1180 RETURN
1190 IF INSTR("Yy", AXIS$) > 0 THEN GOTO 1200 ELSE GOTO 1260
1200 PORT = 888
1210 IF MOTORSTEPS >= 0 THEN MOTORDATA = 48: CLEARPORT = 32
1220 IF MOTORSTEPS < 0 THEN MOTORDATA = 16: CLEARPORT = 0
1230 GOSUB 1350
1240 RETURN
1250 ' * * *  ERROR in Data Entry  * * *
1260 PRINT : PRINT : PRINT : BEEP
1270 PRINT TAB(30); "ERROR IN DATA ENTRY"
1280 PRINT
1290 ' * * *  Program Repeat Prompt  * * *
1300 PRINT : PRINT
1310 PRINT TAB(5); : INPUT "Run again (Y/N)? <Y> ", A$
1320 IF INSTR("Yy", A$) > 0 THEN GOTO 790
1330 END
1340 ' * * *  MOTOR STEP and DELAY ROUTINE  * * *
1350 OUT PORT, CLEARPORT
1360 FOR STEPLOOP = 1 TO ABS(MOTORSTEPS)
1370   OUT PORT, MOTORDATA
1380   FOR DELAYLOOP = 1 TO STEPDELAY: NEXT DELAYLOOP
1390   OUT PORT, CLEARPORT
1400   FOR DELAYLOOP = 1 TO STEPDELAY: NEXT DELAYLOOP
1410 NEXT STEPLOOP
1420 OUT PORT, 0
1430 RETURN
```

COMPUTER PROGRAM
"MOTOR 1. BAS"
Figure 12

Build Your Own CNC

by Roland Friestad

FLASHING LED'S

The March/April 1992 column in *HSM* presented the pin connections and a short program, MOTOR1.BAS, which would use the parallel port of your computer to flash LED's. Now, we'll take this further and begin construction of a complete, build-your-own CNC controller. The MOTOR1.BAS program will be used to test the CNC controller when construction is completed.

THE NEW PROJECT

Now for the new project! These instructions will allow the home hobbyist to assemble a 3-axis CNC controller capable of operating motors of up to 3 amps per phase. The controller will use standard G-Codes for programming. Construction will use standard components that can be readily purchased, and I'll indicate sources.

During the ten years I have been building CNC machines of one kind or another, I have primarily used some kind of commercially available CNC components which were available as a packaged unit. This was the only way to go short of designing and building your own system completely from scratch, including electronic logic, motor drivers, electronic packaging, and operating software. As your basic mechanical-type, with a working but not expert knowledge of these areas, I never felt that the time required to design "from scratch" was worth it, particularly since many of the clients for whom I design and build special machines have no problems with the cost of commercial CNC packages.

However, the costs for commercial units have been a little steep for the home hobbyist, at least until recently. As an example, about ten years ago, commercially available components that could be purchased for building your own CNC controller cost in the $20,000 to $25,000 range. More recently, it has been possible to buy components for around $4,000 to $5,000. This is a significant improvement but still a little high for many of the home hobbyists. I have worked on nearly 300 machines using these controllers and feel that the $4,000 to $5,000 price tag is definitely still the most "bang for the buck" available in CNC controls.

There is a strong demand for a much lower cost CNC controller for those who want to experiment with this fascinating technology. Consequently, I have pulled together some commonly available components and, with the help of Jerry Berns of Cardinal Engineering who wrote the G-Code software, designed a simple, low powered, and low cost CNC controller that can be built by anyone who has the ability to read and follow a schematic. It isn't necessary to be an electronics genius, and with careful shopping it can be built for under $1,500 using all new components.

If you are particularly lucky and can find some older scrapped CNC or NC controller from which to rob parts, it might end up costing a lot less. I have several of these old controllers from which I have salvaged components and this can cut the cost down to the $500 range.

The information published in this column will specify new components since the availability of used or surplus is sketchy at best. The types of components normally seen on the surplus market often come from "distressed merchandise" originally intended for consumer products such as appliances, television sets, etc. There are a few industrial surplus houses around the country, but CNC equipment was never exactly a consumer-type of item and surplus CNC equipment is therefore sort of rare. If anybody out there knows some good sources for industrial and CNC surplus equipment, let me know and I'll be glad to publicize the addresses in this column.

There are several critical parts to a CNC controller. Until recently, a few key items were not readily available without a high level of technical electronic and software ability to integrate the components. The two "missing links" that were most difficult (or impossible) to get were 1) motor drivers and 2) operating software. Now these items are available and at reasonable cost!

The recent development of packaged drivers that can take step and direction signals at computer voltage levels and generate high power output to the coils of stepper motors has solved some major electronic design and construction problems. These are available in different power levels and packages ranging from integrated circuit chips to chassis mounted units with complete power supplies.

In an earlier column, I also mentioned a computer program, *Indexer LPT,* that generates step and direction signals using an MS-DOS computer. This software is manufactured by Ability Systems, Inc. These two elements, combined with the G-Code software written by Jerry, make it relatively easy to build a simple CNC controller. The software, as it currently is written, allows programming using the four most-used G-Codes, G0, G1, G2, and G3. Both software packages together cost $300.

Most MS-DOS computers have one or two parallel ports, but it is possible to have more. Two parallel ports, LPT1 and LPT2, are required to run three motors for the X, Y, and Z axes. The software can generate linear and circular step patterns at computer voltage levels (known as TTL). These signals appear on the parallel port connectors and can be used to run any stepper motor driver that uses what is known as "step and direction" signals. The cables between the computer and the CNC controller are standard cables with male-to-male DB-25 connectors and straight through connections. They should be approximately 6 feet long. These cables are readily available from a computer store.

"Step and direction" drivers use a method that relies on only three wires to control the travel of a stepper motor. The driver has one connection that is the step control line. When this line is pulsed between 0 volts and approximately 5 volts, the motor will move one step for each pulse. The driver also has one connection that is the direction control line. When this line is at 0 volts, all of the step pulses will cause the motor to rotate in one direction. When the direction line is at 5 volts, all of the step pulses cause the motor to rotate in the opposite direction. The third wire is the ground or reference signal to which the voltages of the other two wires are compared.

"DRAGON DRIVERS"

I have chosen to use packaged drivers that are commercially available rather than to build them "from scratch." The driver I chose, the Model 29A55 *Dragon Driver* from General Control Electronics, Inc., will provide 3 amps to each coil of the stepper motor and will generate all of the correct sequences of coil switching to rotate the motor in either direction. This is all controlled by the "step" and "direction" signals described above. The *Dragon Driver* also has a built-in power supply so the only voltage required is an AC voltage of approximately 30 volts. There is even a built-in oscillator for self-testing the driver so a motor can be run without using any software. The folks at General Controls also claim that the drivers are impossible to short out and will shut themselves down in an over-temperature condition. Sounds like just what I need! These drivers cost $200 each in small quantities.

The transformers are *Stancor* filament power. This type of transformer will normally have a center-tapped secondary coil. The center tap is not used, and should be suitably insulated to prevent short circuits. The transformers are available from Newark Electronics for about $35 each.

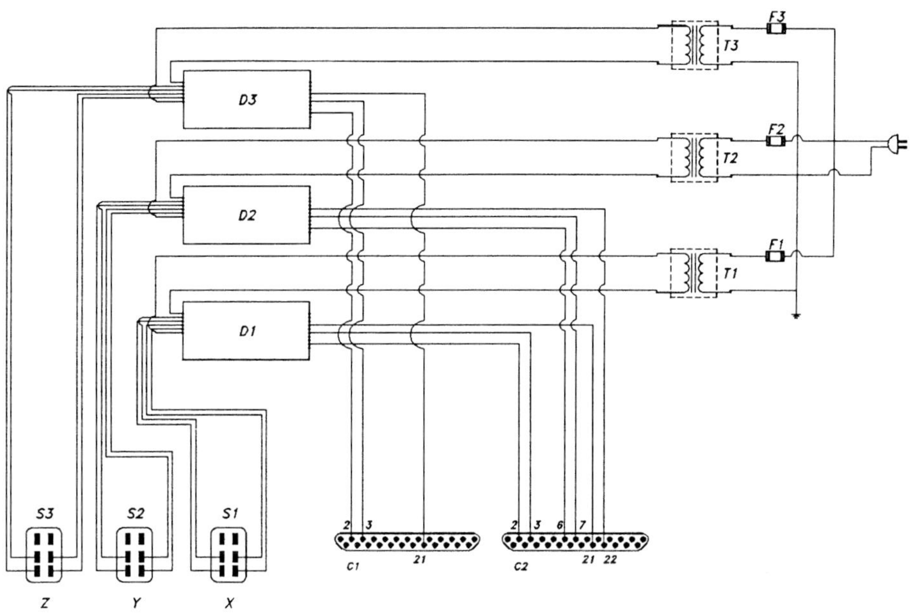

Figure 1

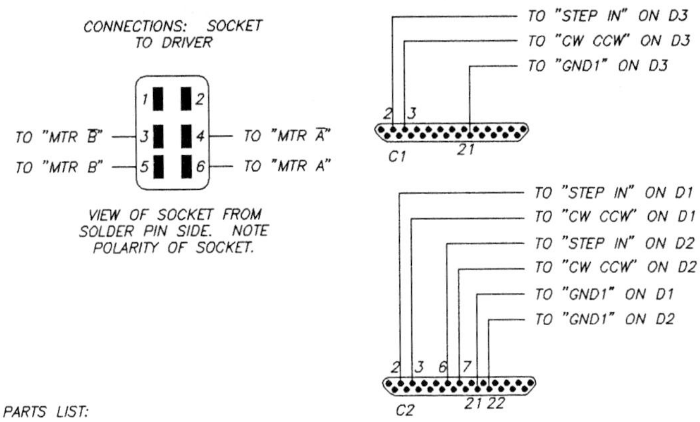

PARTS LIST:

D1, D2, D3:	DRIVER – GENERAL CONTROLS ELECTRONICS, INC. MODEL 29A55 "DRAGON DRIVER" BIPOLAR CHOPPER
F1, F2, F3:	FUSE – STANDARD 1 AMP FUSE
S1, S2, S3:	SOCKET – CINCH MODEL S-306-AB 10A, 125V 6-PIN FEMALE CONNECTOR
C1, C2:	CONNECTOR – DB-25 FEMALE SOLDER CONNECTOR
T1, T2, T3:	TRANSFORMER – STANCOR MODEL P-8669 117V, 28V, 4A, 50/60HZ TRANSFORMER

Figure 2

The driver has terminal strips on each end of its circuit board and they are clearly identified. Wiring is just a matter of running connections as shown in Figures 1 and 2. The schematic of Figure 1 shows the basic wiring of the components. The bill of materials and details shown in Figure 2 provide more specific information regarding wiring.

The CNC controller was laid out on an aluminum plate of 14 × 20". Aluminum bar stock was used to make feet to elevate the plate, and the feet were fitted with rubber crutch tips available from your local hardware store. Figure 3 shows the physical layout I used for this project. There is nothing critical about the lengths and positions of wiring, so builders who want to use a different arrangement do not need to adhere to the layout I used.

Now, wire the stepper motors you are planning to use. The literature provided with the *Dragon Drivers* gives information on wiring different types of motors. The sockets specified as S1, S2, and S3 are available from Newark Electronics. Select a matching plug from Newark for the motors and wire them per the *Dragon Driver* instructions. Pre-wired and tested motors are available from Cardinal Engineering.

Sometimes it is possible to obtain used or surplus stepper motors at very attractive prices. A good source for these bargains is C&H Sales. The drawbacks with used or surplus motors is the possibility that the motor will not be well matched to the characteristics of the driver or the machine upon which it will be used, and you will almost never get wiring information on the motors. However, the low cost of surplus motors allows some experimentation to be done without too much of a loss if it doesn't work. Don't try to put too much current into a stepper motor. It will only cause overheating and possible damage to the electronics or motor.

Once the motors are properly wired, it is time to test the system. Use the BASIC language program, MOTOR1.BAS. This program allows you to drive the motors and test the controller. First, set the switches on the driver so switches 1 through 8 are all off. Next, plug in the cables connecting the computer to the CNC controller. These cables are connected to the parallel ports of the computer and to the connectors labeled C1 and C2. Plug the motors into the sockets labeled S1, S2, and S3.

When wiring the CNC controller, be sure to use a grounded, 3-conductor cable. The ground (green) wire should be connected to the baseplate. Also, be sure the socket into which you plug the cable is properly grounded. Because the drivers are isolated from the 110-volt mains by the transformers, the danger of shock is greatly reduced. However, transformers have been known to short out and, should this happen, it is possible to have the full 110 volts applied to the chassis. A properly grounded plug and socket can deal with this eventuality. Proper size fuses should also be used in the primary circuit of the transformers. The secondary circuit is current limited by the electronics on the *Dragon Driver* and is advertised as burnout-proof. I have not shown a switch in the circuit. You might want to put one in the 110 volt supply line. As always, when working with electricity, BE CAREFUL!!

Now that everything is connected, apply 110 volt power, boot up your computer and run

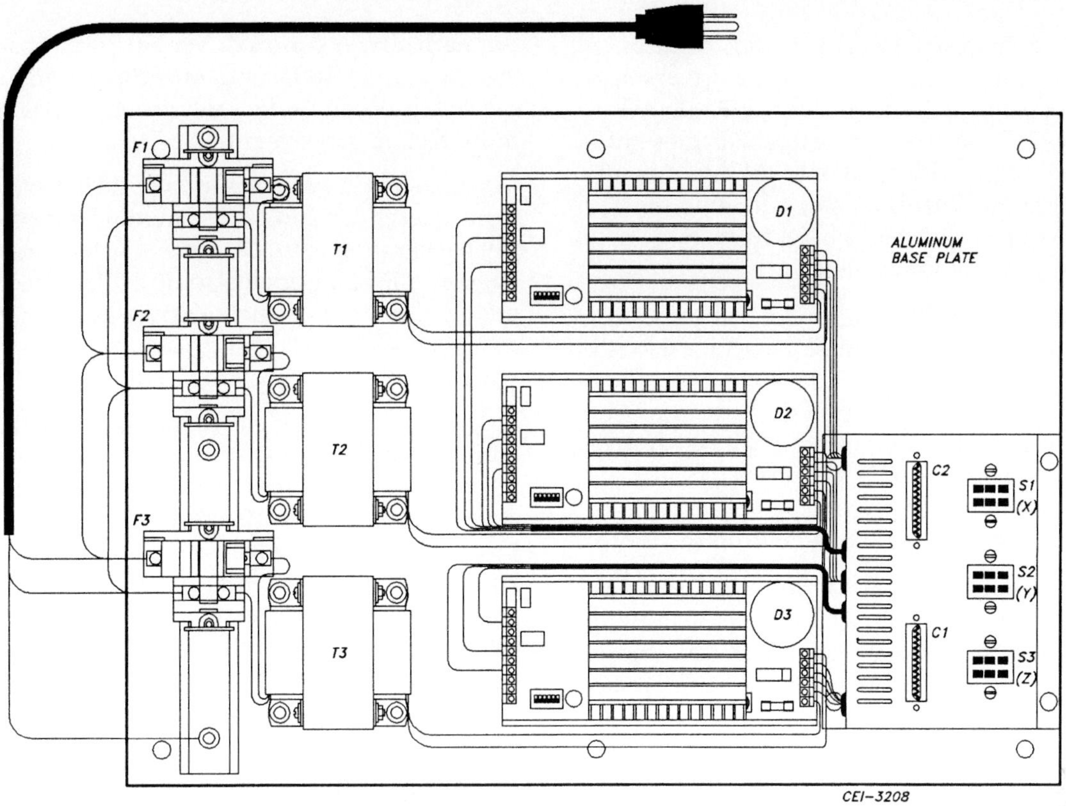

Figure 3

MOTOR1.BAS. This program should work just as it did with the LED's in the March/April 1992 issue, except that instead of flashing lights, you should now have rotating motors! Experiment with the speeds and number of steps. If all motors do not work properly, carefully check your connections and try again. One way to troubleshoot these systems is to change the motor plugs to different sockets. If the problem moves with the motor, then it is in the motor or its wiring. If the problem stays with the socket, then it is somewhere "up the line."

THE MARKETPLACE STRIKES AGAIN

When I first started working with build-it-yourself CNC projects, there was very little to choose from. This situation has changed a lot in the last two years. This project is just the first of several similar ones you will be able to build as CNC building block components become readily available. The software and hardware I have used are not the only choices; they are just the ones I happened across first, and that met the guidelines I was looking for. I have noticed an increasing number of ads for CNC products in *The Home Shop Machinist* and other technical publications. This means other companies and individual developers are waking up to the interest in this technology, and will be creating products hobbyists can use and afford. The marketplace is working in your favor.

CREDITS

Thanks to Jerry Berns for programming and Rick Morrell for drawings.

SUPPLIERS:

Ability Systems, Inc.
1422 Arnold Avenue
Roslyn, PA 19001
(215) 657-4338
(Indexer LPT software)

Cardinal Engineering, Inc.
RR 1, Box 163
Cameron, IL 61423
(309) 342-7474
(Kits and components)

C & H Sales Company
2176 East Colorado Blvd.
Pasadena, CA 91107
(213) 681-4925
(Surplus components)

General Controls, Inc.
2350 Brickvale Drive
Elk Grove Village, IL 60007
(708) 595-2152
(Stepper motor drivers)

Newark Electronics
4801 North Ravenswood Avenue
Chicago, IL 60640-4496
(312) 784-5100
(Industrial electronics, branches nationwide)

Referring back to my previous column, you will remember that the drivers used were capable of delivering about 2-1/2 to 3 amps at 28 volts to the motors. For those of you who are thinking about building this unit, be aware that this is an adequate amount of power to run the Sherline, Unimat, or Hobbymat sized machines.

I have used this CNC controller and successfully operated the benchtop milling machine described in the March/April 1990 through September/October 1990 columns previously printed here. However, the maximum speed of travel available with this power level is "leisurely" at best. If you want to run a larger machine, you will need to have more power available from the motor.

Stepper motors are sort of funny in the sense that power ratings do not really apply quite the same way they do with the normal AC motors with which we are all familiar. The manufacturers' rating charts don't generally rate the motors in horsepower, but instead provide a graph of torque and speed at a given current value. Many people have asked me what sort of horsepower is required to drive a machine table, for instance. I usually answer that these motors aren't really rated in horsepower, which is true.

However, some general statements can be made based on the experiences I have had over several years of using this type of motor. If the motors are considered in their simplest form as having a power output equal to the operating voltage multiplied by the current draw, then some comparisons can be made.

In using these motors I have found that the typical benchtop milling machine table, like that listed above, will require at least 120 watts (4 amps times 30 volts) in order to run effectively. Also, I have found that a floor model milling machine of the Bridgeport size will require approximately 240 watts to function acceptably. I am assuming the voltage to be as measured and the current either as measured or set by selector switches on the driver electronics. Incidentally, 750 watts is approximately equal to 1 horsepower.

The drivers I have used in this series are rated at 2.5 amps and 28 volts. This equals 70 watts which is a little weak for a larger milling machine or lathe. However, as mentioned earlier, it is adequate for the small desktop machines like the Sherline, etc.

The company that makes the Dragon Driver listed in the last issue also makes some higher powered drivers but the prices go up quite dramatically. A 6 amp driver (30 volts) costs approximately $600; and with three being needed, it doesn't take long to come up with quite a large investment.

Now, it should be possible to build a driver from scratch for a LOT less money, but since my background is mechanical and aeronautical engineering and not electronics, I don't feel qualified to develop such a driver from scratch, at least not with all of the other projects to which I have already committed.

So, if there are any electronics types out there who would like to assist in designing a good homebuilt stepper motor driver capable of delivering about 8 to 10 amps at 30 to 40 volts, I would be glad to publish schematics, addresses, etc. This is one of the missing links to really affordable CNC control. Contact me through the publisher if you are interested.

The second missing part is the software. This has been partially answered by the Ability Systems "Indexer LPT" software. As mentioned in an earlier column, this software generates the step and direction pulses to do linear and circular interpolation using an MS-DOS computer's parallel ports. The software has a lot of other useful functions also, but for the present discussion, the capability to generate the pulse sequences for straight lines and circles is of primary interest.

Indexer LPT has its own language which allows the software to do many things besides lines and circles. If you are interested in experimenting with motion control, I can highly recommend it. For my own interests it is essential that a CNC system respond to standard "G-Codes" which Indexer LPT does not do.

Indexer LPT is loaded into your computer first as what is known as a TSR (Terminate and Stay Resident) program. This means the software remains in memory and is sort of "watching" your computer to see if a command it recognizes is given. If so, Indexer LPT will intercept that command and execute it.

Jerry Berns at Cardinal Engineering has come up with a G-Code interpreter that works with Indexer LPT so it can be programmed using G-Codes. This interpreter, known as "CNCLPT" is written using QuickBASIC, a standard Microsoft dialect of the BASIC programming language.

CNCLPT allows CNC programs to be written with the four most-used G-Codes. They are G0 (Rapid Traverse), G1 (Straight Line), G2 (Clockwise Arc), and G3 (Counterclockwise Arc). All programming is done in what is known as the

Absolute or G90 mode, since this seems to be the easiest for most people to understand. If you refer back to the January/February 1990 issue of *HSM*, you will find a more complete explanation of these G-Codes and how they work.

If you load Indexer LPT and CNCLPT into your computer, and have connected up the hardware as specified in my previous column, you should now have a functioning CNC controller.

Now it's time to take a sample part and step through the process of programming and making it using your new CNC controller (or any other CNC controller that uses G-Codes).

Figure 4 is a sketch of a wooden moulding. The problem will be to make a set of blades for a jointer or thickness planer so we can duplicate this moulding. This is the sort of project at which CNC excels. Only three pieces need be made, but they must be identical and they are complex enough that it is difficult to make them manually.

Figure 5 is a drawing of the knife that would be used to cut this moulding. The program to machine this part will use only the four basic G-Codes, G0, G1, G2, and G3. A CNC controller with more than these four G-Codes would allow more efficient code to be written, but these four are enough to do the job. We will limit the program to these codes since these are the four that have been implemented in the CNC controller you have just finished building.

Notice the end view of the knife drawing (Figure 5). The angle on the back side of the cutting edge will be machined by using a dovetail cutter. These tools are available in different angles and sizes. As an alternative, you could make your own cutter.

The blanks for the knives should be made first. The overall size should match the particular cutter head you are using. However, for this example, I will assume the knives are 1/8" thick, 6" long, and 1-1/4" wide. They are made out of tool steel, machined in the annealed condition, and heat-treated to the proper hardness. The three holes shown are used to hold the blanks while machining the contour in the edge. If you are making blades for a particular machine, then you can adjust your dimensions to suit.

Workholding is one of those areas of machine shop practice I find to be similar in nature to heat-treating and plating. They appear to be part science and part art form. Consequently, I will propose a clamping fixture to hold the knives for machining, but I make no claims that it is the only or best way to get the job done. A design of such a fixture is shown in Figure 6. This fixture clamps into the milling machine vise and allows the dovetail cutter to have adequate clearance along the cutting edge of the knife.

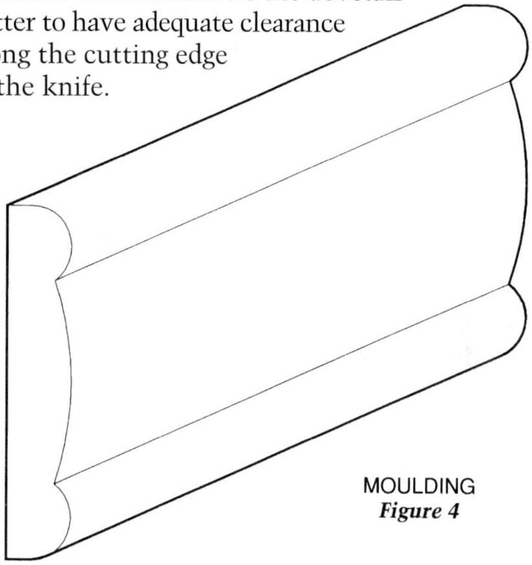

MOULDING
Figure 4

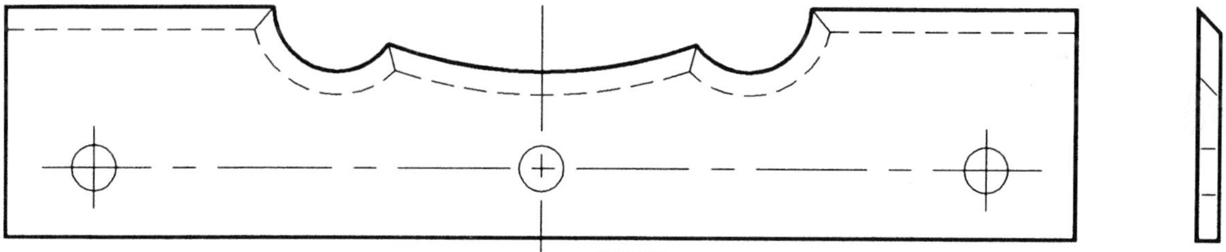

KNIFE
Figure 5

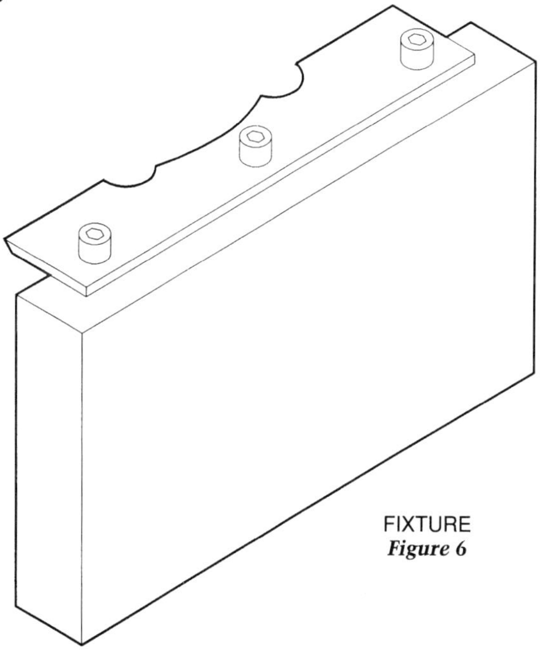

FIXTURE
Figure 6

As mentioned previously, I used the *Indexer-LPT* software from Ability Systems to produce the pulse trains used to drive the stepper motors. Also, since Indexer-LPT has its own native language, which is different from the normal "G-Codes" used in industry, Jerry Berns of Cardinal Engineering has written a program, "CNC-LPT," that allows the user to program using the four most fundamental G-Codes: G0, G1, G2, and G3.

For a more complete discussion of these G-Codes and the Cartesian coordinate system used for programming, please review this column as published in the November/December 1989 and January/February 1990 issues of *The Home Shop Machinist* magazine.

A detailed drawing of the part to be made is shown in Figure 1. The material used for such a knife blade should be a hardenable tool steel. I would suggest 01 tool steel as it can be hardened quite satisfactorily using home shop methods. The biggest problem with hardening a part such as this is the difficulty of keeping it suitably flat. Since the topic of this column is CNC programming, we'll ignore this part of the problem.

Figure 2 shows the milling cutters to be used, and illustrates how the dovetail cutter is used to machine the angle on the knife edge. Note that the center line of the cutter path is shown as a dotted line 0.1875" (3/16") away from the finished edge of the part. This is half the diameter of the 3/8" end mill. If the dovetail cutter is also set to the proper depth to machine the bevel, then it can use the same program.

Note also that the coordinates of the program are taken to be measured from the lower left corner of the part as viewed in Figure 2. This means all dimensions of the cutter path must be referenced to the lower left corner of the part as shown.

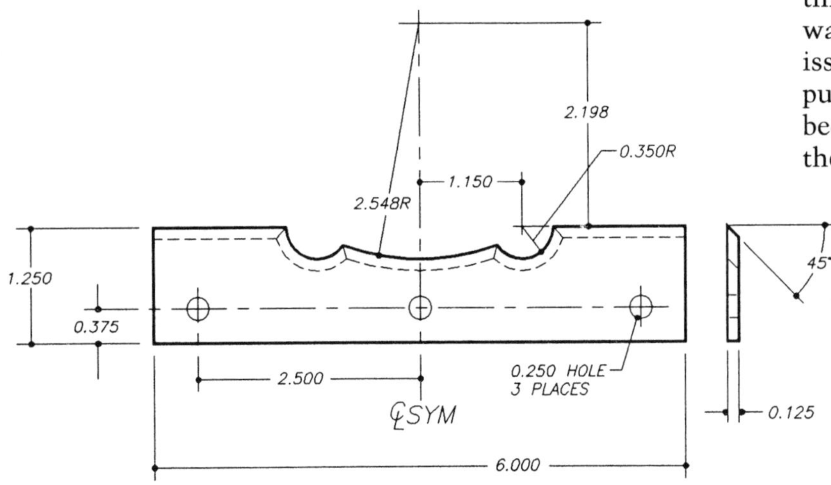

DETAIL DIMENSIONS
Figure 1

In order to write a CNC program for a contour such as this, it is necessary to know the end points and center points of each individual arc and line that make up the motion. This information must be calculated from the dimensions given in Figure 1 and Figure 2. When writing CNC programs, I have found preparing the manuscript is really the easiest part of the job. The more difficult part is the calculation of the coordinates, generally using basic principles of trigonometry and geometry.

Another method of calculating the coordinates is by using a CAD system to make a scale drawing of the part. CAD systems are generally looked upon as means to make drawings on paper. What is often overlooked is the fact that any decent CAD system contains a very powerful trigonometry and geometry problem-solving "engine" that may also be used to make any calculations required.

However, if you do rely on a CAD system to make a scale drawing of the part in question, it must be accurate in every way. When making a pencil drawing, it is easy to make a part "look right." But when working with CAD, just because a drawing on the screen of the computer appears to be correct does not mean it is mathematically accurate. On the other hand, the capabilities of the CAD system make it easy to make mathematically correct drawings.

I use AutoCAD for most of my own drawings but have tried nearly all of the commonly available CAD systems. Most of them have the facilities for making drawings that can be used for CNC programming. The main requirement needed is the ability to make geometric constructions and to extract the numerical values of the end points and centers of lines and circles with respect to any given point on the drawing. As I say, most CAD systems have this capability, and a brief review of systems was published in the January/February 1991 issue of this magazine. The conclusions published will probably still apply; however, because of the rapidly changing nature of the CAD market you should check before you buy. CAD software is continually being upgraded and improved with the result that increasingly powerful systems are available all the time.

Referring to Figure 3, note that the dashed line representing the path of the center line of the cutters has been reproduced and broken down into its component parts in two different ways.

The top part of Figure 3 shows the cutter path broken down into its individual straight line and arc segments. Each line and arc is further labeled with the G-Code that applies to it. Note that we have assumed the program will start on the right end of the path and proceed towards the left. If you trace along this path in that direction, you will see that those arcs labeled as G3 are counterclockwise and those labeled as G2 are clockwise. The straight line segments are, of course, labeled as G1.

An interesting exercise at this time would be to trace the path in the opposite direction, starting from the left end and proceeding towards the right end. You can see the direction of the arcs will then be reversed and all of the G2 and G3 codes would be reversed also.

Now look at the bottom portion of Figure 3. This has each line and arc segment marked with a sequence number or "N-number." This sequence number is used with most CNC programs as a means of identifying each "Block" of code for purposes of editing and "debugging" CNC programs. In more advanced programming, the sequence number is also used as an identifier for "loops" and when a program must be started at a place other than the beginning.

In its simplest form, CNC programming can be described using two common rules: 1) The computer knows where it is, and 2) the programmer tells the computer where to go and what path to follow.

Each segment along the cutter path is identified by its own sequence number. Each sequence number represents a separate G1, G2, or G3 code and its associated coordinate information. The G-Code tells the computer the path to follow from one segment to the other. It is the coordinate information that provides the computer with the "where to go" data.

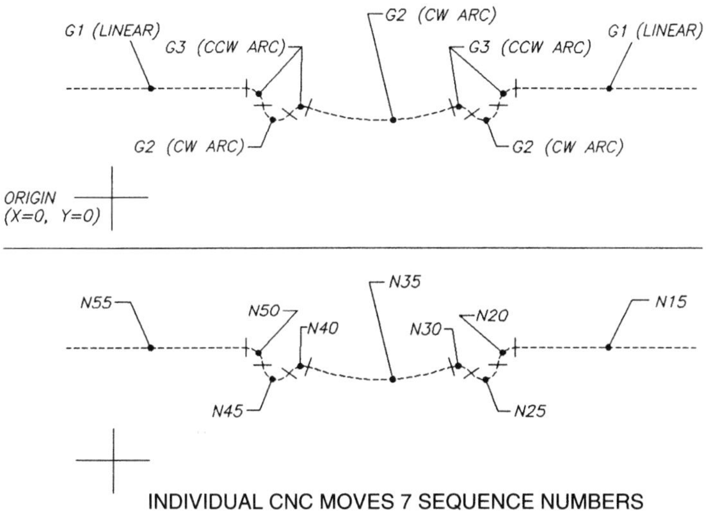

INDIVIDUAL CNC MOVES 7 SEQUENCE NUMBERS
Figure 3

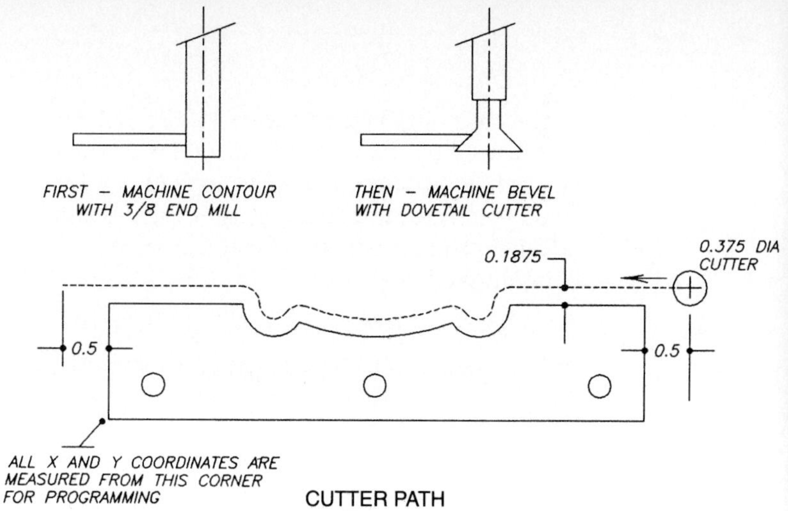

CUTTER PATH
Figure 2

The end points of each line and arc must be calculated or extracted from the CAD drawing. In the case of arcs, the center of the arc must also be calculated. This information is written in each block of CNC code using a few simple rules.

Before proceeding, I should state that all of the programming in this example is considered to be measured in inches and in what is called the "absolute" system. As mentioned earlier, this means all measurements are taken relative to an arbitrary origin. I have selected this origin as the lower left corner of the part as detailed in Figure 2.

Another bit of information that should be mentioned here is that arcs can be programmed up to a full circle in a single block of code. This is mentioned because not all CNC controllers have this capability and some of them must have a separate block of code for each quadrant of the circle, thus requiring as many as four blocks to describe a complete circle and five blocks to describe an arc that is not a complete circle but crosses into all four quadrants.

Refer to Figure 4 for a listing of the CNC program to follow the path shown in Figures 2 and 3.

Now for details on the programming of each block. The first element of each block of CNC code is normally the sequence number. The value of this number is somewhat arbitrary, but many programmers, myself included, elect to start with N5 and increment by five units for each subsequent block. This allows for the insertion of new blocks between the blocks of the initial program while still retaining an ascending numerical sequence, and still not duplicating sequence numbers. This is useful when corrections or additions need to be made to an existing program. Each sequence number is followed by the G-Code for that particular block.

In the case of G0 (rapid traverse), the G-Code is followed by the X, Y, and Z values of the desired destination. If a move in only one or two of the axes is desired, it is not necessary to repeat the coordinates for the axes not being moved.

G1 is programmed in the same way as G0 except that the coordinates are followed by a feed rate designated in block "N15" as "F50." This means that the CNC controller will cause the milling machine to travel at a feed rate of 5.0" per minute.

Note that a feed rate should be established in the first block of a program that calls for either a G1, G2, or G3 move, since these moves are all intended to be used for cutting. Obviously, when cutting, it is essential that the speed of travel be specified. Once a feed rate is specified, it does not need to be repeated in the program unless it is necessary to change it to another value. G0, on the other hand, does not require a feed rate since the purpose of that code is to move from one point to another as rapidly as possible.

Arcs G2 and G3 are programmed similarly to G1 and G0 with respect to the X and Y coordinates. You will note, however, that an additional set of coordinates is provided, labeled I and J. This is used to provide information about the center of the arc to the computer. The I value is the X coordinate of the center of the arc and the J value is the Y coordinate.

The last thing that should be mentioned in this case is the use of the semicolon – ";" – followed by descriptive text. Most of the newer CNC controllers allow some kind of rmarks to be inserted into the text of a CNC program to make it easier for we "mere humans" to understand. The use of the semicolon is quite common for this purpose and any text following a semicolon is ignored by the CNC controller. If you should find yourself programming other types of controllers, the colon –":" – parenthesis – "()" – slash – "/" – and backslash – "\" – have also been used for the same purpose.

When running the program to make this part, the corner of the part should be first "picked up" using an edge finder, wiggler, or any other standard method. The CNC controller is then set to zero at that X and Y location. The selected cutter is installed and set to a location equal to 1/2" above the top surface of the work. This position is then set to zero for the Z location.

```
; PROGRAM TO MILL KNIFE CONTOURS
N5 G0 X6.500 Y1.438 ;RAPID MOVE TO START POINT
N10 G0 Z-.75 ;RAPID MOVE TOOL TO PROPER DEPTH
N15 G1 X4.500 Y1.438 F50 ;FIRST CUT MOVE - LINEAR
N20 G3 X4.312 Y1.250 I4.500 J1.250 ;COUNTER-CLOCKWISE ARC
N25 G2 X4.016 Y1.157 I4.150 J1.250 ;CLOCKWISE ARC
N30 G3 X3.799 Y1.227 I3.862 J1.050 ;COUNTER-CLOCKWISE ARC
N35 G2 X2.201 Y1.227 I3.000 J3.448 ;CLOCKWISE ARC
N40 G3 X1.984 Y1.153 I2.138 J1.050 ;COUNTER-CLOCKWISE ARC
N45 G2 X1.688 Y1.250 I1.850 J1.250 ;CLOCKWISE ARC
N50 G3 X1.500 Y1.438 I1.500 J1.250 ;COUNTER-CLOCKWISE ARC
N55 G1 X-0.500 Y1.4375 ;LINEAR MOVE TO END OF CUT
N60 G0 Z0.0 ;RAISE TOOL CLEAR OF WORK
N65 G0 X0.0 Y0.0 ; RETURN TO ORIGIN
```

CNC CODE
Figure 4

I would suggest when making a part like this that the program first be run with a smaller diameter cutter (say, about 5/16" diameter) and with the Z depth set so the cutter will just machine a path about .010" deep in the surface of the part. Do this with all parts and then saw most of the stock out on a bandsaw. The 5/16" diameter cutter will allow for a finish cut of 1/32" when using the 3/8" cutter for which the program was written.

Place the parts back in the fixture and again run them with the 5/16" cutter, thus cleaning up any areas where too much material was left, and making the load on the 3/8" end mill more even when making the final cuts.

Then replace the 5/16" end mill with a 3/8" end mill, reset the Z axis, and run the program at least twice on each part. The first time the program is run, most of the material will be removed. The second (and possibly third) time the program is run, a very fine shaving will be removed. This fine shaving represents the amount the end mill deflected when making the previous pass. When no more material is removed, the part has been machined to its finished size.

Now replace the 3/8" end mill with the dovetail cutter, reset the Z-axis again, and run the CNC program to machine the bevel on each of the knives. It would be advisable to set the depth of the dovetail cutter so the bevel is machined a small amount at a time to reduce the burr that will be produced at the cutting edge.

Heat-treat the part, stone or grind the surfaces to sharpen the knives, and you are ready to make some moulding. This example was chosen because it was reasonably simple and would be fairly difficult to do manually. Whether it is the proper design for a set of moulding knives is a question I have chosen to ignore in the interest of producing an example that would illustrate some of the features of CNC programming.

On another topic, I have just recently returned from the Biannual International Manufacturing Technology Show. It is held in September on even-numbered years at McCormick Place in Chicago, and is known by industrial people worldwide as the "Tool Show." I can highly recommend that if you are at all interested in seeing just about anything you can imagine in the line of machine tools, this show would be worth your while. I live in a rural area about 180 miles from Chicago and am never really happy to fight my way through the Chicago traffic, but the Tool Show is one of those things I never miss.